P9-ECJ-202

Zulfi Bhutto of Pakistan

To the Sayeds
(all of them, including in-laws)
– but especially to Noori.

David

Zulfi Bhutto of Pakistan

His Life and Times

STANLEY WOLPERT

New York Oxford
OXFORD UNIVERSITY PRESS
1993

Oxford University Press

Oxford New York Toronto
Delhi Bombay Calcutta Madras Karachi
Kuala Lumpur Singapore Hong Kong Tokyo
Nairobi Dar es Salaam Cape Town
Melbourne Auckland Madrid

and associated companies in
Berlin Ibadan

Published by Oxford University Press, Inc.
200 Madison Avenue, New York, New York 10016

Library of Congress Cataloging-in-Publication Data
Wolpert, Stanley A., 1927–
Zulfi Bhutto of Pakistan : his life and times /
Stanley Wolpert.
p. cm.
Includes bibliographical references and index.
ISBN 0-19-507661-3
1. Bhutto, Zulfikar Ali. 2. Prime ministers—
Pakistan—Biography. I. Title.
DS385.B45W65 1993
954. 9105′092—dc20
[B] 92-30044

1 3 5 7 9 8 6 4 2

Printed in the United States of America
on acid-free paper

To the Memory of Our Mutual Friend,

Piloo Mody

and for

Khalid Shamsul Hasan

and

Ardeshir Cowasjee

Preface

Since 1980, when I visited Pakistan to do research on my *Jinnah of Pakistan,* I have been fascinated by the mercurial and seemingly self-conflicting life of Zulfikar Ali Bhutto. Most Pakistanis I met either loved or hated Zulfi Bhutto, the People's Party prime minister, who was arrested by his own commander-in-chief General Zia ul-Haq, and hanged after two years in prison. Millions of Pakistanis still hail Zulfi Bhutto as their *Quaid-i-Awam* ("Leader of the People"), even as they do Mohammad Ali Jinnah as Pakistan's *Quaid-i-Azam* ("Great Leader").

For most of its brief history since its birth in mid-August 1947, Pakistan was ruled by unpopular generals who seized and held power using martial force. Bhutto seemed different, the almost uniquely popular founder-leader of Pakistan's People's Party, who had just swept the polls throughout Punjab and Sind only months before his arrest. How then could he be hanged without inciting mass riots throughout Pakistan, if not a revolution, I wondered?

This book is the product of a decade-and-a-half of my reflection on that question, the last five years of which have been devoted to research into the life of Zulfi Bhutto as a microcosmic mirror of Pakistani society and its troubled history in his times.

I am indebted to many people for kind assistance in completing this book. First of all, it was thanks to Prime Minister Benazir Bhutto and her mother, Begum Sahiba Nusrat Bhutto, that I was given full and free access to Zulfikar Ali Bhutto's Library at his 70, Clifton home in Karachi. Prime Minister Benazir, now Leader of the Opposition in Pakistan's National Assembly, graciously granted me free access to all of her father's books and papers preserved in that Bhutto Family Library and Archive, providing that her mother, who uses 70, Clifton as her Karachi residence, had no objection. Begum Sahiba Nusrat kindly welcomed me to carry out my research in that Library whenever I came to Karachi over the next three years. Prime Minister Benazir and Begum Sahiba also took time from their busy schedules to permit me to interview them at length.

I interviewed more than 100 of Zulfi Bhutto's colleagues and family, and am grateful to all of them for their patient cooperation and frankness in answering my questions. Most helpful were Zulfi's closest collateral relative, Mr. Mumtaz Ali Bhutto, and his only living sibling, Begum Manna (Bhutto) Islam, both of whom shared their intimate recollections with me.

Foreign Minister Sahabzada Yaqub Khan was most encouraging when I informed him of my intention to embark upon this study, and Ambassador Jamsheed K. Marker was singularly instrumental in launching my research. I am deeply grateful to both of those gentlemen-diplomats for their more than "diplomatic" kindness and consideration. Special thanks to Mrs. Marker for her gracious hospitality and help. I thank Ambassador Abida Hussain and Minister Fakhar Imam for their illuminating insights. Thanks to Ambassador John Kenneth Galbraith, Ambassador Henry Byroade, Ambassador Robert Oakley, and Ambassador William Clark, for sharing so many recollections and so much South Asian wisdom with me. I am also most indebted to Deputy Assistant Secretary of State Teresita Schaffer and Deputy Assistant Secretary of State Nancy Ely-Raphel for their frank responses to my many questions.

Thanks to Fellowship support from our excellent American Institute of Pakistan Studies (AIPS), I was able to start my research in Pakistan early in 1989. I am grateful to all members of the AIPS Board for their scholarly confidence, but must specially thank my good friends, Professor Hafeez Malik and Professor Ralph Braibanti, without whom AIPS would never have been born. Hafeez helped convince Prime Minister Bhutto of the importance of starting this premier Institute for U.S.-Pakistan scholarly exchange in 1973, and Ralph presided over the first Board meeting in Washington that year. Thanks to Dr. Peter Dodd and Dr. William Lenderking for their warm hospitality in Pakistan, and to Mr. Ali Imran Afaqi for his kind and efficient help, and to Consuls Kent Obee and Helena Finn, and Consul-General Joe Melrose.

UCLA's Gustave E. Von Grunebaum Center for Near Eastern Studies also supported my research, and I thank my good colleagues, Director/Professor Georges Sabagh and Professor Richard Hovannisian, for their kind assistance. I also thank Dean John N. Hawkins, who heads our Institute for International Studies and Overseas Programs at UCLA, for his unfailing help and encouragement, and my old friend and colleague, Vice-Chancellor Richard Sisson, for his warm support.

Piloo Mody first "introduced" me to Zulfi Bhutto with his memoir, *Zulfi, My Friend* (1973). Piloo hoped someday to "complete" that inchoate biography of Zulfi, but he too died young. I remain, however, singularly indebted to Piloo for his insightful work, and to his wife Vina, who generously granted me full access to her and Piloo's fine collection of Bhutto photographs. Thanks to dear Joyce Hundal for introducing me to Piloo and Vina. I also thank Mr. J. J. Mugaseth, Zulfi's oldest school-buddy, for so kindly assisting me, and also sharing his early photos of Zulfi. Thanks to Omar Qureshi and Al Cechvala for their time and hospitality, and to Husna Sheikh for kindly agreeing to allow me to interview her, and to Begum Mehru Rahim Khan for her gracious assistance. Special thanks to General Gul Hasan, and to Minister Roedad Khan for their interviews.

Many friends in Pakistan opened their homes as well as their hearts and minds to me, too many to list all of them, but I must particularly thank Brigadier Noor Hussain and his good Begum Hushmat; Rizvan Kehar and his good wife; Suhail and Yasmeen Lari; Dr. Haye Saeed and his daughter Salima; Yahya Bakhtiar and his daughter Zeba; Begum Shaista Suhrawardy Ikramullah; Khalid Shamsul Hasan and his family; and Ardeshir and Nancy Cowasjee.

Khalid-Sahib knows I can never adequately thank him, and Ardeshir, like Khalid, is a true gentleman-scholar and a friend, with whom I join in mourning the death of his wife and father.

My warmest thanks to my splendid editor, Nancy Lane, who had now been midwife for many of my books. I thank Jane Bitar for typing the final manuscript.

To my darling wife, Dorothy, I can but inadequately express my growing admiration and devotion that has had only 40 brief years in which to blossom, with much love.

Los Angeles
August 1992

S. W.

Contents

Zulfi Bhutto of Pakistan

1

Sindhi Roots

(pre-1928)

No individual in the history of Pakistan achieved greater popular power or suffered so ignominious a death as Zulfikar Ali Bhutto (1928–79). Zulfi Bhutto's political rise and fall were, indeed, so meteoric as to make his name a legend in the land over which he presided for little more than half a decade prior to his hanging. A full decade after his death, Bhutto remained popular enough to ensure the election of his daughter, Benazir, to the premier position he once held. Wherever she campaigned in Sind and much of Punjab, the popular roar that greeted her was "*Jiye Bhutto!*"—"Bhutto Lives!"—by which millions of Pakistanis meant and still mean, *Zulfi* Bhutto.

Zulfi Bhutto roused such diametrically opposed passions and has left such divergent images among his disciples and adversaries that it remains virtually impossible to reconcile them as reflections of any single personality. Much like the nation he led, and in many ways came to epitomize, born irreconcilably divided, partitioned into East and West, torn from the subcontinental fabric of Mother India by the Islamic faith of his fathers, Zulfikar Ali Bhutto, microcosmic reflection of Pakistan, was never a simple personality. His unique charisma and the deep-rooted failings that brought him to an early and violent death emanated from his schizoid personality, the strengths of one part of which were matched by weaknesses of the other, the depths of its dreadful darkness mirroring the brilliant heights of its most powerful peaks. Torn apart by his inner conflicts, never able to reconcile his romantic dreams of glory with the mundane realities and misery so prevalent all around him, Zulfi hoped to the bitter end of his brief but flamboyant sojourn at the top of Pakistan's slippery pole to save himself and his land from a destiny of diminution, death, and fragmentation, vaingloriously viewing himself as an Islamic Napoleon, the "Shah-in-Shah" of Pakistan.

"For over a year and a half I have been kept in solitary confinement," Zulfi wrote from his acrid death cell in Rawalpindi's now-demolished prison, that bitterly cold January of 1979. "My family has owned not thousands of acres of land but hundreds of thousands, for generations. . . . My genesis to political fame is written in the stars. . . . If I am not a part of Pakistan . . . Sind is not a part of Pakistan . . . my roots in the soil of this land are very deep."[1] So it is in the subsoil of Sind we must search for Zulfi's roots, in the shifting treacherous sands of that ancient southernmost province of Pakistan.

Sind means "river," named for its Indus lifeline. The Bhuttos migrated to Sind in search of water, leaving their parched Rajputana desert homes farther east, in what is now India's Rajasthan, near the dawn of the eighteenth century. Zulfi's paternal ancestors had before his clan's long trek been Hindu Rajputs, but during the seventeenth century Central Asia's great Mughals conquered virtually all of India, and many Rajput warriors, like Zulfi's progenitor, Sheto,[2] converted to Islam, reaping a harvest of tax exemptions and other benefits from the devout Mughal Padishah (Emperor) Aurangzeb (r. 1658–1707) and his Muslim amirs, and claiming for himself, as clan leader, the regal title of khan. Thus, as far back as history traces them, Bhuttos have been adroit at seizing whatever opportunities life offered, equally ready to move on or to change their faith if they deemed it expedient for survival's sake, determined to overcome drought, famine, or the fiercest of adversaries in struggling to advance themselves and the fortunes of their clever family. Sheto Khan found the fertile land he sought in Sind's Ratodero, just a few miles north of Larkana, the rural town in which Zulfi would be born two centuries later. In the fecund watershed of the Indus River, the silt around Ratodero and Larkana proved rich enough to raise as much rice and sugar cane as all the Bhuttos could eat, and cotton enough to clothe that fast-growing clan as well. Though unknown to Sheto Khan, less than ten miles south of Larkana was the still-undiscovered site of one of the world's greatest ancient cities, Mohenjo Daro.[3]

The untilled growth and woods around Ratodero and Larkana were full of game, especially wild boar, which Bhutto men always enjoyed hunting. Hunting would, in fact, become Zulfi's favorite sport, and though he was considered only a fair shot, for almost two decades, whenever he was in Sind, he annually hosted a regal hunt at Larkana to celebrate his birthday. Zulfi probably had the best and largest private collection of hunting rifles and ammunition in Pakistan, and his guests would include the sheikh of Abu Dhabi and the shah of Iran, as well as Pakistan's presidents Ayub Khan and Yahya Khan.

Mughal governors ruled Sind from its bustling port city of Thatta (some fifty miles east of Karachi), which had attracted English merchants early in the seventeenth century by reports that it was "as big as London" (with then close to a quarter million population) and that "its port filled with ships laden with every kind of merchandise, which arrive by passing down the navigable Indo river, on whose bank it stands . . . a most wealthy and most vicious spot."[4] The breakdown of Mughal power soon after the death of Aurangzeb in the early eighteenth century, left Thatta and Sind in the hands of the martial Kalhoras, under whose rule British East India merchants established their first factory in 1758. Sheto's great-grandson, Pir Bakhsh Khan Bhutto, fought and defeated the Arbos, a tribe related to the Kalhoras by marriage, in "many battles in Larkana," where he established Bhutto power. Before the end of that century of violent conflicts and rapid change, however, the Kalhoras themselves were defeated by fiercer Baluchi invaders, the amirs of Talpur, who took control of

Sind, moving their capital almost a hundred miles north along the Indus to Hyderabad. Wadero Pir Bakhsh Khan swore allegiance to the new Talpur rulers of Sind, and was confirmed in return over his vast tracts of Bhutto land around Larkana, Sukkur, and Khairpur. In 1821, however, Wadero Pir Bakhsh was "invited" by His Highness Mir Ali Murad Talpur to send his eldest son, Allah Bakhsh Khan Bhutto, to Khairpur, where he was kept "as an honourable hostage" at the Talpur court for five years, "to ensure that my family did not revolt."[5]

The British had been forced to abandon their first factory in Thatta less than two decades after opening it, and tried without initial success to negotiate a new commercial treaty with the Talpurs. East India Company servants continued, nevertheless, to seek fresh footholds in the region that had once proved profitable, and soon after 1830 British ships sailed up the Indus to Hyderabad, eliciting the prescient prophecy from observant Sindhis that "our land is lost now that the British have navigated our river." With the more powerful and ever-threatening Punjab kingdom of Sikh Maharaja Ranjit Singh to its north, Sind's amirs first hoped to ally themselves with the British company in order to keep the one-eyed "Lion of Punjab"[6] at bay. In April 1838, Colonel Henry Pottinger concluded a treaty with the Talpur amirs, which permitted a British minister and his "escort" of troops to reside in Hyderabad. Two months later the British concluded their infamous Tripartite Treaty with Ranjit Singh and the exiled former amir of Afghanistan, Shah Shuja, who had lived as a pensioner of the British in Ludhiana, and was to serve as the company's puppet-prince on the throne of Kabul after he was returned to it by a mighty British army that invaded Afghanistan in 1839. That doomed army of the Indus would, moreover, invade through Sind, and Sind's amirs were expected to pay all military bills as part of the "debt" Shah Shuja claimed from them, which the British company was pleased to support, despite repeated Talpur disclaimers. Sind was thus used as the launching pad for the tragic First Anglo-Afghan War.[7]

Sir Charles Napier led the brutal conquest of Sind in September of 1842. Three Talpur amirs had divided their kingdom, the major portion ruled from Hyderabad, the others farther north from Khairpur and Miani. Those armies joined forces in a last desperate stand against the British invasion at Miani, the Talpurs' Waterloo, in March 1843, when ten thousand Sindhis were slaughtered. Napier established his court in Karachi, then just a tiny port on the Arabian Sea, where he remained as the company's first governor. He was honest enough to confess to his own diary: "We have no right to seize Sind, yet we shall do so, and a very advantageous . . . piece of rascality it will be."[8] Napier personally reaped some £70,000 in Sindhi plunder. His most famous subordinate was a brilliant young Canal Department servant named Richard Burton, who wrote two cogent volumes on *Scinde, The Unhappy Valley,* which would be published soon after he returned to London in 1851.[9]

Pir Bakhsh's son, Dodo Khan Bhutto, ruled the clan during Napier's overlordship of Sind, and he was, indeed, appointed by Napier's lieutenant as "the sole arbitrator" to give an award on the boundaries of the Chandio Jagir; his liberality to the Chandios would long win their support for all Bhuttos. His hot-blooded grandson, Mir Ghulam Murtaza Bhutto (c. 1869–99), Zulfi's grandfather, fell in love with the beautiful Sindhi "mistress"[10] of Larkana's British collector-magistrate, Colonel Mayhew. Mayhew was then "an old man . . . over sixty." Handsome Mir Murtaza, of course, was much closer in age to the young Sher beauty, who soon reciprocated his love. The wily old colonel suspected the affair and "laid a trap to catch the sinners," Zulfi's father recalled. The colonel pretended to leave on tour, letting word of his departure reach Larkana from his Sukkur residence, luring Mir Murtaza as fast as his horse could gallop to the Sher mistress's bed. The colonel returned that same night to catch the naked lovers, lashing out at Mir Murtaza with his horse whip. Young Bhutto managed to grab the whip, however, knocking the old man to the floor and giving him "a few lashes" before fleeing with his young love, depositing her with "her parents . . . near Napierabad," then galloping home. That was early in 1892. Zulfi's father, Shah Nawaz Bhutto (1888–1957), was only four, but the passionate feud that started between his father and the British colonel evolved into a vendetta, whose memory and impact affected his entire life, teaching him much about "politics," especially in Sind, and how to behave with the British, who would one day knight him, yet never win more than outward displays of his loyalty.

Sir Shah Nawaz reported how the colonel had been careful to "keep . . . secret" the "unpleasant incident" with his father, noting that as an administrative officer he could have lost his position if the "scandal had leaked out." But he never forgot "such an insult," patiently waiting almost a year, to "take revenge." Then, according to Shah Nawaz, a Sind police inspector, Sher Mohammad Shah, was almost murdered "by his enemies" in a village near Larkana. Mayhew promised Inspector Shah promotion and property if he would assert that he was attacked by Mir Murtaza, which he did. The Sessions judge acquitted Mir Murtaza, however, though his defense proved very expensive. A year later the colonel tried again, this time securing "eyewitnesses" to charge young Bhutto with the murder of a Hindu. Now the legal fees were much higher, for two British barristers were engaged, a Mr. Anwerty of Bombay at Rs. 1,000 a day, and a Mr. Rottigin of Lahore, who charged Rs. 1,500. The trial was long, but the barristers earned their high pay, winning acquittal. Still the vendetta continued, and the colonel won a retrial motion he filed in Bombay. Mir Murtaza's father had no more funds to waste in court, and advised his hot-blooded son to flee. The young man "disappeared," first to Bahawalpur State, where his British nemesis discovered him, then to Kabul, where no British officers could follow.

Zulfi and his own elder son, Mir Murtaza's namesake, would both inherit that first Mir Murtaza Bhutto's fiery nature. In her *Autobiography,* Zulfi's oldest child, Benazir, recalls how her father "loved to recount" that particular family story to his children. But Zulfi's version was significantly different from that which his father had preserved and passed down to him. Reflecting perhaps his own passions and pride, as well as some personal experiences he had in California, London, and Karachi, Zulfi amended the tale, insisting that "all the women in Sind were in love with him (Mir Murtaza), including a young British woman. In those days, it was haram—forbidden—to marry a foreigner but he couldn't prevent her feelings for him."[11] Zulfi's version, then, had the colonel "send for" his great grandfather and try to "lash" him because of his "brown skin" for daring to "encourage the affections of a British woman!" Mir Murtaza, however, seized the whip and lashed the colonel, then fled, with British cavalry in "hot pursuit." They almost caught up to the "British woman" who loved Bhutto too much to live without him, so instead of leaving her to "face . . . dishonour" Mir Murtaza ordered his own "retinue" to kill her, then galloped on alone into Afghan exile. "I loved hearing those family stories," Benazir would write, unaware of the amendments in family lore made by her ardently imaginative father, who inspired his children to form "our own moral code, just as my father had intended. Loyalty. Honour. Principle."[12]

For Zulfi's father, Mir Murtaza's young son, the fierce vendetta that forced his father into prolonged exile proved equally instructive, initiating him into "the domain of politics," as he called it, at a remarkably precocious age. "In those feudal times the politics of our environment was essentially personal and tribal," wrote Sir Shah Nawaz, adding with unfortunate accuracy that the "weaknesses of the system" long continued to hold sway over Pakistan. In the Sindhi language the word for politics is *siasat,* an ancient term derived from the feudal code of honor whose most important attributes were loyalty and vengeance. "Our feudal environment draws us to Siasat from the cradle to the grave," noted Shah Nawaz, concluding with the melancholy reflection that "it covers the whole spectrum of our backward society." Mumtaz Ali Bhutto (b. 1933), Zulfi's uncle, who was more like a brother to him and his closest male confidant, recently reflected that Zulfi "was a born politician. I think there was no blood in his veins, it was all politics."[13] Benazir's mother, Begum Nusrat Bhutto, said that her elder daughter, then the prime minister of Pakistan, was "just like her father—in so many ways."[14]

Mir Murtaza's Afghan exile left his father, Khuda Bakhsh Khan, to bear the brunt of the colonel's anti-Bhuttoism. His grandson wrote that Colonel Mayhew's police "induced" criminals to ambush Khuda Bakhsh one evening as he was returning home after inspecting some of his distant property in Jacobabad district. The highway robbers hurled a hatchet at the old man's horse, sending the wounded animal off at so swift a gallop that he threw Khuda Bakhsh from his saddle, and though Bhutto *haris* (peasant serfs) found their *wadero* (landed baron) lying unconscious in the field and carried him to what was then the main village of the Bhutto haris, Amirabad, Khuda Bakhsh never recovered. Two weeks after the old Khan's death, since his only son and heir was "an absconder," Colonel Mayhew confiscated all moveable Bhutto properties, including invaluable jewels, heirlooms, and gold-embossed saddles and guns, ordering kerosene poured over the guest houses and godowns to set fire to the furniture, carpets, and durries. "We saw the fire ablaze at night," Shah Nawaz recalled. "In the morning we saw the ashes."[15] That was 1896, when 8-year-old Shah Nawaz and his aged grandmother, mother, and brother were forced by the superintendent of police to leave their once-luxurious home in Garhi Khuda Bakhsh Bhutto village of Sind's Ratodero Taluka, "with only the clothes on our bodies." They took shelter with their poor *haris,* and for the first time in his life Shah Nawaz was obliged to walk barefoot in Sind's blazing heat some ten miles a day to the vernacular school at Naudero, taking his crust of dry bread along for lunch, he who had always been served on "silver plates."

When Mir Murtaza learned of his father's death and its tragic aftermath, he appealed to Kabul's "Iron Amir" Abdur Rahman Khan, in Kabul, for help to reclaim his confiscated fortune. The Afghan amir provided the young Bhutto Khan with "sufficient funds not only to return to Sind but to enable him to fight out his case in Court."[16] But Mir Murtaza lost all his Afghan gold when the boat he had hired to take him from Peshawar to Karachi capsized in a storm at Panjnad. Ever-resourceful, he managed to borrow more money from an official in Bahawalpur State, where he was well known, and so completed his journey to Karachi.

Once he reached Karachi, Mir Murtaza disguised himself as "a Sikh labourer," to avoid detection by the colonel's police, then went to visit his friend and former pleader, Rais Ghulam Mohammad Sheedi. The rais first refused to "believe his eyes," such was the perfection of Mir Murtaza's disguise, but once he heard the "poor labourer's" voice and several of the names he so swiftly dropped, he recognized it was Bhutto-*sahib* himself. These two wealthy Sindhis then put their heads together till they figured out how best to get Bhutto inside the carefully guarded residence of the British commissioner of Sind, Karachi's highest official, so that Mir Murtaza might personally plead his case, without running any risk of having it shot down in advance by Colonel Mayhew. They knew better than to attempt a direct approach to Commissioner James's office, for that gate was guarded by the commissioner's *mir munshi* (head clerk), who not only required much "consideration" but would at best end in sending Mir Murtaza to the commissioner's political assistant, a young Englishman who would probably never agree to "waste" any of the commissioner's precious time on so questionable a matter. The plan they devised was worthy of a tale from the Arabian Nights, and most instructive as to how important meetings have long been arranged in Karachi and Sind.

Rais Ghulam Mohammad knew the mir munshi well enough to appeal to him "to permit a poor Sikh labourer from the north to get employment on a construction job that was going on in the Commissioner's residence." Police clearance was, of course, required for all such jobs, but the rais assured his friend that this "labourer" was "hardworking" and because it was only *one* job there was no problem. So Mir Murtaza was smuggled into the house of the most powerful Englishman in Sind, disguised as a poor worker. But the wily Bhutto Khan and his clever pleader carried their conspiratorial plot much further, the rais having secured the list of guards to learn that one of the commissioner's many *chowkidars* (gate-keeper guards) was a Sindhi Khokar from *Larkana!* Once Mir Murtaza got inside the great house, therefore, he had a natural ally, and from this Larkana informant learned that the commissioner came out to inspect "the progress of the construction" twice weekly, and on Saturdays "invariably spoke to the labourers and the foreman."[17]

The following Saturday morning, as the commissioner, Sir Avan James, made his rounds of the construction, Mir Murtaza stepped forward to announce his true identity, boldly pleading, "I have a story to tell and I want you to do justice by hearing me." Sir Avan "did not show surprise or anger," only stared "more closely" at this strange man before turning him over to his aide-de-camp to take into custody. The aide listened to Bhutto's story, and reported what he had learned to the commissioner, who had by now heard of the notorious vendetta between Mayhew and Mir Murtaza, and probably even knew about the lovely mistress they had shared. So he had Mir Murtaza brought to him, and found him reasonable and courteous as he pleaded for "a fair trial" before the commissioner's own court in Karachi rather than under Mayhew's jurisdiction in Sukkur. That seemed just enough to Sir Avan, who immediately dictated an order of transfer, and shook Mir Murtaza's hand before he left, sagely advising him "to leave his office without bitterness."

More than a quarter century later, when Sir Shah Nawaz himself became a minister in the government of Bombay, he looked up Sir Avan James's reports, and took note of one the wise old commissioner had written about the "great influence" of the Bhuttos, ending with a warning: "They have to be watched."

After producing two sureties and raising bail, Mir Murtaza was released, his Karachi pleader seeing to the transfer of all trial documents from Sukkur to the Commissioner's Court. Free though he was to move about Karachi, Mir Murtaza could not leave that city till his trial, another lengthy one, in which he was defended by barrister Moti Ram Advani, ended in his vindication. All the confiscated Bhutto property was now to be restored to Mir Murtaza. It was 1899, the year the Black Plague first reached Karachi from China, via Bombay. Mir Murtaza swiftly left that city of death, galloping north to Sehwan to give thanks at the shrine of Sind's greatest Sufi, Saint Lal Shahbaz ("Red Falcon") Qalander, who first came to Sind in the fourteenth century and whose shrine at Sehwan, thirty miles south of Larkana, would often be visited by Zulfi for worship, as well as by his mother and grandfather.

Mir Murtaza stayed with his good friend Ali Mohammad, son of Khan Bahadur Hasan Ali Affendi Majidi, who was posted by the British in Dadu (just north of Sehwan) as civil judge, and who gave him happy news of Colonel Mayhew's imminent retirement. Bhutto's old adversary was, in fact, just then on his way down from Sukkur to Karachi for his last voyage home. They met in Dadu at the affendi's house and Mir Murtaza asked the colonel to expedite "restoration of all his immovable properties including his houses" in Larkana. The old colonel, who said he still "loved Sind and her people," agreed to "let bygones be bygones," and signed the order, prepared in the civil judge's house, to restore all such properties. With that powerful document in his hand, Mir Murtaza returned to Larkana, to a royal welcome.

"There were celebrations all over Ratodero," Shah Nawaz recalled. From the village of Garhi Khuda Bakhsh Bhutto to the town of Ratodero people lined both sides of the road, bright carpets spread at every intersection. Accompanied by three faithful servants who had remained with him throughout his arduous exile, Mir Murtaza Khan Bhutto rode on horseback to be feted at the grandest reception ever held in that district. He publicly assured everybody that he "harboured no grievances" toward anyone who had "given evidence" against him, and thanked them all for such heartwarming hospitality. There was, he said, "no greater happiness than to return to the land of birth and burial and to be with one's children, with one's friends and people." It was the high point of his short turbulent life. One month later Mir Murtaza was poisoned to death by the machinations of some "influential zamindars," his son suspected, who had "falsely deposed against my father" to please Colonel Mayhew, and now feared the hot-blooded Bhutto Khan's "revenge." Siasat!

Shah Nawaz, though only ten at the time, was a full *day* older than his half brother, Ali Gohar Khan, and thus found himself suddenly "in charge of the family and the state." Those same "influential" landlords of the district who had conspired to murder his father were, of course, "jealous" of Shah Nawaz and "hatched" what he called "dangerous intrigues" against him and his young brother. The new British collector, Mr. Mules, took personal care and responsibility for protecting the boys, however, under his Court of Wards, while their great-uncle, Wadero Ellahi Bakhsh Khan Bhutto, Mumtaz Ali's grandfather, took charge of the management of their estate fortune and family until they reached maturity. Young Shah Nawaz had fond memories of his good guardian, Mr. Mules, and his assistant collector, Mr. Tarapat, who founded a "very good" middle school in Larkana, modeled after the excellent Chief's College of Lahore, with English curricula designed to prepare the children of local gentry for higher education in Bombay or London, if not Oxford or Cambridge. The boys were encouraged to take up boating, for "prosperous" fast-growing Larkana, with its ample mango gardens and orchards, had been built on the banks of the huge Gharwah Canal, and was by then commonly called "the garden of Sind." Shah Nawaz and his friends often played along the banks of the canal after school, and though he had not learned how to swim, decided one hot afternoon to "take a dip." He suddenly found himself carried away by swift currents into the deep center of the wide canal. Luckily, his friends' cries roused several nearby fishermen, who jumped in to save him. He had lost consciousness in the canal, waking up much later in a hospital. For many years he did not dare to enter the water again.

Shah Nawaz's new Zamindar school was named the Larkana Madressah, and he completed his 5th Standard in English there before being sent off to Karachi, where he was enrolled in the higher Sind Madressah, whose principal was an English "gentleman" named Vines. Shah Nawaz lived in the principal's home, where he had two of his own rooms on the ground floor of that bungalow. He remembered Mr. Vines and his wife as "kind" people, who were "sympathetic to Muslims." Hockey had recently been introduced by the British to Karachi, and young Bhutto got to "like it" and soon became "a keen player." He did not, however, take any of his studies very "seriously" and was anxious to finish school and get back to Larkana. Principal Vines tried to advise "patience," but the young Bhutto was "in an agitated frame of mind, anxious to burst out," tempted to leave almost daily for home "by the first available train." A deep current of restlessness flowed through the Bhutto gene pool.

The British government of Bombay, which still included Sind, decided in this early part of the twentieth century to recruit loyal young zamindars and waderos into its civil service and police cadres, much as they had earlier enlisted the cooperation and loyal support of Indian princes and chiefs. One of Shah Nawaz's friends became subinspector of police and went off to Nasik for training at the Police Academy. Shah Nawaz hesitated but finally opted to apply, receiving strong support from his principal and others in high places. He was offered the post of deputy superintendent of police and instructed to join the Nasik Police Training School by the end of December 1908. With that post assured, he took leave of school, not bothering about his matriculation examinations, and returned home, stopping one night at Garhi Khuda Bakhsh Bhutto, then moving on next day to Garhi Pir Bakhsh Bhutto, ancestral home of his great-uncle–guardian, Ellahi Bakhsh, and his family. Shah Nawaz delivered some valuable gifts of silver and gold furniture and ornaments he'd had made for his uncle in Karachi, and that evening they rode together to Ratodero to meet the new deputy collector of Larkana, Ellahi Bakhsh Shaikh, his uncle's namesake, who was "encamped" there. They returned home late and dined together.

Shah Nawaz was still in bed next morning when his uncle's servant came "in a perplexed state" to report that the wadero had been locked in his bathroom for "well over an hour and that this was contrary to his normal routine."[18] They broke open the bathroom door and found Ellahi Bakhsh lying face down on the ground, dead at the age of 28. "His premature death was a terrible shock to me," Shah Nawaz recounted. "Wadero Ellahi Bakhsh was too young to die in natural circumstances. . . . The medical authorities thought it to be a collapse of the heart, whilst others opined that it was suicide. I could not get myself to believe that a man of Ellahi Bakhsh's promise could commit suicide."

Thus, at the tender age of 20, young Shah Nawaz "found myself occupying the position of a family elder." Overnight he became the guardian of Ellahi Bakhsh's two minor sons and all of his "female dependents" (two widows, a sister, a daughter, mother, and an aunt) and of an estate four times as large as his own father's had been, totaling more than 100,000 acres. The morning after the funeral, with the women's loud "wailing" still ringing in his ears, Wadero Shah Nawaz Khan Bhutto walked around the village outskirts of Pir Bakhsh Bhutto, accompanied by a servant. He had much to worry about, much to ponder, from his "impetuous brother" and "lackadaisical cousin" to "the legion of womenfolk who had to be looked after."[19] Reflecting on the virtues of his former guardian, he thought, "The man was a jewel. . . . He had the hands of an artist . . . spoke English fluently and strangely enough without an accent. He was a man of letters well versed in Sindhi and in Persian . . . possessed of a sharp intellect and . . . polished in demeanour . . . if he had lived a full season, he would have shone like a star in the politics of undivided India. Fate ordained otherwise."[20] In "a perplexed state of mind," the new young master of all he could see turned to ask his servant for a cigarette, taking the stronger, cruder Sindhi *biri* he was offered instead. "Since that first puff, for many years to come I became a chain smoker," Shah Nawaz reflected, in the twilight of his own life. All thoughts of Nasik and police school had now to be banished, moreover, and though he would still be obliged to remain a nominal ward of the British courts for almost a year, he assumed complete responsibility for the growing Bhutto clan and its incalculable collective fortune.

C. M. Baker, the recently appointed collector of Larkana district, proved more difficult than Mules had been in his dealings with the new Bhutto wadero. Shah Nawaz found him "arrogant" and "sadistic," a petty man, whose head had become "swollen" from the "sycophancy" of the tribal chiefs of Jacobabad district on Sind's frontier, his previous posting. Baker had appointed a retired deputy collector, a Bengali named Kimat Roy, to serve as manager of the Bhutto lands and estate, which Shah Nawaz vigorously challenged. As manager, Roy would have enjoyed "vast discretion and complete authority," and Ellahi Bakhsh's widows wished, therefore, to partition his property and save at least some portion of it for their minor sons, rather than risk losing everything to the new "alien Manager." "Interested persons fanned their fears," Shah Nawaz noted, seeking to reassure the anxious widows and assuage their concerns. Sindhi custom kept all landed property for male heirs, of course, and when Shah Nawaz explained that to Baker, the collector "flew into a rage and thought that I was challenging his authority."[21] It would take some time and not inconsiderable arguing and substantial legal fees before Shah Nawaz managed to convince Baker as well as the widows of the wisdom of keeping things the way they had been when Ellahi Bakhsh was still alive. The former wadero's namesake, Deputy Collector Ellahi Bakhsh Shaikh, had to be recruited to help bring the stubborn Baker round, as was Shah Nawaz's former principal of Larkana Madressah, G. M. Mirza. Shah Nawaz now proved remarkably adept at untangling intricate legal problems and adroit in disarming the opposition he faced from both sexes.

From his deputy collector friend Shaikh, Shah Nawaz learned that every new British officer who came to Larkana was immediately told, by "enemies of the family" of "details of my father's quarrel with Col. Mehyhu (Mayhew) and about the Hindu . . . who was murdered because . . . [he] did not stand up to receive my father when he [Mir Murtaza] went to his office. These officers were told that as I was a chip of the old block . . . I should be kept in my place."[22] In March of 1909, Wadero Shah Nawaz turned 21, and his father's estate was released from the Court of Wards and divided between him and his brother Ali. Shah Nawaz, however, continued to manage the much larger estate left by Ellahi Bakhsh for his widows and sons. He also became actively interested in public affairs, for this was a time of important constitutional reforms in British India.

The Liberal Party's victory at the polls in Great Britain in 1906 had brought John Morley (1836–1923)[23] to the helm of the India Office as secretary of state for India. Morley, one of the leading radical reformers and liberal philosophers of his time, was determined to open constitutional windows throughout British India to winds of popular representative feeling and to bring the autocratic racist government of imperial India into the more liberal mold of England's responsible parliamentary rule, making the stiff-necked bureaucrats of Calcutta and Simla in part responsible to Indian public opinion as expressed through the views of elected Indian leadership. It was a long slow process, but Morley's faith in representative rule and liberalism was second only to that of his philosopher-friend, John Stuart Mill, inspiring a generation of such fine young Indian Liberals as Gopal K. Gokhale and Romesh C. Dutt. As a Lincoln's Inn barrister, moreover, Morley also inspired Mohammad Ali Jinnah (1876–1948),[24] who followed him through the mighty labyrinth of law to the pinnacle of British India's bar and Privy Court powers, before veering off to become the Quaid-i-Azam ("Great Leader") of India's Muslims and Pakistan. The Councils Act of 1909 opened British India's many council chambers to Indian representatives elected for the first time by their own constituents. Franchise qualifications were narrowly limited, however, in this first step of constitutional change, for many British officers still thought of Indians as "savages," dark "natives" of a lower order of humankind. Those who paid enough taxes or who held higher educational degrees or titles were deemed fit to select their own spokesmen, and the Muslim minority community (about one-fourth of India's population) was also now to be specially honored and trusted, given the chance to vote for its own separate Muslim candidates for the limited number of council seats open to election.

Coming in the wake of Bengal's First Partition, in 1905, when the government of India carved its first Muslim-majority province, Eastern Bengal and Assam, out of the Bengali-speaking heartland of what had previously been the raj's premier presidency, India's National Congress leadership viewed this new act as proof positive of "Perfidious Albion's" policy of "divide and rule."[25] British Viceroy Lord Minto (r.1905–10), had been quick to encourage the Muslim aristocracy of India, led by His Highness the Aga Khan, to approach him in Simla as a special deputation to request "a fair share" in any contemplated constitutional change for their minority community, promising from his viceregal throne, long before Morley's introduction of his reform bill in Parliament, that Muslim "interests as a community will be safeguarded." By thus promptly playing his "Green Card," Minto was warmly congratulated by conservative officials for "pulling back" India's 62 million Muslims from joining the "seditious opposition" of a Hindu-majority National Congress. Two months after their encouraging success, moreover, the Aga Khan's deputation, now doubled in size to seventy members, met in Dacca (Dhaka), capital of the new Muslim-majority province, to launch the first nationwide Muslim political organization of British India, the Muslim League, "to protect and advance the political rights and interests of the Musalmans of India." Muslim princes and landed barons, who had never before taken any interest in politics or public affairs suddenly found important opportunities to influence British policy in virtually every way.

"My new responsibilities and my general interest in politics aroused by the Reforms, drew me closer to public affairs," Shah Nawaz recalled. "My services were available to the common man from morning to night. I ran an open house at Garhi Khuda Bakhsh Bhutto without distinction and without motive. I tried to help people who came. . . . I tried to give good advice. . . . I was courteous to the common man and rarely did I lose my temper. . . . I spared no efforts to cultivate and befriend the masses."[26] It was from his father that Zulfi learned politics, including his populist attempts to "befriend the masses." Shah Nawaz was a remarkably shrewd, mostly self-educated man, and taught his only brilliant child most of the tricks he had learned, often by bitter experience. He always tried, as he put it, "to side with justice" in advising his neighboring tribal chiefs and waderos on ways of seeking to mitigate their never-ending disputes.

"Our good family relations with Pir Pagaro (leader of Sind's Hurs) and His Highness the Mir of Khairpur were further cultivated by me," Shah Nawaz noted. "Similarly, we were on very friendly terms with the Jatois of Nawabshah and Dadu. . . . with the Bozdars of Sukkur and the Buledis and Bejaranis of Jacobabad. In Baluchistan my relations were equally cordial with the Bugti Sardar, with the Raisanis and the Khan of Kalat."[27] Zulfi was careful to continue to cultivate the same princes, chiefs, waderos, and sardars, for like his father he well understood that in Sind, Baluchistan, and Punjab the "masses" voted as their feudal lords and tribal leaders told them. Bhutto influence extended to the Princely State of Bahawalpur, with whose royal ruler Shah Nawaz maintained an "intimate" association. The Multan district in Punjab had, moreover, developed many bonds with Sind's Bhuttos. These would remain key bases beyond Sind of Zulfi's power and popular support.

No opportunity for early political advancement eluded Shah Nawaz. He was careful to seize any advantage for his family's fast-rising fortune, and never forgot his loyalty, almost as great as that premier loyalty to the Bhutto family, to his region and its people. The Indian Councils Act of 1892 had given Sind a single seat on Bombay's Legislative Council, but that seat was held by the governor's nominee, Shaikh Sadiq Ali, wazir of Khairpur, who enjoyed premier power among the waderos of Sukkur, Larkana, and Nawabshah, and retained his seat at the council table of the governor of Bombay until 1904, when he returned to govern Khairpur State as its chief minister. His place was taken by influential Sindhi barrister G. M. Bhurgari, and after the 1909 reforms Sind received three more seats on the Bombay council, for until 1935 Sind remained part of Bombay.

Bhurgari was Sind's first British barrister. He had returned to Sind with liberal political ideas, and was wealthy enough to devote himself entirely to politics, joining the Indian National Congress first, and later the Muslim League, as Jinnah had also done. Bhurgari's seat was supposed to represent the zamindars of Sind, most of whom were Muslim waderos, but Shah Nawaz felt that he came more and more under the influence of Hindu managers, who "convinced him that he was a leader of all Sindhis, both Hindus and Muslims."[28] As Shah Nawaz viewed the situation this was betrayal, for "Muslim zamindars had to be saved from the Hindu moneylender," the "educated Muslim needed employment," Muslim lawyers needed "briefs," and more Muslims had to be brought into trade and commerce. How could "Muslim interests" be "safeguarded" without effective political leadership? Such was the growing cry among Muslims, not only in Sind but in other parts of India as well, which would in three decades give birth to the Muslim League's demand for Pakistan.

British law gave Hindus equal rights to acquire auctioned state lands and forfeited Muslim zamindar lands that Muslim landlords alone had long enjoyed, and now many "upstart" Hindus had enough cash in hand to take advantage of such sales. "No Muslim leader," Shah Nawaz bemoaned, was in the council chambers to "protest against the creeping conquest of the Hindus into the rural heartland of Sind."[29] Bhurgari had become a Hindu advocate and was no longer trusted by his zamindar constituents.

In October 1913 the waderos of Sind joined forces to fight this new "threat" posed by Hindu buying of their land, and held a political meeting in Hyderabad, which Shah Nawaz attended, his first public commitment to political action. "It was a marathon session full of confusion," he recalled, but "we were anxious to evolve a united front and a common strategy." Though not a high school graduate and only 25 years old, Zulfi's father was now "fully bitten by the bug of public life."[30] He regularly attended all future zamindar conferences, and said "little," for he considered most of what he heard at such meetings "a lot of hot air." Yet when he spoke, he was listened to, and he tried to reconcile differences, to strengthen wadero common ground, compensating for the "handicap" of his incomplete education by "hard work and a cool head." He believed, moreover, that "life itself is the best teacher," and less than seven years after embarking on his political career, Shah Nawaz defeated overconfident barrister Bhurgari in a hard-fought election to win the one seat from Sind on Delhi's Imperial Legislative Council in 1920.

Shortly after the start of World War I Shah Nawaz's first son, Sikander, fell victim to pneumonia when he was only 7 years of age. Luckily, a second son, Imdad Ali, had been born two months earlier, making the pain of that blow "bearable" to the bereft father. Imdad Ali would remain heir to Shah Nawaz's fortune till his own early death from cirrhosis of the liver at the age of 39 in 1953.

Curiously, Shah Nawaz makes no mention of the birth of Zulfikar Ali Bhutto in his unpublished "autobiography." Zulfi was the son of his second wife, a "beautiful" young Hindu dancing girl with whom he had fallen in love at "first sight" at the home of her elder sister and brother-in-law, the superintendant of police of Sukkur, Mir Maqbul, in 1924.[31] Lakhi Bai, who changed her name to Khurshid when she converted to Islam to marry Shah Nawaz, had come on a visit from Maharashtra's capital, Poona (Pune). She was married to Shah Nawaz in 1925, the year in which he was honored by the British raj with a companionship in the exalted Order of the Indian Empire. The marriage, surprisingly enough, was not in Sind, perhaps in deference to the sensitivity of his first wife and family elders, but at the palatial home of his good friend, the khan of Kalat, in Quetta. Shah Nawaz wore a plain business suit and hat, marching with his petite bride through the line of upraised crossed swords of the khan's bodyguard. At 37 he was more than twice the age of his bride, who was to become the mother of three of his children, two daughters, Manna and Benazir, and his youngest son, Zulfikar Ali.

The strain of fresh married life took its toll, however, on portly Shah Nawaz's health. Before 1925 ended he had suffered a serious heart attack, and had to be taken to a Hyderabad hospital. He recovered strength enough to sire his third son, but his days and nights of galloping through Sind's hinterland were over, as were his weeks of wild boar hunting.

Soon after entering Delhi's Imperial Council Shah Nawaz made the most of his new position and the options for local influence it opened to him. He was appointed by the British raj to preside over Larkana's District Board, giving him virtual administrative as well as judicial control over his home district from 1920 till 1934. He also chaired Larkana's Central Co-operative Bank, and in 1930 he was knighted. As one of only sixteen Muslim delegates chosen from all of British India, Sir Shah Nawaz sailed from Karachi for England late in 1930 to attend the first of three Imperial Round Table Conferences in London.

King Emperor George V himself was to inaugurate the first of those conferences at which fifty-eight Anglo-Indian officials, princes, politicians, and sages of many sorts sat round an enormous table trying to agree upon the next great step for India's constitutional future. India's National Congress had expected full dominion status within the British Commonwealth immediately after World War I, but received instead renewed martial "law" and the terrible massacre at Jallianwala Bagh.[32] Hence, the Congress leadership boycotted that first conference. Mahatma Gandhi (1869–1948) languished in prison at the time, in the wake of his famous Salt March Satyagraha campaign, and some sixty thousand of his loyal disciples had followed him behind bars for having defied British law by picking up untaxed salt from seashores around India's peninsula. Jawaharlal Nehru (1889–1964), Congress's youngest president, had drafted the Purna Swaraj ("Complete Freedom") resolution, which Congress passed by acclamation at its revolutionary Lahore session in December 1929, proclaiming that "the British Government . . . has ruined India economically, politically, culturally and spiritually," and that the time had come to "sever the British connection" entirely.

Barrister Jinnah, however, who was to set up residence in London's Hampstead and his own chambers in the City for Privy Council appeals, joined Sir Shah Nawaz, the Aga Khan and other Muslim leaders that November of 1930, sitting at the House of Lords round table with Prime Minister Ramsay MacDonald, former viceroys Lord Hardinge and Reading, the nawab of Bhopal, maharajas of Patiala, Baroda, and Alwar, Sir Tej Bahadur Sapru, M. R. Jayakar, Sir Homi Mody, Sir Chimanlal Setalvad, and forty more luminaries recently arrived from India. Jinnah startled some of his English as well as Hindu auditors at that table when he insisted with no hesitation in his opening speech that there were "four main parties" in attendance, whose "aspirations of India" needed to be "satisfied" by that conference: "the British party, the Indian princes, the Hindus and the Muslims."[33] He warned his English friends that though most Muslims had "kept aloof" from Congress's platform of noncooperation they might be "tempted" to join Congress unless their just aspirations and demands were now met constitutionally. "India wants to be mistress in her own house," argued Jinnah, who had been the first member of the viceroy's Imperial Council to resign in principled protest over passage of the "Black" Rowlatt acts extending martial law after the war. Brilliant politician and constitutional lawyer that he was, he could "conceive" of no revised constitution for British India that would not "transfer responsibility" over India's central government to a cabinet "responsible to the Legislature." Had the "parties" at that round table taken to heart Jinnah's sage advice, drafting a government of India act in accordance with it, India might possibly have been spared the anguish of partition with its bloody communal prelude of civil war some fifteen years later. Neither the British bureaucrats nor Hindu leaders took him quite seriously, thinking him little more than a "slippery" negotiator who looked and sounded more British than Indian, and hence could readily be dismissed for lack of a mass following. After the third round table conference ended with painfully little accomplished, Jinnah resolved to transform himself into the Quaid-i-Azam ("Great Leader") of the Muslim League, at whose helm in the late 1930s and early 1940s he boldly led his Muslim nation to its promised land of Pakistan.

Shah Nawaz first met Jinnah in Larkana in 1928, the year Zulfi was born. The then brilliant Bombay barrister, himself born in Sind and reared in Karachi, had been invited by Sind's Mohamedan Association, over which Shah Nawaz presided, to settle a legal dispute among several Muslim waderos of Sind. Jinnah stayed with Shah Nawaz in the Bhutto compound's grand house Al-Murtaza, in Larkana, and may well have heard some of Zulfi's earliest cries, though he surely never went near the purdah quarters of Shah Nawaz's younger wife and her children. Nor is it likely that he met Shah Nawaz's first wife, or either of the two other possible wives he was permitted and "may" have married as a good Muslim. Shah Nawaz, like Jinnah, was a heterodox Shi'ite Muslim who believed in the claims of Ali and his martyred heirs to historic succession of leadership over the Islamic community, not as most orthodox Sunni Indian Muslims did, and most Sunni Muslims of Pakistan do. Like many other Sindhis, however, though nominally Shi'i, Shah Nawaz and his formerly Hindu wife both worshipped at the shrine of Sufi Saint Lal Shahbaz Qalander, and would periodically visit Sehwan to pray for favors, as had Mir Murtaza Khan, after whom the Bhutto residence in Larkana was named, and as would Zulfi Bhutto much later.

The crowning political achievement of Sir Shah Nawaz's life was to convince Great Britain's rulers at the Round Table Conferences in London that Sind deserved separate provincial status, thus liberating his home region from Bombay, elevating sleepy Karachi overnight, once the Government of India Act of 1935 took effect, to equal status with other booming provincial capitals like Bombay, Calcutta, and Madras. It was the one public cause to which Shah Nawaz devoted himself wholeheartedly during the early 1930s, having started to lobby on Karachi's behalf as soon as he reached Delhi's Imperial Council in January 1920. "On coming to Delhi the thought crossed my mind on how I could get the Council to pay more attention to the development of Karachi. I believed that with the development of Karachi the hinterland of Sind would not escape the benefits," this single-minded Sindhi recalled. "I therefore chose to concentrate on Karachi. Whenever I went to other places and particularly to Bombay, I tried to learn how to improve Karachi. Bombay was a much bigger city but I looked upon it as a rival of Karachi."[34] It took this patient politician fifteen years, yet he won his point, and thanks to his victory Karachi land values skyrocketed, as did the concomitant "benefits" to Sind's Larkana and other rural hinterlands. For India's Muslim bloc, moreover, Shah Nawaz's victory was almost as important because with Bengal reunited after 1910 and the Frontier districts yet to be raised to full provincial powers, Sind became British India's only full blown Muslim-majority province.

It was the single-most-important economic-political coup won by an Indian Muslim since the founding of the Muslim League, eclipsed only by Quaid-i-Azam Jinnah's subsequent victory more than a decade later in winning his suit for separate nation-statehood for Pakistan. After the Round Table Conferences, Shah Nawaz remained in touch with Jinnah, and toward the end of his career became Jinnah's emissary to the nawab of Junagadh State, whom he served as prime minister in 1946, convincing his royal ruler to opt for Pakistan in August of 1947, though Junagadh was surrounded by India's Gujarati territory and most of that state's population was Hindu. Sardar Vallabhbhai Patel (1875–1950),[35] the "Iron Man" of Congress, Nehru's deputy prime minister, would never permit such "treachery," however, surrounding the tiny state with Indian troops, who took it virtually without a shot that September. Junagadh's nawab and his family, together with his diwan, Sir Shah Nawaz, fled by sea from the port of Veraval to Karachi. For Shah Nawaz it was a welcome homecoming, after almost a quarter century's political "exile," first to Delhi, then to Bombay, as Governor Lord Brabourne's leading Muslim confidant and cabinet minister after 1934, and finally to Junagadh.

"Politics," Zulfi's father reflected, required "diverse talents," perhaps the most important of which was an ability to strike "a pose" and develop a special "style." Shah Nawaz proved singularly successful at doing both. So would his youngest son. "It requires one individual to change the destiny of people," Shah Nawaz wrote in the personal memoir, which his talented son must have studied. "Gandhi and Jinnah were such individuals." Napoleon was another hero of Shah Nawaz, and later became Zulfi's favorite historic role model. "Until the Day of Judgment, everything is temporary," the elder Bhutto cautioned his son, who learned much about diplomacy as well as provincial politics from his shrewd father.

"Britain provided a sophisticated cover to her domination. . . . [I]f international factors had not intervened, the British would have remained eternally in the colonies on the pretext of schooling them. Their subtle policy caused much confusion."[36] But Shah Nawaz was never fooled by the poses or loud professions of self-sacrifice made by his enemies or friendly neighbors, whether in Larkana, Karachi, Bombay, Delhi, or London. "Some of the worst sycophants of the British now unabashedly pose as gallant freedom fighters," he cautioned. Although this knight of the Order of the British Empire dined and drank at more elegant tables of British governors and viceroys than most of his contemporaries, he never forgot that he was a Sindhi Muslim wadero named Bhutto. His precocious son would also dine at the homes of presidents, prime ministers, generals, and kings, but he too trusted none of them.

2

From Larkana to Bombay

(1928–1947)

Zulfi's mother was careful to note the precise minute and hour of her precious son's birth, for though she had converted to Islam before marriage, Lakhi Bai never lost her Hindu faith in astrology. Begum Bhutto had her son's horoscope cast by an old Bombay Brahman astrologer as soon as she could after Zulfi's birth on 5 January 1928. A close family friend, who much later saw the horoscope,[1] reported that the astrologer had "predicted everything," Zulfi's "marriages," his singular "success," and remarkable "power, up until the age of fifty," beyond which the old man refused to say any more, vaguely muttering to Zulfi's mother, "I don't know what I see!"[2]

"Bupa," Zulfi's baby name, was a frail child, "always ailing," his elder sister Manna recalled, frequently down with head colds during Larkana's brief but often bitter winters. At the age of 3 he ran so high a fever from influenza, possibly compounded by malaria, that "we almost lost him."[3] Zulfi's mother rushed off in desperation to pray at the shrine of Lal Shahbaz, and was later convinced that her son's swift recovery after she returned home was thanks to Sind's greatest Sufi saint.[4] More than forty years later, as prime minister of Pakistan, Zulfi prayed at the same shrine, and ordered a pair of solid gold doors to be installed there.

No familial bond in South Asia is closer than that which ties a son to his mother, yet Zulfi's deep emotional attachment to his young, often lonely mother was even stronger, it seems, than the norm among families as large as the Bhuttos. Not only was Khurshid the younger bride, which naturally strained her relations with Shah Nawaz's first wife and her tight family coterie, but her "outcast" birth and religious differences compounded her sense of ostracism in that clannish Larkana family. Her only local friends, indeed, were the young Sindhi wife of Larkana's superintending engineer, Mrs. G. N. Mirchandani, and a Hindu neighbor, Mrs. Shivdasani. Her first child was a comfort to Khurshid, but daughters rarely raised a rural mother's prestige or family power because with marriage a daughter must leave the home of her birth and generally take a substantial dowry with her, rarely returning again, except when in need of help or care. A son, however, could be a prince! He might one day become Wadero Khan Bhutto, even though his elder half brother, Imdad Ali, was the designated heir. Zulfi's mother, moreover, knew that her son was destined for greatness! Was he not the brightest, sweetest, dearest child ever born? She adored, clung to, and hopelessly spoiled him as only a desperately lonely, virtually outcast insecure young mother can spoil a darling son. Wherever Khurshid went "Bupa" Zulfi went with her, first at her breast, then on her hip, and once he could walk, holding her hand, never beyond her reach. It was as if she feared that losing touch with him would sever her own lifeline to the Bhutto world. Her busy husband rarely saw her any more, and most of his clannish family viewed her as an intruder from a hostile, remote world, neither Sindhi nor Muslim by birth, who had shamelessly "seduced" the old wadero, dancing her way into his heart.

Zulfi's mercurial temperament, the strange sudden shifts in his moods, his suspicious ambivalence toward most people, even his closest "friends" and colleagues, may in some measure be traced to his doting mother and aloof father, and to the inherent incompatibilities of their worlds and natures, one basically marginal and insecure, the other overconfidently arrogant. His genetic code was at any rate a curious blend of polar qualities: Shah Nawaz's feudal pomp and power, Lakhi Bai's low-caste fears, lack of familial fortune, deep-seated doubts as to her very worth, her own tenuous toehold in the Bhutto world she had entered by a side door of passion left ajar. How many of her darkest hidden anxieties had this mother conveyed to her frail, sensitive child? No records survive of her whispered words to him in the long days and nights of their unbroken intimacy, yet Zulfi's mother surely communicated more than her mystic faith in astrology and the power of prayer to immortal saints to her only son. All the superstitious, traditional caste fears of Hindu India, her ingrained belief in rebirth and karmic pollution, in the impact of every deed, and the dark dangers of mixing forbidden human fruits and duties, dangers worse than death, for death itself could mean release from rebirth, the highest form of happiness, escape from the traumas of life and all the hateful curses of jealous "enemies." Surely she must have conveyed through her small body's mother's milk itself her deepest fears of strange, foreign "enemies" everywhere, within the very walls and shaded rooms of the house they inhabited, among his father's closest relatives, his jealous first wife and her kitchen coterie of "spies" and servants.

All the clashing pluralisms of South Asia itself, a riverine blend of Hindu-Muslim, highborn and outcaste passions, conflicts, harmonies, and hatreds flowed in Zulfi's blood, an inharmonious mix of the light and dark, a microcosmic reflection of Islam's violent assault upon and unstable conquest of the body politic of Hindudom. And added to that impossible blend, of course, was the superimposed external world of an English Christian West, whose patina of polite privilege and imperial arrogance had been donned with knighthood and knee breeches by his proud, pompous prince of a father, magistrate-wadero of Larkana, Sind's Mughal on the Imperial Council, a red-cheeked English governor's boon companion and right-hand man. Was it any wonder Zulfi should be so strangely complex, inherently incompatible within his many psyches, forever unsure of which Bhutto he was? Hindu outcaste? Sufi saint? Or the *Zul-*

fikar—Sword—of Islam? The Sword—*Zulfikar*—that first belonged to Caliph Hazrat Ali, and was used to liberate Arabia's poor, converting them to Islam.

Zulfi attended the convent kindergarten school at Bishops High School in Karachi for just a few months, after he turned six, before Bombay's governor, Lord Brabourne, invited Sir Shah Nawaz to join his cabinet in 1934. Sind's provincial separation had been agreed upon in London but would not take effect till elections could be held under the new Government of India Act in 1937. Lord Brabourne knew Sir Shah Nawaz well enough, however, to want his immediate advice on all matters affecting the Muslims of Sind as well as Bombay. So the Bhutto family left Karachi late in 1934 on a P & O liner, steaming for three days and nights across the Arabian Sea to beautiful Bombay harbor, the British raj's cosmopolitan gateway to India.

As the governor's leading Muslim minister, Sir Shah Nawaz was given a splendid home atop Malabar Hill, Bombay's most gracious residential area, overlooking the bay with its sugar-loaf islands. Young Zulfi enjoyed his big new home, playing hide-and-seek in its many spacious rooms with his sisters. His father's third and only other child, a younger daughter named Benazir (meaning "Matchless"), was just 3 at the time, old enough to join in some games. Zulfi adored her, as only an elder brother can feel about his lovely little sister. But his first Benazir's health was to prove more fragile than his own had been. She would succumb to influenza at her boarding school in Pune at the tender age of 14. Zulfi visited his sister's grave whenever he came to India, and when his first child was born a decade later, insisted on naming her Benazir.

For all his power and natural sagacity, Sir Shah Nawaz, who never graduated high school, became keenly sensitive about his lack of formal education after moving to Bombay. Daily contact with Lord Brabourne and His Highness the Aga Khan, both of whom lived nearby, and with his cultured Parsi commercial neighbor, Sir Homi Mody, and Jawaharlal Nehru's charming sister, Krishna Nehru Hutheesing, who lived next door, made him feel acutely conscious of his intellectual "handicap." Shah Nawaz resolved, therefore, to spare no resources to assure that his brightest son should receive the best possible education money could buy. Zulfi's mother, however, was determined to keep her favorite at her side as long as possible.

Sir Shah Nawaz managed to enroll his younger son in Bombay's Cathedral High School in 1937 at the age of 9. Yet though Zulfi had reached that school's eligible age of admission, Colonel Hammond, Cathedral's principal, thought that in view of Zulfi's having been "taught at home" before this, it might be best for him to "join the Girls' School before he could be admitted" to Cathedral Boys' first standard.[5] Young Zulfi, however, "revolted" at the idea, and the British colonel liked his "spirit," patting him on the back, admitting him to Boys' first without further demur. By thus striking a "pose" of bluff self-assertion and self-confidence, Zulfi got what he wanted. It became the pattern of his life in times of crisis. The young man soon entered Savage House at posh Cathedral High and took pride in his handsome green Boy Scout uniform with its broad polished leather belt, short pants, and high black woollen socks held to his calves by shiny green ribbons tied above bright white sneakers.

Zulfi was still painfully slight, "only skin and bones held together by a squeaky high-pitched voice," as his friend at Cathedral High, Piloo Mody (1926–80), recalled, perceptively finding "his whole make-up . . . somewhat incongruous."[6]

Piloo was to finish his secondary education at the more prestigious British Indian Doon School in Dehra Dun's hill station, and would then return to Bombay, to the new palatial mansion his father had built for the family on Carmichael Road, across the way from the Bhutto's Ghia Mansion. Piloo's cousins, Silloo and Jehangir Jal Mugaseth, and Omar Qureshi and his cricket star friend Mushtaq Ali, were Zulfi's companions in Bombay. Those were cricket and film-watching years of "fun and joy," a "carefree life devoted to much pleasure," Piloo reminisced. Zulfi's subsequent passions for fine food and wine, expensive clothes, and buxom women blossomed then. He worried much less about his studies than Piloo, failing the first senior Cambridge examination he took in 1945. By then he had other distractions from schoolwork.

Zulfi was only 13 when he was first married to an older cousin, Sheerin, whose married name is Amir Begum. She was one of three daughters of his wealthiest uncle, Khan Bahadur Ahmed Khan Bhutto, one-third of whose wadero estates would be inherited by Zulfikar after his father-in-law's death. The bride was much older[7] than her adolescent groom, but he took her to Kashmir's Srinagar for their honeymoon and handed all bills for that happy holiday to his father-in-law, who reluctantly paid. Zulfikar assured his second bride, Begum Nusrat, on the eve of their marriage that his first marriage was "purely for the property." He insisted that he felt "no physical attraction" for his cousin-wife, who remained at Al-Murtaza in Larkana, coming only briefly for special family reunions to Karachi. Some close friends, however, recalled that he not only enjoyed marital intimacy with his first wife, with whom he slept during his frequent visits to Larkana, but that shortly after his daughter Benazir was born, Amir Begum also "gave birth to a daughter." That was later, however, after he returned from his university years abroad. Zulfi's closest relatives all deny that he had any offspring from his first wife, who still lives in Larkana, though she now "comes quite often to Karachi."[8]

As president of the Sind Mohammadan Association from 1925 through 1934, Sir Shah Nawaz had led his province's politically conscious Muslims for most of the decade prior to his departure for Bombay, and confidently expected to win election from Larkana to Sind's new Legislative Assembly as soon as elections were announced. He was elected leader of the newly organized Sind United Party, all of whose members were Muslims. The deputy leader was Sir Haji Abdullah Haroon (1872–1942), Karachi's leading entrepreneur. The party wisely offered cooperation to Karachi's Hindus, thereby attracting many more votes than the competing Sind Muslim Party, led by Sir Shah Nawaz's arch rival for control of Sind, lawyer Sir Ghulam Hussain Hidayatullah (1879–1948), who had the support of Larkana's other wealthy wadero clan, the Khuhros. When all the votes had been counted for Sind's new assembly late in February 1937, Sir Shah Nawaz's United Party won a clear majority. As his party's leader, of course, Sir Shah Nawaz should have been named the first chief minister of Sind.

The new British governor, Sir Lancelot Graham, had, indeed, been ready to do just that, but much to the governor's amazement, Shah Nawaz lost his own Larkana seat to a virtually unknown young man, Sheikh Abdul Majid Sindhi, who had campaigned vigorously with the strong backing of Wadero M. A. Khuhro (1901–80). Overconfident Shah Nawaz had hardly bothered to return to Larkana during the campaign. That rejection by his own people came as so shocking a political blow to him that he opted to quit politics rather than accept Governor Graham's offer of appointment to his cabinet without a seat in Sind's assembly.

"I thought all was going well," Graham reported from Sind's Government House in Karachi to Viceroy Lord Linlithgow in Delhi that March of 1937, "but it fell to pieces in two or three days over the division of the spoils; and I have decided to drop Shah Nawaz as he said that he could not be happy with Sir Ghulam Hussain as Chief Minister. . . . Hussain who is easily the most competent man in the Province has agreed to advise me. . . . There's a lot of intrigue going on about the speakership."[9] Siasat! Young Zulfi may have taken much too much to heart the lesson of his father's election defeat, resolving even at his tender age never to risk losing an election, no matter how high a price need be paid to insure victory.

Sir Shah Nawaz, however, by then too tired for public contests, accepted defeat philosophically. Never again would he subject his fate to the will of a fickle majority. Bombay's governor liked him so well that he offered him the plum of membership on the prestigious and powerful Public Service Commission for Bombay and Sind, a post he held and would chair in comfort till 1946, when he left Bombay to become *diwan* (prime minister) of Junagadh. Ironically, the father of Sind's independent provincial status thus found no seat in the newly elected government of Sind convened in Karachi in 1937, just a decade before that fast-growing port city became the capital of the new nation of Pakistan.

Since 1930 at least the idea of a separate Muslim Nation, called Pakistan (Land of the Pure), had been mooted by Muslim leaders of British India, most important among whom was the great Urdu poet-philosopher, Muhammad Iqbal (1877–1938).[10] Iqbal, who presided over the Muslim League's meeting at Allahabad in December 1930, called for the amalgamation of Punjab, the North-West Frontier Province, Sind, and Baluchistan into a "single State" as the "final destiny" of "Muslims, at least of North-West India." A few years later several "students" in Cambridge, whose leader was Rahmat Ali, started a Pakistan national movement, and wrote widely circulated pamphlets about Pakistan as the "Muslim Fatherland," demanding its immediate creation, "a demand based on justice and equity."[11]

It was not until 1937 that Mohammad Ali Jinnah finally lost faith in the possibility of reaching a political agreement with Mahatma Gandhi, Jawaharlal Nehru, and their National Congress, which seemed to him more arrogantly "Hindu" in the wake of election victories. That October of 1937 he donned his Persian lamb Jinnah cap and long black sherwani coat for the first time, in Lucknow, to urge more than five thousand Muslim delegates to unite and work "for the cause of your people and your country." Jinnah assured his followers that with 80 million Muslims they had "destiny in their hands" and "nothing to fear" from a "Hindu Congress." Two and a half years later, in March 1940, Jinnah addressed a much larger audience in Lahore, where the famous resolution, known to history as the Pakistan Resolution was adopted. "The Musalmans are not a minority. The Musalmans are a nation," thundered the League's Great Leader. "The problem in India is not of an inter-communal but manifestly of an international character. . . . The only course open to us all is to allow the major nations separate homelands, by dividing India into 'autonomous national States.'" For the next seven years, till the Dominion of Pakistan would be carved out of once-united British India with partition of Bengal as well as Punjab, this Pakistan demand was to remain the single platform of Jinnah's Muslim League, which after the end of World War II spoke for India's Muslim majority.

During the war years Shah Nawaz met Jinnah often at the office of their Bombay physician, Dr. Jal R. Patel, on Hornby Road. Shah Nawaz's heart and various stomach ailments, including ulcers, kept him an almost daily visitor to Bombay's most sociable "chronic-bachelor" doctor, whose "lavish" afternoon teas were as renowned as his medical reputation. Jinnah had fatal spots on both lungs before war's end. But for Dr. Patel's professional silence about the X rays he had taken of them, Pakistan might never have been born, at least not in 1947. If Nehru had realized how precarious the health of his arch adversary was, he might well have waited, instead of agreeing to Lord Mountbatten's proposed partition plan. But the urbane, diminutive Dr. Patel, a lecturer in Bombay's Medical College, had taken his Hippocratic Oath to heart. He was, moreover, specially indebted to Sir Shah Nawaz, who as Bombay's minister of health, "accelerated" the young doctor's "promotion." Patel's patients included most of Bombay's great names of the time, Parsi industrialist, J.R.D. Tata, and Sir Cowasjee Jehangir, as well as Jinnah and Sir Shah Nawaz Bhutto. Zulfi's Cathedral High School was nearby, on Outram Road, and if he was not watching cricket or going for a swim after school at four, he would walk over to Hornby Road and climb the stairs to join Dr. Patel's clinic tea party, which started daily at 4:30 and usually lasted till 6:00 P.M.

Zulfi generally sat with "my mouth shut," he later recalled, ears wide open, however, as he listened to Sir Cowasjee and others press Jinnah "to define Pakistan in territorial terms," or when wealthy Hindu banker S. D. Schroff argued that a separate Muslim nation could never be "economically viable." Jinnah's answers to such questions, Bhutto remembered, were "the kind of replies that made me feel proud of him."[12] Zulfi wrote his first letter to Jinnah in 1945, while he was on holiday with Piloo at posh Mussoorie Hotel. At 17 he may have felt twinges of political guilt for going off during school break to eat "enormous quantities of food" and take "lessons in ball room dancing"[13] while so many of his less fortunate contemporaries, who had taken a stand for Indian independence throughout the war, were still languishing behind British prison bars, or had suffered broken bones from police *lathi*-charges.

"Musalmans should realize that the Hindus can never and will never unite with us, they are the deadliest enemies of our Koran and our Prophet," wrote young Zulfi to his Great Leader on 26 April 1945. "You Sir, have brought us under one platform and one flag, and the cry of every Musalman should be 'onward to Pakistan.' Our destiny is Pakistan, our aim is Pakistan. . . . Nobody can stop us, we are a Nation by ourselves and India is a sub-continent. . . . You have inspired us and we are proud of you. Being still in school I am unable to help the establishment of our sacred land. But the time will come when I will even sacrifice my life for Pakistan."[14]

The war's end in Europe that June precipitated a change of government in London. Winston Churchill's Tory-led cabinet was displaced by Clement Attlee's Labour leadership, whose cabinet mission flew to India determined to find an acceptable constitutional formula through which to transfer British power to indigenous hands. The three-tiered federal constitution devised by Sir Stafford Cripps, Lord Pethick-Lawrence, and Admiral A. V. Alexander was after hard negotiations accepted by Jinnah for his League, as well as by the Congress High Command under Maulana Abul Kalam Azad's (1888–1958) presidency. Azad then made what he subsequently called "a mistake" of "Himalayan dimension,"[15] nominating Jawaharlal Nehru to succeed him as Congress president, for that July, Nehru at his first press conference in Bombay announced that Congress would enter the forthcoming Constituent Assembly "completely unfettered by agreements," rejecting the basic premise on which the cabinet mission's delicately constructed plan had been erected. Nehru's remarks to the press undermined Muslim confidence in Congress's willingness to work the plan without prejudice. Jinnah lost the last shred of faith he had in the "higgling *banyas*" of Congress and most regretfully felt obliged to say "good-bye to constitutions and constitutional methods." He called upon his League and the Muslim Nation behind it to take "Direct Action" a month later, launching a tragic full year of communal civil war in mid-August 1946, South Asia's bloody prelude to Partition and the birth of Pakistan.

Zulfi was invited to Jinnah's Malabar Hill home with a "small group" of Muslim students to advise their Quaid-i-Azam about how best to launch direct action in Bombay. "Every one talked in circles and vague language," Zulfi Bhutto recalled thirty-two years later in his death cell. "I remarked that Bombay was a Maharashtrian stronghold and Elphinstone College was a student fortress of the Maharashtrian militant students. Some kind of a strike in Elphinstone College would have tremendous psychological effect."[16] He knew the son of Elphinstone's principal, Arun Seal, his classmate at Cathedral High, so together they rounded up about two hundred local students, who staged a sit-down inside, blockading the college entryway. The police were called, but the principal, anxious to avoid hurting his son, closed down the college instead of asking the police to clear a passageway. Jinnah was "very pleased" because all Bombay newspapers carried reports of the "Direct Action protest," giving Zulfi's baptism in a long career of political action "publicity . . . quite out of proportion to the achievement."

Sir Shah Nawaz also enlisted in the Pakistan movement in 1946, resigning his Public Service Commission chair to move to Junagadh State. Veraval, that princely state's port capital, is some 400 miles south of Karachi on the Arabian

Sea. Jinnah and Shah Nawaz had agreed that it was worth trying to wrest wealthy Junagadh from Indian control, especially because it was so close to Mahatma Gandhi's birthplace. The Muslim nawab accepted his new diwan's advice, but Sardar Patel, himself born in Gujarat, had not the slightest intention of ever letting Junagadh leave the Indian union.

Zulfi remained alone in Bombay after his father and mother moved with their daughter to Junagadh. He then lived in "The Nest," their home on Carmichael Road, and watched over another Bhutto property in Bombay's busy Fort area, the multistoried Hotel Astoria Building, net rental from which was more than 110,000 rupees, a not insubstantial sum in 1946–47.[17] There were many securities, other valuables, and cash left in the hands of the bright and precocious 18-year-old Zulfikar, who was a decade later to become sole heir to his father's fortune. With countless cricket matches to watch by day, and several posh clubs to visit nightly, Zulfi had his hands full. He was also obliged to retake his senior Cambridge examination, having failed it in 1945. He hoped to continue his education abroad, and though that initial failure made it impossible for Zulfi to enroll in any college of Cambridge or Oxford immediately, he might still manage to finish at one of those most prestigious schools, if he did well for a year or two at some easier-to-enter college in the United States.

With the war over in Europe and the havoc of pre-Partition communal riots spreading across South Asia, Zulfi, Piloo, and their wealthy Bombay friends all resolved to go abroad for higher education. Piloo decided to study architecture and felt that the United States would prove more fertile ground for that than England. He applied to Harvard, the University of California at Berkeley, and the University of Southern California (USC). He was admitted to all three but chose Berkeley because Harvard was "too big" and USC was "a school for playboys."[18] Zulfi applied to the same schools but was admitted only to USC. On 5 September 1947, less than a month after Pakistan's traumatic, blood-scarred birth, Zulfi Bhutto flew west with Piloo's cousin, Jehangir Jal Mugaseth.

3

Brief California Interlude
(1947–1950)

"When the skyline of New York became visible I was filled with excitement and expectancy," Zulfi recalled. "We saw the Statue of Liberty and took . . . pride in it . . . because it personified liberty. And for our naive minds, freedom had the connotation of unrestricted and unfettered right of the individual to live free and unmolested."[1] Zulfi reveled in the nightlife of New York in the movie houses, restaurants, and bars around Times Square and along Broadway, but after a week he headed west, to Los Angeles.

By the time he reached Los Angeles, Zulfi sorely missed his servants. He shared a room with his cricket-loving buddy, Omar Qureshi, in Bess Jones's rooming house on South Flower Street, just a few blocks from the USC campus. Al Cechvala, a composer and now professor of music living in Glendora, also lived there, where room and board cost just $13.50 a week. Cechvala vividly recalls Zulfi's first cry from the top of the stairs, outside the bathroom door: "Who down there is going to draw my *ba-a-ath?*"[2] Much of Zulfi's daily life in his new home came to this spoiled scion of Sindhi aristocracy as a rude awakening. Though Nazism and fascism had been defeated abroad, racial prejudice and Jim Crowism were still alive throughout the United States. Zulfi had always considered himself to be a member of the "top crust" of Indo-Pakistani society: suddenly he found himself being treated, on campus and off, as a "Mex" or "nigger." The only interracial off-campus fraternity at the time was the night school's Kappa Alpha Xi, which Zulfi joined under Cechvala's brotherly sponsorship. Because there was no cricket being played at USC, Zulfi and Omar signed up for tennis, at which both excelled. The abundant nightclubs along the Sunset Strip and their risquê girlie shows provided much more entertainment than the relatively tame clubs of Bombay.

Zulfi dated many women in California. He was by nature singularly passionate, and finding himself so liberated from all social and familial constraints, quickly lost whatever inhibitions he may have felt in Bombay or Larkana. None of his new friends had any idea, of course, that he was married. Slim and handsome, he had grown a Clark Gable-type moustache, was endowed with a full head of wavy hair, and always dressed nattily. He was quite the dapper rake.

Zulfi dated Margaret, an art major who later would become Cechvala's first wife. She subsequently confided to her husband that she had found Zulfi much "too impatient with women." He'd always left her "dissatisfied," and she soon came to feel that their relationship was "meaningless" to him. It was a not uncommon female complaint about the swift sexuality of adolescent males, hardly unique to Zulfi, yet what was unusual perhaps is how often, how long, and how late in Zulfi's life that pattern of instant passionate desire and equally sudden loss of interest in its object seems to have persisted. Did that never-resolved ambivalence toward women reflect perhaps the instant gratification of love his mother's ever-present breast and tender hand had always provided during his purposely prolonged infancy, clashing with his mature consciousness of guilt at still somatically needing such succor?

Cechvala also introduced Zulfi to Mary Ellen, a hometown friend who was a widow in her early thirties, with two small girls, when Zulfi dated her. He liked coming to her apartment for dinner, dressed in Indian costume, feeling perhaps as if he were almost in Larkana during family evenings, especially because Mary Ellen was "so much older," like Amir Begum. Zulfi, who was just over 20, tried to pass himself off to Mary as "twenty-five or twenty-seven."[3] Thanks to his mustache, physical precocity, and English accent, she readily believed him. Anne Reynolds was another of his girlfriends, as was Leili Bakhtiyar, with whom he loved to walk on the Santa Monica beach. "He said he would like to marry me," Dr. Bakhtiyar, now an eminent physician recalled, "and it was a very unusual kind of proposal, because there had not been any kind of intimacy."[4] To demonstrate his sincerity, Zulfi gave her a "very beautiful" watch and a sari. He seems to have felt more tender toward her because she was Persian and "he had this fervor of pan-Islamism." He often talked of his dream of turning a unified bloc of all Islamic nations into a "major force against the great powers." Bakhtiyar was only 18 at the time, and "it scared me," she remembered; he was so "fanatical about it."

Zulfi never declared a major during the two years he spent working toward a bachelor's degree at USC, but he took as many political science courses as possible and joined the debating team. He argued in a debate in early February 1948 that "as a citizen of the world, I consider it my solemn duty to try and convince people that our salvation lies in one world."[5] He was impressed by Wendell Willkie's global consciousness, and his bright mind tried to integrate all the new things he was experiencing with the image worlds of Larkana, Karachi, and Bombay that he carried in his genes and head. "For the sake of humanity," Zulfi reasoned, "let us re-examine without prejudice the norm of our politics. Let us close the bloody chapters of war and engage ourselves in harmonising our people. Let us erase from our minds the crazy nation-centric notion leading to fanaticism and intolerance." He passionately prayed that "the weapon of our one world will not be the atomic bomb, but the weapon of love and . . . the creed should be that of simplicity—the simplicity which Prophet Mohammad expounded so effectively in the deserts of Arabia."

On 1 April 1948 Zulfi spoke at USC of "the Islamic heritage" as one of "my own accomplishments, for I genuinely consider any accomplishment of the

Islamic people as a personal feat, just as I consider a failure of the Muslim world as a personal failure. There is something binding about the Muslim world in spite of the fact that it is torn by dissension. . . . I am not a devout Muslim. . . . My interest is soaked in the political, economic and cultural heritage of Islam."[6] He then gave a brief exposition of the early history and doctrinal development of the Islamic faith and its spread the world over. "Our position is pathetically unstable," Zulfi continued, reflecting perhaps his own condition. "Imperialism has sapped our vitality and drained our blood in every part of the globe. . . . The young generation of Muslims, who will be the leaders of a new force, of an order based on justice, wants the end of exploitation." Under an Islamic confederation, the future world, at least "the disciples of Mohammad," would be more secure. "Destiny demands an Islamic association, political reality justifies it, posterity awaits it, and by God we will have it. Courage is in our blood; we are the children of a rich heritage. We shall succeed."[7]

Zulfi often told his roommates that when he went back to Pakistan he would start a pan-Islamic league. "Then it dawned on me later," Cechvala reflected, that "this guy really likes power . . . he enjoyed the thought of being in charge." Zulfi often spoke of Jinnah to his friends, saying, "That's *my* man! That's my idol, the man I respect." Not that any of them, except Omar, had ever heard of Pakistan or of Jinnah.

Zulfi was a keen fraternity brother, initiated by hanging inside a dark potato sack for half an hour next to other initiates, shouting "Hey, what's going on?" As a new member, he was expected to do the usual dirty work. On the eve of the school's annual football game with the University of California–Los Angeles, Zulfi, Cechvala, and Joe Phillips, who drove the "getaway car," ventured into UCLA's new campus in what was then quiet Westwood Village. Zulfi and Cechvala's job was to paint "OUTHOUSE" on any UCLA walls that they found. Then, racing toward the waiting car, Zulfi tossed his paint can and reddened brush into UCLA's fountain. The three miscreants broke every traffic law as they sped home.

Piloo Mody spent several weekends with Zulfi; the New Year's Day Rose Bowl game and parade of roses in Pasadena were exhilarating to the two students. Zulfi returned those visits, venturing north to Berkeley, where he happened to be just after Mahatma Gandhi's assassination on 30 January 1948. Young Piloo felt "fatherless, countryless and creedless . . . and it was quite some time before I could bring the tears and sobs under control."[8] Zulfi "consoled me and nursed me out of my condition, sharing my sorrow," Piloo recalled, adding that when Jinnah died eight months later, "I did the same for him."

Shortly after Jinnah's death in early September 1948, Zulfi wrote to M. A. H. Ispahani, Pakistan's first ambassador to Washington, one of Jinnah's closest friends: "[W]e have been orphaned at this crucial moment when we needed more than any other force, the torrential magnanimity of our beloved Leader. . . . [T]hough the Quaid is no longer with us, yet his pure and virgin spirit will remain forever fertile in our mind. . . . His entire life was a struggle for the betterment and emancipation of his people."[9]

Zulfi spent some time during his summer holidays as a volunteer in the Pakistan embassy in Washington. One of the papers he wrote there dealt with the man he revered: Jinnah

> is solely responsible for the creation of a State for those whom he led in the struggle for the emancipation of their lives. . . . His dream of creating a Pakistan has been a great dream, the realization of his dream has been nothing short of a miracle, for it has been an achievement carried out single handedly. He has led a people who were thoroughly derelict and disunited and depressed. He has been a God-inspired Man, a man of purity of heart, of an unbelievable audacity and a unique courage and determination.[10]

Attempts would be made by Zia's regime in several of the *White Papers* published after Bhutto's arrest, as they had been by Ayub Khan's martial government a decade earlier, to discredit Zulfi Bhutto by insisting he continued to think of himself as an Indian citizen all the time he spent in the United States and Britain, but the assertions were obviously untrue. He did try to receive fair payment for his father's Bombay property in a High Court petition filed after returning home, but much of his time in the United States was spent seeking to educate his friends and community groups about Pakistan and its significance for the modern world. "I will not forgive you for not knowing where my country is," Zulfi told the Ventura County American Association of University Women when speaking at its International Program. "It is your duty to know where every tiny island in the world is located if you are to play your rightful part in this world." The newspaper account of the talk reported that Bhutto "felt it would have been wiser to unite India, but that the seeds of British imperialism were so deeply inculcated that an honest compromise was impossible. . . . 'There is some strife between India and Pakistan, but all countries go through that. It should be known to all that Pakistan will go along with all nations interested in world peace. We have a great destiny and are proud of it.'"[11]

Most Americans, even USC professors who should have known better, were quite confused about India and Pakistan forty-odd years ago. Diplomatic historian T. Walter Wallbank wrote in October 1948 that "one of my students from India, Mr. Z. A. Bhutto . . . has been in my history classes for a period of one year. I would say that this young man is definitely a superior scholar, perhaps even in the brilliant category."[12] Two months later, the head of the Department of Political Science, Wilbert L. Hindman, wrote in support of Zulfi's reapplication to Oxford's Christ Church College, attesting his "improving academic record" and "superior potentialities as a student."[13] After two years at USC, however, Zulfi's average was "lower than B." He would not be admitted to Oxford until after he completed his degree "with honors" in political science at the University of California's Berkeley campus in June 1950.

For his twenty-first birthday on 5 January 1949, Zulfi received two literary gifts from his father, then living in Larkana: a leather-bound five-volume set of William Sloane's biography of Napoleon Bonaparte, a lifelong role model, and

Karl Marx's *Communist Manifesto* in pamphlet form. "From Napoleon I imbibed the politics of power," Zulfi later reflected. "From the pamphlet I absorbed the politics of poverty."[14]

"We used to kid him and ask if he was a Brahman," Joe Phillips recalled, "and he just laughed and said, 'Of course!' He liked Miller's draught beer, and cold beer. . . . He watched his diet a lot though. . . . Zulfi had a good time."[15]

After two years at USC, in June of 1949, Zulfi moved to Berkeley. In a house on Allston Way, Piloo did the cooking, Yadu Kaul and Madhav Prasad washed the dishes, Zulfi made the beds, and Asif Currimbhoy and Edward Mannon[16] were supposed to clean the house. Inder Chabra and Brij Mohan Thapar later moved in. "We had communal cooking. Nobody was fussy. We all pitched in money for groceries," Inder reminisced. "Zulfi was always politically minded, a born political animal, very intense, very pro-Jinnah, pro-Pakistan."[17] Berkeley political scientist Leo Rose, a classmate of Zulfi's at the time, remembers that he "dated a lot of girls," was "very bright, and loved Professor Hans Kelsen [1881–1973] and International Law."[18]

Kelsen had recently published his Oliver Wendell Holmes Lectures as *Law and Peace in International Relations* (1942) and his most famous text, *The General Theory of Law and State* (1949). He was called by some "the jurist of our century," and had disciples at the UN as well as in Berkeley. Kelsen's "value-free science" of the law and his oft-reiterated question "What is justice?" appealed to Zulfi's passionate nature and fascination with international affairs. Kelsen's legal relativism focused on most major questions relevant to international law, "the justice of freedom, the justice of peace, the justice of democracy, the justice of tolerance." An Austrian Jewish refugee from Hitler, Kelsen championed democratic freedoms as the surest key to justice. His mentor's wisdom and brilliance sparked Zulfi to probe more deeply in sources that held no simple answers, no formula solutions to problems central to political power in the postwar world of swiftly changing alliances and realities. For the rest of his life, Zulfi Bhutto would remember what Kelsen had taught, often quoting him in debates on the Pakistan Constitution[19] and in speeches before the UN General Assembly—and finally in his three-day defense statement during his last public appearance, before Pakistan's Supreme Court.

The only course Piloo and Zulfi took together was a graduate class in the history of political philosophy, taught by Professor Lipski. They started with Socrates and Plato's *Republic* and continued through Hume and Locke to modern times, trying to formulate some "theory of international relations." Piloo recalled that a "keen competition" soon developed between the old friends, each of whom hoped to become prime minister of his home country. Zulfi wrote at this time, reflecting on his own feelings, no doubt, as well as one of Lipski's lectures:

> What is the disease that makes an animal of man on a Monday; glorifies him on the following day; turns him into a barbarian on a Wednesday; inspires him to become a messiah on a Thursday; draws forth from him a glimpse of Plato on a Friday; makes him display a streak of Machiavelli

> on the Sabbath? . . . There are those that hate you, those that hate me; and . . . those that dislike our philosophies. . . . [T]he day for my domination is still to come."[20]

Zulfi returned home for his sister's wedding in September 1949. Stopping over in New York, he invited his USC friend "Selman the Turk" to dinner at the Raja Restaurant, and the following evening they shared a Turkish meal together. They talked all evening about "everything from Pan-Islamism to import regulations, from the nostalgic memories of glittering Los Angeles, to the Young Turks' Revolution, from geology to the morals of American women." About midnight, the headwaiter asked them to leave. They "loitered about aimlessly . . . walked into a movie." Out they came at 3:30 in the morning.

> The city was half dozing . . . yet alive. . . . [E]rect, masculine skyscrapers standing undisturbed and unaffected by the events that went on inside them. . . . They seemed to be terribly human, soaring with emotion. It was a strange feeling, the strangest I have felt. Suddenly their character changed. The mask was withdrawn. . . . They began to symbolise something magnificent—the elevation of man. . . . I wondered if they had . . . become uncontrollable monsters. . . . I imagined they were capable of plotting against their own creator called man. . . . I almost shouted and rebuked the tall stone structures, calling them ungrateful. At the same time, knowing full well that I had made them, they were a source of pride to me. . . . People in primitive times were hospitable and simple. Now they are cold and aloof and complex. . . . I went on thinking . . . till I staggered into my room.[21]

Zulfi, of course, had had too much to drink that night, yet even given his intoxication, his half-hallucinatory description of "terribly human" skyscrapers "soaring with emotion" is startling in what it reveals about the state of his mind and the ubiquitous polarity of his passions and feelings. As soon as the phallic "monsters" were unmasked they became "uncontrollable" and started "plotting against their own creator," who was none other than Zulfi himself! "Knowing full well that I had made them," they were "a source of pride to me," yet "cold and aloof," as were the "people" in this city of strangers, where weary headwaiters ordered him to leave restaurants, and lovely women he smiled at or spoke to warmly were too often "cold and aloof" not "hospitable and simple"—like the women of Larkana or Karachi perhaps? Zulfi's ambivalence toward New York reflected all the polarities he felt about the United States, which had at once warmly welcomed yet coolly rejected him, giving him love, education, and hospitality on the one hand, dished up with doses of racial prejudice, mistrust, and indifference on the other. He would never reconcile those feelings. Zulfi's love-hate relationship with the United States was to remain unresolved, new "uncontrollable monsters" of suspicion looming in his mind till his last days.

That afternoon Zulfi flew home to Karachi, where his family eagerly awaited the prodigal. Manna's marriage to Mr. Islam was a splendid affair: a gala wedding followed by three days and nights of parties, luncheons, dinners, celebrations, and exchanges of gifts and greetings by both families, each visiting the other's homes, forging new alliances, finding potential brides and grooms for future family weddings.

"The first time I saw him was when I had gone to my bank to take out some of my jewelry to wear for the wedding," Zulfi's widow Begum Sahiba Nusrat Bhutto recollected. "We kept our jewels in the safe deposit vault. I went to sign and I saw Manna's mother and a man, a boy, standing in front of me. They wanted to take out some jewels also for the wedding. So his mother said, 'Oh, hello, how are you, Nusrat?' . . . [S]he said to the son, 'Oh, here, meet Nusrat, who is Manna's friend.' And she told me, 'This is my son who's come, Zulfi.' "[22]

Nusrat Ispahani was one of Karachi's most beautiful debutantes. Her Kurdish Iranian parents had migrated to Bombay, where she was born on 23 March 1929. Her two much older sisters had been born in Iran, and were more like surrogate mothers to her, pampering young Nusrat. Her father had founded Bombay's Ispahani Soap Factory, which soon exported such large quantities of soap to Iraq that he later changed its name to the Baghdad Soap Factory. A clever, energetic, enterprising man, he was the first Indian soap manufacturer to make samples of his merchandise, distributing them free of charge in villages in Maharashtra and Gujarat and winning customers wherever he went. Early in 1947 as Hindu-Muslim communal riots intensified, several Bombay Iranian friends were killed, convincing Nusrat's father that it was time to sail to Karachi.

The family moved into the grand Palace Hotel (now the Sheraton) in the heart of Pakistan's fastest-growing city, which was destined to become the nation's capital. Sir Haji Haroon and Lady Haroon invited them to stay at their palatial home, and within a few months Ispahani had bought a place at 23, Clifton, the main residential road leading from the heart of Karachi to the seashore. Nusrat's father started a silicate of soda factory in Karachi, and was able to rebuild his soap business without too crippling a financial loss.

Nusrat joined the Pakistan Women's National Guard, was good at martial drill, and soon learned to drive trucks and ambulances. A tall, slender, dark beauty, she was promoted to captain, with silver pips on her shoulders. One of her best friends, Marri Habibullah, General Habibullah's wife, had glimpsed Zulfi before Nusrat did, and reported to her on the eve of Manna's wedding, "She's got a very handsome brother!"

"So he looked and I looked at him," Nusrat reminisced about their first meeting in the bank, "and to me he didn't look very handsome or dashing that first moment, so in my mind I said I must tell my friend, 'Why did you say he's *handsome?* He's not at all handsome!' . . . Of course, later on I thought he was very handsome and I loved him." They met again at his sister's wedding, and Zulfi "just looked at me and said 'Hello,' and went away," Nusrat recalled, thinking him "shy." But during the reception "he kept dancing with me," and "tried to hold me tight, and I said, 'This is Pakistan, it's not America.'" To which Zulfi "laughed," obviously taken with this new beauty, who looked Pakistani but sounded "cold and aloof."

The reception was held at the Central Hotel and before the buffet ended, Zulfi asked if he could drive Nusrat home. She explained that she had her own car and driver waiting, and wouldn't think of going with him alone. Sounding "deflated" by her response, he asked if she would "like an ice cream." She refused. "Don't you know who I am?" he irately asked. "I've come for the holidays!" She pretended to have thought he was his "brother," but he could see she was "just fibbing," so he sat down beside her and became more keenly interested as he looked at her beautiful face and warm eyes.

"I got to know her when we started acting in the National Guard, arranging a variety program for the Red Cross,"[23] Nusrat's friend Hushmat Habibuddin, the charming niece of Pakistan's second governor-general, Bengali Khwaja Nazimuddin (1894–1964), recollected. They went to rehearsals together. Hushmat's father, Khadesha Habibuddin, was the minister of refugees and rehabilitation in Liaquat Ali Khan's cabinet—a "very open-minded" man, much more liberal then most of his contemporaries. Begum Liaquat Ali, who was another Iranian beauty, asked Hushmat's uncle, "I want my guards to swim. Can they use the governor-general's pool?" Nusrat and her sister, Zinnat, also came to Government House to swim and "sometimes they would leave their jewelry in the aide-de-camp's room and come back again for it." The army ADC was especially tall and handsome, and being a "tall lady," Nusrat was "also interested in him," though Hushmat would marry him, the now-retired Brigadier Noor Hussain of Rawalpindi.

Two days after Manna's wedding, Zulfi arranged for a married friend's wife to invite Nusrat to dinner at one of Karachi's best restaurants. Nusrat found Zulfi waiting with their chaperones, and before the evening ended, he complained of how difficult it would be to "wrack my brains" to find other married friends they both knew whenever he wanted to see her. "Then he proposed to me, and I took it as a joke," Nusrat recalled. She knew that in a week or two he would return to Berkeley to complete his degree. He would neither forget nor retract his proposal, however. Distance and time served only to strengthen Zulfi's resolve to make the lovely cosmopolitan Iranian lady his second begum.

Upon his return to Berkeley Zulfi launched his political career, running for the representative-at-large seat on the Student Council—the only foreign candidate in a field of seven. His populist platform called for more complete "integration of foreign students in student affairs"; opposition to the California loyalty oath, a product of the Cold War witch-hunt mentality then sweeping the United States; and a "Fair-Bear" (a bear is Berkeley's symbol, as it is California's) minimum student wage of a dollar an hour. Zulfi was in his element, badgering all his friends to work "very hard," in Piloo's words. Zulfi Bhutto became the first Asian member of the council (forty years later Chang-Lin Tien would be appointed the University's first Asian-born chancellor). Zulfi's growing political consciousness and liberal democratic ideals were given expression the following year when he volunteered in the 1950 campaign of Democrat Helen Gahagan Douglas for a seat in the U.S. Senate against Republican Richard M. Nixon. It was during that infamous campaign that Nixon was first dubbed "Tricky Dick." Unfortunately, Zulfi learned more about the kind of politicking he later employed from Nixon's deplorable tactics that fall than he did from his contact with the more principled losing candidate for whom he worked.

On the eve of his departure from California for England, in the fall of 1950, Zulfi reflected on what he had learned: to "live by a profitable absence of scruples if we are to be successful politicians. We have to do what others do to us, but we must do it before the others have the opportunity." These were Machiavellian lessons he would soon apply in every corridor of political power he entered in Pakistan. "We have also seen the tragedies of Jim Crowism in Lincoln's America," Zulfi noted, ambivalently concluding: "We have seen the warmth and tenderness of the citizens of this great democracy, and we have seen the crass and naked discrimination against a certain sect of people."[24]

Zulfi continued dating Leili during weekends in Berkeley, whenever she flew up from Los Angeles. He had another special girlfriend (in addition to his many "one-nighter" dates) in the Bay Area: Carolyn. She wrote regularly to him at Oxford for more than a year. Late in 1951:

> It has been obvious to me for quite some time—long before your last letter—that the feeling you once had for me . . . is quite dead. . . . It strikes me as being rather ironical that you were the one to change your mind—the same Zulfi that worried so about the sincereness and durability of my love for him. . . . Perhaps it is for the best, for the problems we would have had to face most surely would be great ones. . . . I would rather not have my cross back—keep it, for when you are old and can do no more than reflect, you can remember how once you captured the heart of an American girl.[25]

"Although Zulfi is perfectly capable of using people to gain his own ends," Piloo noted, in trying to "evaluate the character" of his best friend, "he is almost as willing to allow others and particularly his friends to use him even when he realises it. The candour with which Zulfi can admit having tried to fool someone, and failed, is what ultimately induces people to trust him. . . . He normally sought approval from his friends about the way he looked. Even his casual appearance was studied." As to "speech," he would "revel in language and concentrate on effects and when necessary would exaggerate or overstate a proposition. This was also for effect." Zulfi was ever the master of rhetoric, uniquely able to bring huge crowds to a fever pitch of emotion. He excelled, moreover, in the art of sarcasm, even aiming barbs at his closest friends, including Piloo Mody.[26]

Zulfi called himself a "socialist" during his last year at Berkeley, reading Harold Laski's *Grammar of Politics,* as well as the works of Marx and Engels, and such Fabians as Jawaharlal Nehru, who was one of his early heroes and long a political role model in several respects. He was always a militant advocate of pan-Islam; nationhood for Palestinian Arabs; freedom from French rule for Morocco, Tunisia, and Algeria; and independence for Vietnam. The Student Council was a forum in which he tried to educate fellow students on important contemporary issues. Zulfi was confident of his destiny: a leader of Pakistan, a world statesman. He viewed the UN as yet another, larger forum for his most advanced ideas in international affairs, and was ready, even in his early twenties to take a stand there. He would lead not only his countrymen but all Muslims, indeed, the entire Third World toward a bright socialist future that carried no trace of Western domination.

"We are not brute iconoclasts of systems," honors graduate Zulfi Bhutto wrote as he was about to leave for Oxford.

> Despite some of our disturbed thoughts, in the space of three breezy years, we have grown to love this Country, our University and the many other fine and attractive facets of this immense continent. . . . We are deeply in love with the great good that we have seen all around us. Above all, we shall desperately miss . . . the people. . . . [T]he cute waitress Suzie, . . . the saucy Carolyn. . . . and the many other wonderful friends.[27]

The nostalgia was followed by his vision of the world and its dominant pattern of change, linked to his destiny as a leader:

> The peoples of the World are becoming more outspoken and vehement about putting an end to all the vestiges . . . of destruction and carnage. . . . We have had enough of the destruction of our Londons, Berlins, and Stalingrads. . . . Let us, therefore, unite in thought and purpose and be the harbingers of a lofty ideal—the salvation of Mankind. Let us learn not to think only in terms of America, Anglo-Saxon or Protestant, but in terms of World harmony and co-operation. Then only can we avoid the harlotry of the past and the suspicions of the uncertain future.[28]

4

From Oxford to Karachi

(1950–1957)

Zulfi's first formal meeting in Oxford was with Christ Church College counselor Hugh Trevor-Roper (now Lord Daker), who asked the young man what he wanted to study. "I said, I wanted to study jurisprudence and law," Zulfi replied, upgrading his undergraduate record as he added, "I had a summa cum laude from the United States in first class and am, therefore, entitled to finish my schooling in two instead of the usual three years." Trevor-Roper was not quite as easily impressed by Zulfi's bluff confidence as Principal Hammond of Bombay's Cathedral High had been, simply asking the new boy if he knew any Latin. "I replied in the negative and he advised me to do my course in three instead of two years. Latin was a compulsory subject and if I failed in Roman Law, I would fail altogether. Who would not have wanted an extra year at Oxford?" Zulfi rhetorically asked Pakistan's Supreme Court as he started three days of passionate pleading for his life in December of 1978. His Oxford education proved he had the "character" of "a gentleman."

". . . [W]hen I was leaving, [Trevor-Roper] said to me, 'You know, even the best brains of our own boys would not be able to do it in two years.' I turned back and said that I would do it in two years because of what he had said and I would show him that I had brains as good as their boys, if not better. And I did it in two years and got high honours. . . . I am a cultivated and educated person."[1] Zulfi's ego could never ignore even the slightest challenge or insult. Trevor-Roper's Oxford hauteur and condescension in mentioning "even the best brains of our own boys" was precisely the sort of gauntlet that blazed scarlet in Zulfi Bhutto's Sindhi-siasat consciousness.

At a later time Zulfi revealed to Piloo Mody that he still saw "Latin in his dreams." Zulfi had in fact crammed Latin so intensely that he developed "a repugnance for learning languages that is with him to this day," Piloo reported in 1972.[2] Yet he proved to Trevor-Roper that he had pluck enough to try to match the best British brains, a feat that so favorably impressed the Oxford don that he soon became one of Zulfi's boon companions and remained a lifelong friend.

"I remember on one occasion in early '52," Mumtaz Bhutto recalled,

> when he turned up at London and rang the bell . . . at about 3 o'clock in the morning, and . . . was there with the censor of Christ Church, Hugh Trevor-Roper . . . and Robert Blake, who's now Lord Blake, and they came in and sat down and asked for drink. . . . I distinctly remember the remarks of these two distinguished gentlemen. . . . Trevor-Roper said, "Young man, watch your cousin, he's going very far!" So at that early stage, when he was a student, and in many ways a reckless one, those who were teaching him had this opinion. . . . [H]is tutor at Christ Church, Mr. Grant Bailey, also had a very high opinion of him. . . . His *presence* was there. When he entered a room, all heads would turn, and I felt the same way. I accepted him as a leader, and as a man who would have a say in the destiny of the country.[3]

Zulfi's self-confidence, at least his appearance of assurance and complete command of a situation, whether political, social, or personal, won admiration and support from a remarkably wide spectrum of people the world over. He was witty and charming, handsome, glib if not brilliant, certainly well read, well informed, and passionately opinionated, ready at any time of day or night to drink, argue, laugh, or make love. He was one of those fortunate young people of high birth and great wealth for whom the world becomes a moveable feast of fine foods and rare wines. For a variety of reasons, however, in Zulfi's case, perhaps because of the "color bar" that kept him from the best rooms and restaurants in the United States, or the monumental ignorance and indifference to Pakistan and pan-Islam that he found in the West, he felt it his personal destiny to lead his land and its hungry millions toward a happier horizon and greater world recognition. The incongruous blend of his feudal genetic load covered over with a thin veneer of socialist ideals acquired from his studies and from personal exposure to prejudice of every sort conspired to help Zulfi resolve during his years overseas to devote his life to global political diplomacy at the helm of Pakistan. Had he not gone abroad to complete his education, it is more than probable, given his father's experience and inherited siasat-shrewdness that Zulfi would in any event have embarked upon a political career in Karachi. His ambitions, then, would have been less grand, confined perhaps to the chief ministership of Sind or membership in the Pakistan Central Cabinet. Because of the University of Southern California, the University of California, and Oxford University, the world became his stage, and premier power the pearl he sought in every oyster he pried open.

Zulfi returned to Karachi after the spring term of 1951. He immediately told Nusrat that for "all these two years I've been thinking of you."[4] He had dispatched a friend the year before with a message from "Zulfikar," but she had feigned ignorance of who the sender might be: "Which Zulfikar?" Indeed, he was not the only Pakistani named "Sword," but the report of her studied indifference fired Zulfi's resolve to marry her. He would prove his worth to her, as he had to Hammond and Trevor-Roper; he would conquer the Iranian beauty as he was going to conquer Latin.

Nusrat herself had completed only senior Cambridge exams at the Convent

School of Jesus and Mary in Bombay, never starting college. Yet she was worldly wise. "I was happy, but I tried not to show it," she confessed when she thought back to the first time he proposed marriage. "I went to the door to see him out and said, 'I thought you were joking.' He got mad and said, '*Dammit,* I'm serious! I'll have to go back to study, and I want to finish it off now. I've no time, not enough time!' . . . So then, of course, I had to tell my sister, and my parents [father] wouldn't agree. 'How can you marry him? We're Iranians and they're Sindhis!' . . . My mother had already died when I was fifteen years old. We were modern, our parents were more broad-minded. They had taught us what was good and bad and they trusted us. I never stayed out after midnight. . . . But he wanted to get married, so I finally agreed."

Now it was Zulfi's parents who protested. Sir Shah Nawaz was shocked, though not Zulfi's mother, who knew Nusrat and liked her but nonetheless thought her a bit "too modern-minded" for her precious son. Would any woman have been good enough for him? Finally, in frustration and rage, Zulfi gave up arguing with his parents, bought a ticket for Nusrat to accompany him to London, and came to her one evening, bags packed. "But I said, 'How can I elope? I can't come like that!' We both cried, and we were feeling so sad," Nusrat remembered. Then a friend of Zulfi's "turned up" with a Jeep and told them, "'Look, I married a Hindu girl, and then we went to the mosque,'" to give them "courage to act." He drove them out toward Malir, where there was a mosque, and said, "Why don't you two go and marry with the Moulvi?" Nusrat "got such a fright," that she cried, "'I can't do that! My parents, my father! Turn back, I'm going home!'" She wanted a proper wedding. They took her back, and Zulfi "spent the night at that friend's home." When his parents heard nothing from him next day, they were frantic with fear for his life. "Finally his mother came crying to our home, '*Key,* where's my son? He can marry anyone he wants!'" Nusrat wasn't sure where he had gone, suggesting that he might have flown back to London because he had shown her the tickets. Knowing how stubborn and emotional their son could be, his parents accepted reality. They rushed home and then returned to Nusrat with "a ring and jewelry, and said, 'Get married!'" Within the week Zulfi and Nusrat were man and wife.

It was a small wedding for so important a Karachi couple, celebrated on 8 September 1951. "Nusrat wore an embroidered bridal 'gharara'—traditional long skirt—and jewelry." Zulfi "rejected the traditional bridegroom's 'achkan'—coat—and 'turra'—turban, wearing instead an elegant black suit with a folded handkerchief in his pocket."[5] They were married at the Bhutto home on McNeill Road in Karachi, and the following days passed in a whirl of luncheons and dinners. By week's end Zulfi, his mother, and his bride left Pakistan for London, stopping over in Turkey to visit Zulfi's half sister, who was married to Colonel Mustafa Khan, Pakistan's military attaché in Istanbul.

"Mother-in-law came with me," Nusrat recalled, for though this was his second marriage, Zulfi's mother could not as yet let go of her darling son. "So we three traveled together to Istanbul. . . . Then from there to Rome." Zulfi's mother had at last turned back, "so we were alone in London for the first time." Their suite at the Dorchester, where they stayed for a week, overlooked Hyde

Park. One evening they saw Mary Martin in *South Pacific* and "loved it." Zulfi was fond of musicals but liked classical music, especially Beethoven, as well as popular music. He enjoyed acting of every variety, and was himself an excellent mimic. One of his oldest Sindhi friends, Muzaffar Husain, thought "he should have been an actor"; "he would have been the greatest actor the world had ever known."[6] Others, who admired Zulfi less than Muzaffar did, said that, in fact, he was!

Zulfi and his new begum reached Oxford just in time for the fall term. He moved Nusrat into a small hotel near Christ Church, but because he had to return to the college every night by eleven, when the gates were locked, theirs was not the usual first year of married life. "I didn't like [the Mitre Hotel] because the floors used to creak, it was dark, and I had to go alone to my room. So then he took me to a more modern hotel." One night they went to a club and stayed so late that Zulfi could not make it back to Christ Church in time for bed check. The next day Trevor-Roper asked the reason, and when Zulfi told him, he exclaimed, "I can't believe you're married," Nusrat recalled. "So I was brought, and then he let Zulfi off, and we became friendly."

Nusrat lived in Oxford for almost three months. Everyone was still on short rations in England; there was virtually no meat available, and few eggs. Nor could Nusrat cook, for she had never been obliged to learn. Zulfi, who had lived with Piloo, of course, knew quite well how to cook, but had little time for it. Letters from both families urged Nusrat to return to Pakistan. Her father and sisters missed her very much, and his parents wanted her to let their son study more. "They didn't realize that if I'm there I would *make* him study," Nusrat noted—especially Latin, on which she kept drilling and questioning him. By late November she headed home, to live with her in-laws on McNeill Road, in a house they had named Be-Nazir (Without Equal; Incomparable). That was a trying time for the newlyweds. Zulfi wrote regularly, and though it was more difficult in those days, he occasionally phoned: "Why don't you tell my father to send you here?" Nusrat was "too scared," of Sir Shah Nawaz, whose health had deteriorated considerably by then, making him more cranky. He agreed with his wife, moreover, that their son would "study more" if alone. Finally Nusrat's father saw how unhappy she was, how much she missed being with her husband, so he bought a ticket for her to fly to London in mid-1952.

Zulfi by then had completed residence requirements at Christ Church and was permitted to move to London. He still had to work on Latin, which he continued to study while he was enrolled at Lincoln's Inn, reading for the bar and eating the requisite number of dinners there, even as Jinnah had done sixty years before. Nusrat shared his furnished flat on the sixth floor in Chatsworth Court, Kensington for a month while he commuted between London and Oxford. "He used to do some cooking," she remembered. "I learned to go to the butcher. I told him, 'I don't want pork!' so he would give me some eggs or whatever." Nonetheless, most nights they ate out. Zulfi enjoyed French cuisine, but Nusrat was indifferent; when "one is so much in love, you don't know what you're eating, what you're doing." Zulfi confided to his wife his ambition "to be Foreign Minister." He talked about going into the Foreign Office, after he finished his studies and returned home.

After the month in London, Zulfi and Nusrat moved to Oxford, where she stayed with him for two more months. It was there that a doctor confirmed her suspicion that she was pregnant. Zulfi "was so excited he was running in the streets, picking me up and sort of carrying me, dancing in the streets." To every stranger they met, Zulfi exulted, "You know I'm—I—I'm going to be a father. I'm going to have a child!" He could scarcely contain his excitement and pleasure. Nusrat returned to Pakistan, alone.

Zulfi had only the bar to worry about once he finished Latin and was awarded a master of arts degree with honors. There were "bolts" for him to speak at, and "moots" to attend in Lincoln's Great Hall. Rhetoric and debate were Zulfi's best subjects, but by 1953 Lincoln's Inn bar examinations required a grounding in constitutional law, jurisprudence, evidence, criminal law, remedies, land law, civil procedure, conflict of laws, criminal procedure, partnership, divorce, master and servant law, construction, and torts. Zulfi "worked up" all those subjects, from among any of which he might be questioned during that final multiday ordeal. He drew up lengthy, precise outlines on every legal subject, the high points underlined in red, green, or blue, each filed in its labeled folder.[7] Less than a year after being enrolled at Lincoln's Inn, Zulfikar Ali Bhutto was called to the bar there, as William Pitt, Lord Canning, John Morley, and Mohammad Ali Jinnah had been before him. What an exalted fraternity he'd joined, that band of barristers, who had dined in the hallowed Great Hall and passed through the weathered stone portals of the inn named for Thomas de Lincoln and founded in the fourteenth century.

Barrister Bhutto was apprenticed in the chambers of Ashe Lincoln, Q. C. Christmas (Toby) Humphreys was head of chambers at the time, and Barrister Eric Hurst, then a Labour member of London's County Council, who had earlier "spent many hours" with Sir Shah Nawaz, "of whom I was fond," was Zulfi's mentor in chambers.[8] Hurst said some thirty-five years later, "We shared a certain proclivity for being rather less concerned with doctrinaire politics than with practical solutions to palpable evils."[9] He made no "claim" to having played "even a small part" in Bhutto's "stupendous ascent to greatness," but did note that "when he had time to spare for our discussions on law" Zulfi's "mind was among the brightest, most perceptive, and most analytical I have known." It appears, however, that Zulfi did not often find "time to spare" for the law. Hurst attributed his distraction when in chambers to what Zulfi told him was work he'd been doing "on a proposal for a new agrarian policy for Pakistan." Zulfi once showed Hurst a draft of that scheme, trying to explain perhaps why he had been remiss in his legal labors. "He was driven," Hurst concluded, "perhaps sometimes . . . impetuously, but always, I believe, sincerely, by compassion. . . . All, I think, the young Zulfie I knew ever wanted was to procure for his country what he perceived to be its true manifest destiny: peace with honour and a sufficiency for all."

Nusrat was living at Be-Nazir in Karachi when her first child, Benazir, was born on 21 June 1953. The choice of name was "my in-laws'," Begum Nusrat recalled. The doting grandparents usually called her "Nazir, Nazir," but Nusrat always used the full Benazir. Her aunt called her "Pinkie," because she had been very pink at birth, and that was to remain her family nickname. She was "a beautiful child," but felt surprisingly "heavy" to the new mother.

Zulfi must have been pleased at the good news, but even so continued his work in London till year's end. His father's friend, Professor S. N. Grant-Bailey of Christ Church, had just appointed Zulfi a part-time assistant lecturer on the law faculty of Southampton University. Grant-Bailey wrote to convey that information to Sir Shah Nawaz from London in August, adding, "I greatly regret that the necessity of his returning home at the end of this year will confine his lecturing life here to one term. I am confident that his lecturing will be of value from many angles."[10] Zulfi was, however, "unable" to start lecturing at Southampton before steaming home from that port in November, contrary to the note on his life in the *Biographical Encyclopedia of Pakistan*.[11]

Barrister Bhutto returned to Karachi on the same P & O liner that future Justice Dorab Patel, who had been called to the bar after finishing his studies at London's School of Economics, had also booked a berth. "He looked me up on board," Dorab Patel said thirty-six years later.[12] Soon after reaching home, the two young barristers shared an office on the top floor of renowned Solicitor Ram Chandani Dingomal's chambers in the old Indian Airlines building facing Karachi's Municipal Courthouse. Sir Shah Nawaz had known Dorab's Parsi "Uncle Mody" in Bombay. The eminent justice of Pakistan's Supreme Court, who would be one of a minority of three judges voting to acquit Zulfi Bhutto of "conspiring to murder" a political opponent, recalled that "we did not have much in common." Nor did he see very much of Zulfi in Dingomal's chambers because he rarely arrived until early evening by which time Patel had usually gone home. Their working hours must have overlapped enough, however, for Patel to feel the need to have a glass partition installed between their cubicles. "I like to have privacy," the soft-spoken justice explained.

Benazir was almost six months old when her father glimpsed her for the first time as he came down the gangway to his waiting family. "I had the baby, but he ignored it," Begum Nusrat remembered. "He was too shy. But when we went home, he said, '*Now* let me see my child!' And of course he loved it. So I said, 'Why then did you ignore her?' He said, 'In front of my parents? They'll think I'm too weak! They'll think what am I doing, looking after, carrying the baby?' But in the house he kept looking at the face, the hands. . . . She was a beautiful child." He soon went off to Larkana to visit his first wife and to check on his lands. He liked to spend three or four winter months in Larkana, inviting many foreign as well as Pakistani friends there for the hunting season, and to celebrate his birthday.

He returned to Karachi to practice law in Dingomal's chambers. His "first case," according to Begum Nusrat, was to defend a fellow Sindhi charged with murder. Bhutto won the case, but because his client had no money, he gave Bhutto his little girl. "So he brought the child home to me, and said, 'Look what he gave me! How could he give his own child?' So I said, 'Take it back!' So he did."

Zulfi had met Dingomal's son at Oxford, where he had been a student at Oriol College. Young Dingomal, now a famous lawyer in Karachi, was to represent the Bhutto family in a number of "personal matters." He remembered the first time he heard Zulfi speak in Oxford at a Rotary Club meeting: "He said he was going to be the foreign minister of Pakistan, and all of us said, 'What an ambitious young man!' "[13] Zulfi's ambition had long focused on politics and diplomacy, which helps perhaps to explain why he took so little interest in his legal practice.

Barrister Fakruddin Ibrahim, later chief justice of Pakistan's Supreme Court and after his retirement governor of Sind (1989–90), also practiced in Dingomal's chambers, where he met Zulfi. Zulfi turned many of his cases over to Ibrahim, who had been called to the bar from the Inner Temple, and was most diligent in his daily application to the career from which he never deviated till after retirement. He remembered Zulfi as "very generous," but always sensed "a streak of violence" in him, a certain "mean" or "vindictive" quality.[14]

Zulfi joined the bar in Karachi and became a member of the prestigious Sind Club, where he liked to drop in for a drink or a meal with his friends in the old high-ceilinged Victorian palace. His ambition to run the Foreign Ministry naturally predisposed him toward issues of constitutional law; he occasionally lectured on the subject in Karachi's Sind Muslim Law College, and wrote papers on constitutional questions that at the time were being vigorously argued in the clubs and assemblies of Karachi and Lahore. By 1954 Pakistan's Constituent Assembly had been wrestling for seven years with its still unwritten constitution in its search to reconcile fundamental regional and tribal differences within its complex, divided body politic.

The elusive constitution would have to be acceptable both to representatives of some 56 percent of Pakistan's Bengali population, who lived in the East, and to the 28 percent in the West's largest province, Punjab, which also provided almost 85 percent of the nation's recruits for the army and higher civil bureaucracy. Sindhis, Frontier Pathans, and Baluchis clamored, moreover, for special provincial considerations of their own. Seven years had not been long enough to reconcile the many differences among all these discrete cohorts. Nor had the preceding seven-year interlude—from the Muslim League's adoption in March 1940 of its famous Lahore (Pakistan) Resolution to the traumatic birth of the Dominion of Pakistan in mid-August 1947—sufficed to permit Quaid-i-Azam Jinnah and his advisers to formulate clear guidelines for the constitutional character of the nation they sponsored.

Jinnah himself was dying of lung cancer and tuberculosis by the time he returned to Karachi as Pakistan's first governor-general. What little he said about the constitution he envisaged for his nation indicates that he hoped it might be "democratic", based on Islamic foundations of "social justice . . . equality and brotherhood of man." Two days after returning to his birthplace, Karachi, Governor-General-designate Jinnah told assembled followers in the

Karachi Club: "Let us trust each other. . . . With the help of every section—I see that every class is represented in this huge gathering—let us work in double shift, if necessary, to make the Sovereign State of Pakistan really happy, really united and really powerful."[15] Such inspirational generalizations were all he ever articulated, though his first address to the new Constituent Assembly on 11 August 1947—shortly before he died—was his longest constitutional legacy, and the closest Jinnah ever came to explaining how he hoped the Pakistan of his dreams might one day function:

> The first duty of a Government is to maintain law and order, so that the life and property and religious beliefs of its subjects are fully protected. . . . The second thing . . . is bribery and corruption. That really is a poison. We must put that down with an iron hand. . . . Black-marketing is another curse. . . . The next thing is . . . the evil of nepotism and jobbery. This evil must be crushed relentlessly. . . . If we take our inspiration and guidance from the Holy Quran, the final victory . . . will be ours. . . . [B]e prepared to sacrifice . . . all, if necessary, in building up Pakistan as a bulwark of Islam and as one of the greatest nations whose ideal is peace. . . .
>
> Now, if we want to make this great State of Pakistan happy and prosperous we should wholly and solely concentrate on the well-being of the people, and especially of the masses and the poor. . . . If you change your past and work together in a spirit that everyone of you, no matter to what community he belongs, no matter what relations he had with you in the past, no matter what is his colour, caste or creed, is first, second and last a citizen of this State with equal rights, privileges and obligations, there will be no end to the progress you will make. . . . We should begin to work in that spirit and in course of time all these angularities of the majority and minority communities, the Hindu community and the Muslim community . . . will vanish. . . . We are . . . all citizens and equal citizens of one State. . . .[16]

Prime Minister Liaquat Ali Khan, an aristocratic Oxford-educated barrister who had helped Jinnah revitalize the Muslim League and achieve Pakistan during the last hard decade of struggle with India's National Congress leadership, tried his best to carry on in the footsteps of his Quaid. But Liaquat lacked Jinnah's charismatic powers and was weakened by ulcers, the intractable weight of Partition's burdens, and the ancient legacy of tribal and regional, religious, racial, and linguistic conflicts that daily threatened to tear Pakistan apart. Liaquat drafted the Objectives Resolution that was accepted by Pakistan's Constituent Assembly in early 1949. That resolution stated what were generally agreed to be the basic goals of the still-to-be-expanded constitution of Pakistan, "wherein the principles of democracy, freedom, equality, tolerance and social justice as enunciated by Islam, should be fully observed." Fundamentalist Muslim mullahs insisted that the as yet unwritten constitution sounded too secular; Bengali Hindus argued that it was too Islamic. Liaquat lost the support of Pathan and Punjabi militants as well, for seeming "too soft" toward India over

Kashmir; he had agreed to accept a UN-monitored cease-fire line, pending a plebiscite—which has never been held—instead of pushing harder for what might have been the more popular military "resolution" of Pakistan's major continuing conflict with India. On 16 October 1951, as Pakistan's first prime minister rose to address a mostly military audience in Rawalpindi, he was assassinated by a hired gunman, who was in turn instantly shot dead. None of the other conspirators was ever arrested or charged for that public murder.[17]

Pakistan's second governor-general, Khwaja Nazimuddin, stepped into the prime ministership, which in the other dominions was more important and powerful than the usually ceremonial governor-generalship, though Great Leader Jinnah's choosing to serve in that position had exalted its status in Pakistani eyes. Nazimuddin, however dedicated and devout a Pakistani patriot he might have been was Bengali by birth. That "stigma" alone sufficed to make many West Pakistani Punjabi leaders mistrust him. Theirs were languages separated by more than the thousand miles of India that stood between the two provinces or wings of Pakistan. Millennia of cultural distance made them suspect, if not hate and intuitively mistrust, each other. Each inhabited a world more than strange to the other: polluted, corrupt, dishonest, incomprehensible. Nazimuddin knew that he could do virtually nothing to win the loyal respect of the military high command or civil bureaucrats, who treated him hardly as their Quaid-i-Azam but more as a mere shadow head of state, less important than most dominion governor-generals. He hoped, nonetheless, that in the prime minister's office he might at least muster popular political support as head of a cabinet that was meant "responsibly" to govern the parliamentary dominion. Nazimuddin believed in the system the British had bequeathed to Pakistan, as he believed in democratic governance, the rule of law, and popular sovereignty. But Punjabi bureaucrat Ghulam Mohammad (1895–1956), who had vacated the office of minister of finance to replace Nazimuddin as governor-general, believed only in the primacy of Punjabi power, and had nothing but contempt for the little man he had succeeded and would soon dismiss in the first of many such militarily backed coups that became Pakistan's favored technique for expediting political change.

The military secretary at this time was Shi'ite Colonel Nawabzada Syed Iskander Ali Mirza (1899–1969), a scion of Persian aristocracy, whose ancestors had been nawabs of Bengal. He was an old friend of Sir Shah Nawaz's. Mirza was a clever man who recognized the unscrupulous ambitions of Ghulam Mohammad. He encouraged the governor-general to declare martial law in Lahore to stop fundamentalist Muslim rioters there early in 1953. Those Ahrar rioters had launched attacks against the peaceful wealthy cosmopolitan Ahmadiyas, a "heretical" sect of Islam who virtually worshipped their founder, Mirza Ghulam Ahmad (1835–1908) of Qadian, believing he was endowed with prophetic powers. Hundreds were killed in the riots. Soon after the army restored order, Governor-General Mohammad, appreciating how easy it was to act on his own, ousted the prime minister. Nazimuddin was, of course, shocked and outraged. He desperately appealed to the courts and to friends abroad for help in remaining at what he had thought to be the pinnacle of power. But judges have no weaponry, and the world was either indifferent or approved the coup as a change that might help make Karachi's ever-quibbling politicians more competent the next time they had a crack at running a government. Without military support, in fact, no government of Pakistan could stay in power for a day, yet prior to Ghulam Mohammad's "civil coup" of April 1953 that now-familiar truism had never been so clearly demonstrated.

Ghulam Mohammad brought Pakistan's ambassador to Washington, Mohammad Ali Bogra (another Bengali, but much younger and more pliable than Nazimuddin), back to Karachi to serve as prime minister. His extensive U.S. contacts would prove most useful when it came to integrating Pakistan into a number of cold war alliances devised by Secretary of State John Foster Dulles as a diplomatic chain of steel to contain Soviet imperialism in the early 1950s. Pakistan joined the Middle East Treaty Organization (METO), or Baghdad Pact, later renamed CENTO, as well as the South East Asia Treaty Organization (SEATO), and signed a mutual defense treaty with the United States, becoming our "most allied ally" by the mid-1950s. Major General (promoted promptly for his coup services) Iskander Mirza was made minister of the interior in Bogra's cabinet, and the commander in chief of the army, General Mohammad Ayub Khan (1907–74), became minister of defense. Bogra's Bengali birth was supposed to have helped mollify, if not wholly pacify, Bengali feelings of betrayal and alienation in the wake of the coup that had removed the Bengalis' leading spokesman from power in Karachi, but provincial elections held in Dacca early in 1954 proved a disaster for the Muslim League. Bengali public opinion was much more sophisticated than West Pakistani bureaucrats had believed. There was at any rate a fast-growing sense throughout East Pakistan of treachery on the part of the ruling West, and more and more Bengalis felt themselves to be an alienated populace, an exploited "colony" of Punjab rather than the majority province of a single nation.

The 1954 Dacca elections brought the United Front—led by H. S. Suhrawardy's (1893–1963) increasingly popular Awami (People's) League, and A. K. Fazlul Haq's (1873–1962) more radical Krishak Shramik (Peasants and Workers) Party—to power in the East, routing the moribund Muslim League. Prime Minister Bogra made a last-ditch effort to win back the Bengali support he and his party had lost by proclaiming that Bengali, the language of more than 90 percent of East Pakistan's population, would after mid-1954 enjoy equal national-language status with Urdu, hitherto Pakistan's only national language (together with English) for testing and other official government purposes, despite the fact that less than 20 percent of the nation's population spoke it as a mother tongue. But at this very time riots rocked East Pakistan's Adamjee Jute Mills, and police fired at Bengali workers there, killing between five hundred and a thousand in the worst civil explosion since Partition. Governor-General Mohammad dismissed Chief Minister Fazlul Haq's United Front ministry, and sent General Mirza to Dacca as martial governor of that increasingly uncontrollable remote region.

Prime Minister Bogra's various attempts to keep Pakistan within a single body politic were doomed from inception, even as Quaid-i-Azam Jinnah's early attempts to translate the 1940 Lahore Resolution into a single-state rather than a two–nation-state resolution had been doomed. Fazlul Haq, after all, had chaired the subcommittee of the Muslim League that had drafted the ambiguous Lahore Resolution. Bengali Fazlul Haq and a majority of his committee meant from that resolution's inception *two* states, not one, in specifying that the areas of British India in which Muslims were "numerically in a majority, as in the North-Western and Eastern zones of India, should be grouped to constitute Independent States in which the constituent units shall be autonomous and sovereign." Lahore's newspapers the next morning, however, blazoned banner headlines that read "Pakistan Resolution." And when Jinnah was asked by reporters whether the resolution meant one or more than one Pakistan, he unhesitatingly replied, "One!" It was unthinkable to him that there should be more than one Muslim Nation carved out of British India; that would have meant more than one Quaid-i-Azam. Nor would the British, who still ruled India, have taken the time or trouble to carve yet a *third* nation-state out of the domain they were weary of managing and so eager to leave by the end of World War II. They almost rejected the single Partition, refusing absolutely to consider separatist claims by either the Sikhs in Punjab or Suhrawardy, who wished to preside over a separate unified "Nation of Bengal" (Bangladesh) in 1947.

Bogra's last gasp as premier was to convene the new national Constituent Assembly in West Pakistan's hill station at Murree in July of 1955, to which the newly elected East Pakistani delegates came. His own Muslim League Party rejected Bogra's leadership, choosing instead Punjabi bureaucrat Chaudhri Mohammad Ali, who was then still minister of finance. That slap elicited Bogra's resignation in August. Old Ghulam Mohammad was now too sick to function in any capacity, and turned over his governor-generalship to Iskander Mirza. Mirza chose the Chaudhri to serve him as prime minister, sending Bogra back to Washington. The Punjabi Mohammad Ali managed to survive just over one year, less than half as long as his Bengali predecessor had clung to the top of Pakistan's most slippery pole. A new formula trick was devised to try to break the constitution-fashioning deadlock, integrating all the provinces and principalities of West Pakistan into "One Unit," which would then elect exactly as many members to a new National Assembly as would East Pakistan. This clever idea would, in fact, disenfranchise almost 20 percent of East Pakistan's population, thus stripping that region of its national majority. The smaller provinces of the West, however, also felt cheated, for they would now become political cyphers of the Punjabi majority under whose unitary umbrella they were to be integrated. The first governor appointed to head this new Western Unit was the Pathan leader, Dr. Khan Sahib (1882–1958), Mirza's trusted Frontier colleague, elder brother of the "Frontier Gandhi," Khan Abdul Ghaffar Khan (1891–1990), whose son was Abdul Wali Khan.

Zulfi Bhutto had been repelled by the One-Unit scheme when it was first mooted in 1954. He wrote publicly to denounce it as "insidious," arguing that "implementation of this notorious proposal will be tantamount to the liquidation of the smaller units. . . . Thus, the land of ancient Sind which took shape in the civilization of Mohenjo Daro and enjoyed independence up to the days of the Talpur's Miani will wither into a mere pensioner of the Punjab."[18] His devotion to Larkana and its lands made him denounce the scheme as a "relentless effort" to write "the last chapter of Sind's history." His father's feudal arch rival from Larkana, Wadero Mohammad A. Khuhro, whom Shah Nawaz blamed for his 1937 electoral loss, was then chief minister of Sind. Zulfi attacked Khuhro for trying to "ram" the scheme "down our throats."[19] The Bhutto-Khuhro vendetta continued for the rest of Zulfi's life, as it had throughout most of Shah Nawaz's. "He wanted to dominate the political scene," Khuhro's politically active daughter, Hamida, insisted, reflecting on Bhutto's motivation, "hence he felt Khuhros must be eliminated. . . . He had no convictions. He was totally unscrupulous in attaining his ends, crushing his enemies."[20] *Siasat!*

Zulfi's protest against the One-Unit scheme became a launching pad for his entry into Pakistani politics. He was too bored with the law ever to consider devoting his full time and bubbling energies to it. Nor was his really a legal mind. He was too impatient, too mercurial, too impulsive to focus all day on the details of legal abstracts and codes, to comb depositions meticulously, to draft and redraft briefs, each of whose citations had to be minutely checked for accuracy. His mind was too busy for such things, his dreams too lofty. "Politics is the end in view, the objective one strives to attain," Zulfi wrote at this time, "law on the other hand, is one of the many means to achieve the political end."[21] In a longer constitutional critique, "Pakistan: A Federal or Unitary State," he argued:

> In Pakistan, the problem of constitution making has been *sui generis*. The men concerned with Government have waltzed in and out of the labyrinth of casuistry. . . . In view of the ethnic, linguistic and cultural differences . . . in view of the chasm of one thousand miles between East and West Pakistan, only a federal government could foster the *solidarite sociale* of the people. . . . The synthesis of Islamic culture has been essentially a federal process . . . the 1940 Lahore Resolution stipulated that Pakistan would have a general government. The Objectives Resolution . . . reiterated the pledge. . . . In the face of such imperishable circumstances, antagonism towards a federal constitution was not to be expected. The compulsion of logic, however, has rarely been a safe guide in assessing the requirements of the Homo sapiens.[22]

Having thus "waltzed in and out" of the question, did Zulfi's agile mind suddenly reflect on his own Homo sapiens "requirement" that often defied "logic" and its less powerful "compulsion"? "Federalism is not a voluptuous damsel over whose charm and desirability men differ," barrister Zulfi Bhutto continued. "There can be no blind obsession for a legal concept." Proponents of the One-Unit scheme insisted it would eradicate provincialism and reduce expenditures, but Zulfi argued that it would instead only "augment disintegration."

In October 1954, almost three-quarters of the elected members of Sind's provincial assembly, who had been polled by the then chief minister, Pirzada

Abdul Sattar, were "opposed" to One Unit for West Pakistan. Governor-General Mohammad replaced Sattar a few weeks later with M. A. Khuhro, and within a month One Unit was approved by a vote of 100–4 in the assembly. That vote was characterized in the next Constituent Assembly debates as "Khuhroism": "members of Legislative Assemblies shall be arrested; their relatives will be put under detention; officers will be transferred . . . elections shall be interfered with and members . . . shall be terrorized."[23] Similar practices would later be called "Bhuttoism." Both were, of course, subsets of *Siasat.*

Twenty-seven-year-old Zulfi Bhutto, now committed to a political life, assessed his moves carefully, knowing how swiftly political fortunes could fall in Pakistan's volatile state. His elder brother, Imdad Ali, was dying of cirrhosis of the liver, and his father "could hardly walk." Sir Shah Nawaz "used to go to Germany once or twice a year . . . for [medical] treatment there."[24] Hence, for all practical purposes Zulfi owned the vast lands that would officially be his after his father's death in November 1957. His first wife's huge Larkana estate had already come to him after the death of her father, Wadero Ahmed Khan Bhutto. His legal practice was enhanced by a substantial retainer he received in 1955 from the Hunt Oil Company, which embarked on a prolonged exploration for oil along Baluchistan's Makran Coast. Zulfi continued to represent his friends against criminal charges, which he viewed as nothing but "vindictiveness" on the part of Chief Minister Khuhro's police and magistrates.

In October of 1955 General Mirza replaced ailing Ghulam Mohammad as governor-general. Zulfi now had a close family friend in power. Mirza and Sir Shah Nawaz were old friends, thanks initially to Mirza's engineer uncle, who had worked for the government of Bombay when Sir Shah Nawaz was minister there. Mirza had been a regular guest for the annual hunt in Larkana, staying at Al-Murtaza, getting to know Zulfi Bhutto as a clever, rising young man, one who could be trusted. Soon after taking over as governor-general, Mirza reappointed his other good friend, General Ayub Khan, as commander in chief for a second five-year term, and that winter of 1955–56 he brought Ayub along to Larkana for the hunt. Despite his rank, Mirza's entire career had been spent in the civil bureaucratic service rather than as an active military commander; Ayub therefore distrusted him, or at least never considered him more than an "outsider" to the army. Both generals enjoyed the generous hospitality of the Bhutto family, however, whose grand feasts and lavish parties were appreciated by all lovers of good food and fine wines and whiskies. Zulfi and Nusrat were charming, delightful hosts. His quick wit and endless anecdotes of Oxford, London, Los Angeles, and Berkeley, and his sharp grasp of politics and diplomacy impressed the generals. Few Pakistanis of Zulfi Bhutto's class concerned themselves with diplomacy or sought solutions for constitutional problems. Begum Nusrat, moreover, one of the most beautiful women of her day, was enchantingly modest in manner, as soft-spoken and demure as her husband was flamboyant and loud. Governor-General Mirza and General Ayub were captivated; the latter would long remain so.

Pakistan's second Constituent Assembly approved the new Constitution in February 1956, and on the sixteenth anniversary of the Lahore Resolution, 23

March 1956, the Islamic Federal Republic of Pakistan was born. That republic's head was a president, who "must be a Muslim . . . at least 40 years of age." Iskander Mirza was the first and only president under the 1956 Constitution, which remained in force just over two and a half years. During that brief interval no fewer than four prime ministers tried in vain to reassert political dominance over the increasingly martial polity. The second of those prime ministers, Bengal's Suhrawardy, was a brilliant politician, whose Awami (People's) League, would later launch East Pakistan's Six Points, demanding virtual autonomy, under the lead of his disciple Sheikh Mujibur Rahman (1922–75). Suhrawardy became prime minister in September 1956, and he too was invited to Al-Murtaza in Larkana to enjoy the warm Bhutto hospitality. He immediately recognized Zulfi's political talent, and tried to lure him into the Awami Party, but Zulfi was not ready for an affiliation, and certainly had no intention of lining up behind a Bengali, no matter how solid or shrewd. Zulfi bemoaned the swift decline and virtual demise of the Muslim League, which in December 1956 he attributed to its having "lost contact with the masses, with their feelings and problems."[25] Though born as the party of Jinnah and Liaquat, the Muslim League no longer served "the common man," Zulfi argued. Its aging leaderships' "conceit and inertia replaced humility and dynamism," till "the soul oozed out of the body that was once beautiful."

Governor-General Mirza encouraged Dr. Khan Sahib to found the Republican Party, whose cabinet members formed a coalition government with Suhrawardy's Awami League. Zulfi was urged by Mirza to join the new party, but he argued that it too lacked grass-roots support. He stood aside, saying that "as long as Pakistan remains a democracy, the resuscitation of the Muslim League cannot be ruled out." He noted how important political parties were to a modern state, democratic or dictatorial: "In the Soviet Union, the Communist Party [was] the vanguard of the toilers" thanks to which it became the foundation of "centralism and discipline." Similarly, Zulfi argued that Adolf Hitler, embodied the "will of the German people" through his National Socialist Party. In dictatorial states, however, the prime object of a party was its "quest for power," which once seized" would never be relinquished.

Zulfi then went on to argue that in "a true democracy" elections "reflect the will and the sentiments of the people. The party closest to [their] desires and aspirations" wins a mandate to govern. The out-of-power opposition party then seeks to attract "sympathy and support" by offering "even more than that promised or achieved by the ruling party." He understood the potential disadvantages of such a course, "especially in backward countries," where "irresponsible promises are made to capture votes." Such promises could "create a sense of cynicism and frustration" once enough people realized it was all "a political stunt."

The truly successful democratic party, Zulfi explained, is one that sustained "a permanent ideological purpose." After the founding of Pakistan in 1947, the Muslim League lost its purpose and collapsed. Nor did it ever redefine its ideology to revitalize itself. As the afterglow of victory faded, the league virtually died. The Republican Party, though born "in dubious and inauspicious circumstances'' was welcomed, but time alone would tell whether it could serve the cause of democracy. If its only purpose was power, that narrow and selfish ideology would never win popular or permanent support. "Let us see if the Republican Party will become dynamic enough, revolutionary enough to attract fresh blood,'' young Zulfi advised, expressing his own reluctance to join it. "There is reason for doubt and suspicion. . . . Has the Republican Party come into existence to challenge the Muslim League's monolithic control, as a harbinger of democracy, or has it come into being as an opportunistic force? . . . [S]o far the Republican Party is treading the path of its predecessor.'' What was needed, Zulfi advised, was for responsible party elements to frame a manifesto reflecting the "genuine desires of our people.'' Otherwise, the "nebulous utterances of the good old man who leads the party,'' might simply lead it to "disaster.''

Here Zulfi was addressing Mirza, cautioning him not to rely too much on that "good old man'' Khan Sahib, whom Zulfi never trusted, and who was to be assassinated in 1958. More than the Pathan Khan, however, Zulfi feared and disliked Bengal's Suhrawardy. "There is no real need for a coalition Government at the centre,'' realpolitik Zulfi insisted, unhappy to find that Suhrawardy had been offered the premiership "on a silver platter'' by Mirza. Zulfi never trusted Bengalis, neither Suhrawardy nor his foremost disciple, Sheikh Mujibur Rahman. He knew, moreover, that if a Pathan-Punjabi-Bengali coalition government was able to remain in power, it would not only receive majority support from both wings of Pakistan but leave him out in the cold, for the Sindhi member of that cabinet was Khuhro. So he cautioned Mirza that the Awami League was "much better organised'' than the Republicans and could "steal a march'' on the West.

"The disease of coalitions pock-marked the face of the Weimar Republic,'' Zulfi wrote. "Papen boasted that Hitler was his prisoner, tied head and foot by conditions he had accepted. . . . There is always an element of danger in drawing analogies. . . . The Republican Party coalesced with the Awami League at the Centre. The calculation is that Mr. Suhrawardy . . . can be checkmated by the strength of the Republican Party in the National Assembly and also by the Republican representatives in the Central Cabinet. . . . This country is fortunate in that the Awami League of Mr. Suhrawardy, unlike the Nazis, is a party dedicated to democratic principles.''

Khuhro was invited by Prime Minister Suhrawardy to serve as his minister of defense. Zulfi, therefore, not only was an outsider but viewed the Suhrawardy government as totally antipathetical to his Larkana interests as well as his political career, and thus did his best to convince Mirza to withdraw his support of it. *Siasat!* "The Republican Party has a choice either of giving this country's politics an unprecedented turn for the better or of destroying democracy altogether,'' Zulfi warned. "In either event, it has to act promptly. What matters in the final analysis is not the attainment of power, for a defeat with honour is infinitely better than victory with dishonour. . . . It is better to be wrong than to pave the way for a military dictatorship. The Republican Party does not appear to be conscious of this living threat . . . that we are on the threshold of dictatorship.''

Shortly after Zulfi's article appeared, President Mirza, in Larkana for the annual hunt, offered Zulfi the mayoralty of Karachi. But Zulfi was not interested in trying to manage a municipal corporation, an impossible task for anyone, young or old. It was the second offer Mirza had made. A year earlier he had suggested a posting to the United Nations, something Zulfi would have loved, had not Prime Minister Mohammad Ali vetoed him after a brief, acerbic first meeting. Now, of course, the Chaudhri was out. Suhrawardy was more astute, recognizing young Bhutto's potential. That September 1957 his name was added to the delegation to New York, which was led by the Malik Sir Firoz Khan Noon (1893–1970), Pakistan's Republican foreign minister, soon to become prime minister.

Suhrawardy's coalition government collapsed in October, just one month after Zulfi had left Karachi. Suhrawardy's Muslim League Punjabi successor, I. I. Chundrigar, held office for only two months. Sir Firoz Khan Noon stayed on as foreign minister in Chundrigar's cabinet, but soon enough Chundrigar lost the support of his Republican Party bloc on the issue of separate electorates. After several days of indecision President Mirza called upon Sir Firoz Khan, as leader of the largest parliamentary group in the assembly, to form a new ministry.

"He would have preferred to take over Government himself under a new form of Constitution,"[26] Sir Firoz Khan Noon confided to his memoirs, but hadn't "sufficiently" prepared the "ground" with the Army. Noon never really trusted his president, of whom he wrote, "His great quality was easy contact with people of all shades of opinion. While he worked with the Awami or Republican Ministry, he played bridge with the Opposition leader of the Muslim League. . . . Mirza received my advice with great attention and usually agreed with me except on important matters, as for example on whether a democratic or other form of Government was the more suitable for Pakistan. Once he had made up his mind to abrogate the Constitution he talked to me no more . . . planned things and kept me in complete ignorance. . . . Suhrawardy knew that if my ministry also went through lack of a majority, Iskander Mirza would declare Martial Law and assume dictatorial powers."[27] As Zulfi suspected, Mirza by then had resolved that this interlude of political haggling was almost over.

Flying to New York just one decade after his first visit to the United States, Zulfi Bhutto had ample reason to feel proud. He was not yet thirty but soon would address a committee of the General Assembly of the United Nations on an important subject.

Yusuf Buch, who worked then for the *Far and Near Eastern Review* and was head of the Free Kashmir Council, met Zulfi for the first time at the UN, and found him a "very bright young man, rather flamboyant, ambitious, self-confident . . . who had the appearance of superficiality."[28] Buch, now deputy secretary-general at the UN, later worked closely with Zulfi as one of his press secretaries and came to admire how "genuinely concerned" Bhutto was "not only with improving the political situation but also the social situation in Pakistan." Zulfi's grooming and self-confident manner made him look arrogant, more like a dandy than a reformer, at first sight. His was never a simple personality, often projecting one persona and hiding another, usually the more sober side of himself. He had no "One-Unit" psyche.

"The task of defining . . . aggression is indeed a gargantuan one," Zulfi began, addressing the UN's Sixth Committee on 25 October 1957.[29] He recapitulated earlier Assembly resolutions on the subject, spicing his *obiter dicta* with other favorite Latin phrases he had so painfully memorized.

> If man's ingenuity is limitless and if his resources and capabilities know no frontiers, then he is, and indeed must be, ingenious enough not only to define aggression but also to circumvent, subvert, and abuse it. A definition, under these circumstances, would literally mean the presentation of our civilisation on a uranium platter to a would-be aggressor, to a twentieth century Chengez Khan or Attila; a would-be world dictator who would most certainly find the means to distort and mutilate the definition for his own wicked and gruesome ambitions.[30]

Less than two decades later several Western diplomats had begun to fear that Zulfi Bhutto himself might perhaps become that very world dictator he now so innocently warned against.

Zulfi quoted Hans Kelsen at length, putting what he had learned at Berkeley to good use. Kelsen had told his students that there were no general rules of international law valid for the whole community. "My delegation deeply respects the Charter of the United Nations. However, my delegation believes that no disrespect is shown or intended to this august organisation if reality is mirrored accurately," Zulfi stated in an affirmation of Kelsen's realpolitik; it was not possible to have "a legal definition of aggression."[31]

Despite his pessimism, Zulfi insisted that "economic aggression" or "indirect aggression" must be included if a definition were eventually to be adopted. He explained that with the Indus and all Punjab rivers rising in India, or passing through Indian-controlled Kashmir, Pakistan was vulnerable as a lower riparian to Indian interference in the assured supply of irrigation waters. A reduction would "turn green alluvial and fertile fields into a scorching desert," create "famine, frustration and fear," even "compel civilised human beings to resort to cannibalism." He ended on a more optimistic note: "[W]e must strive tirelessly and continuously" to achieve "perpetual peace." That was "a duty we owe not only to our own war-sick generation, but to our progeny." He and Nusrat had three children by now: Benazir; their elder son, Murtaza, born in 1954; and a second daughter, Sanam, born that year, 1957. Their fourth child, another boy, Shah Nawaz, would arrive in 1958. Zulfi had reasons to think of "our progeny."

A month later sixty-nine-year-old Shah Nawaz died in Larkana. Zulfi flew home to bury his father in their ancestral village soil, at Garhi Khuda Bakhsh Bhutto. He remained for ten days in mourning at Al-Murtaza, while relatives, friends, servants, and neighbors came from near and far to offer their condolences and pay their respects, bowing low to the new wadero khan of this fast-rising branch of the Bhutto clan. On the eve of his thirtieth birthday, Zulfi was now heir to many more thousands of acres of Sind and all his father's invaluable property, and godfather guardian of all the women, children, and servants living within the Bhutto compound, a leader who for the remaining two decades of his life would search "tirelessly and continuously," but never find "perpetual peace."

Shah Nawaz Bhutto and his wife, Khurshid (née Lakhi Bai), Zulfi's mother, c. 1925. *(Courtesy of Piloo and Vina Mody)*

Sir Shah Nawaz Bhutto, c. 1930. *(Courtesy of Piloo and Vina Mody)*

Top and bottom: Young Zulfi in Bombay, c. 1940. *(Courtesy of Piloo and Vina Mody)*

Top: Cathedral High School Cricket Team, Bombay, c. 1944. Zulfi Bhutto second from left in first row; Jehangir Jal Mugaseth at his side, third from left. *(Courtesy of Jehangir Jal Mugaseth)*

Left: Zulfi Bhutto in California, 1948. *(Courtesy of Piloo and Vina Mody)*

Top: Zulfikar Ali Bhutto in London with his bride, Begum Nusrat Bhutto, 1953. *(Courtesy of Piloo and Vina Mody)*

Left: Zulfikar and Begum Nusrat Bhutto's children, c. 1963. Left to right: Benazir, Shah Nawaz, Mir Murtaza, and "Sunny" Sanam. *(Courtesy of Piloo and Vina Mody)*

At Simla, President Z. A. Bhutto, Prime Minister Indira Gandhi, and Benazir Bhutto, summer of 1972. *(Courtesy of Piloo and Vina Mody)*

Chairman Bhutto and Chairman Mao, Sino-Pakistan summit, May 1974.
(Courtesy of Piloo and Vina Mody)

5

Apprenticeship to Power
(1958–1963)

After celebrating his thirtieth birthday with a month of festive hunting, Zulfi flew to Geneva early in March 1958 to chair the Pakistan delegation to the United Nations Conference on the Law of the Sea. His boon shooting companion, President Mirza, had personally chosen him for that prestigious assignment. Zulfi was grateful, writing, shortly after he reached Switzerland to reassure Mirza of his "imperishable and devoted loyalty to you." "I feel that your services to Pakistan are indispensable," Zulfi fawned. "When the history of our Country is written by objective Historians, your name will be placed even before that of Mr. Jinnah. Sir, I say this because I mean it, and not because you are the President of my Country," he added, trying to convince himself that it wasn't gross sycophancy. "I do not think I could be found guilty of the charge of flattery."[1]

Mirza was still promising to hold elections, and Noon was still hanging on as prime minister, but none of Pakistan's basic problems had been solved. Frustration and popular discontent were rampant, east and west, and the economy was rapidly deteriorating. General Ayub Khan blamed Mirza for much of the political chaos that followed adoption of the Constitution of 1956: "Shrewd as he was, he could see how the Constitution could be used to promote political intrigues and bargaining . . . yesterday's 'traitors' were tomorrow's Chief Ministers . . . everyone connected with the political life of the country [is] utterly exposed and discredited."[2] Ayub soon confirmed Noon's theory that Mirza never intended to hold general elections but was "looking for a suitable opportunity to abrogate the Constitution. Indeed, he was setting the stage for it." As army commander in chief Ayub would be first to know.

"We are conscious of the failures of the past," Zulfi assured the Conference on the Law of the Sea, "but are also poignantly aware of the dictates of this thermonuclear age that give us the ultimatum to either embrace peace with the arms of law, or perish forever in the graveyard of a world Carthage."[3] Zulfi had not forgotten his lessons in rhetoric at the Oxford Union:

> Times have undoubtedly changed. Institutions and values of yore have become effete. Invincible states that controlled the destinies of teeming

> multitudes are now weak and vulnerable. Those held in bondage are now free and, with that freedom, have changed the path of history. . . . Pakistan is deeply concerned with all the Law of the Sea. . . . Both wings of Pakistan have fairly large coastlines. Its fisheries are of considerable economic importance. . . . Most important of all, it is the sea that connects East and West Pakistan and through this mighty force of nature we maintain the geographical indivisibility of our state.

Then,

> The High Seas are free to all. Every nation, large and small, old and new, has the right to take fullest advantage of the resources provided by this freedom. . . . Nations that have faith and confidence in their intrinsic strength must have the vision to think of their interests in terms of centuries. After all, what are fifty years or even a hundred in the histories of countries that hold the legacy of civilisations dating back to Moen-jo-daro and Pompeii? If the United States of America could subscribe to the doctrine of the Freedom of the High Seas at a time when she was not able to take full advantage of that freedom. . . . why cannot the other young and virile nations do the same?[4]

Zulfi spoke again in Geneva before flying home, winning admiration and kudos from many delegates. His ringing reaffirmation of freedom of the seas elicited praise from Secretary of State John Foster Dulles, and though most of his audience in Geneva may never have heard of Moen-jo-daro before, they now knew at least that Pakistan's antiquity rivaled that of Pompeii. World diplomacy was the setting for Zulfi; he enjoyed every moment in the international limelight. All of his higher education had trained, indeed, primed him for this fascinating arena of power and intrigue. It could hardly have been put to better use. Unlike the practice of law, flying halfway around the world to address a room full of equally learned leaders from every corner of the globe was pure adrenaline to Zulfi's spirit. He reveled in his own wit and charm, his clever turn of phrase, declamatory skill, the ring of rhetoric that came so naturally to him. Perhaps most of all he was exhilarated by the social conviviality, grand luncheons, and sumptuous banquets in gracious rooms overlooking Lake Leman and Lake Success. He was filled with the sense of virile power that freedom brings to young men as well as young nations.

In 1958 Zulfi and Nusrat's youngest son, Shah Nawaz—destined to die of an overdose in Southern France little more than a quarter century later—was born. "I was reading a book by Gogol," Begum Nusrat recalled, "so I called him 'Gogol,' because my mother-in-law said, 'You can't call him Shah Nawaz, because when you scold him it's being disrespectful to his grandfather, so it became Gogol. Then the little ones called him Gogi." Benazir had just finished Lady Jennings's nursery school, and was enrolled in Karachi's Convent of Jesus and Mary private school. Mir Murtaza, four, was still at home, as was Sanam, named for the Goddess of Love but called "Sunny" by the family because of her cheerful disposition. The elegant new Bhutto home at 70, Clifton in Karachi was in those days always bustling with life and activity, filled with beautiful children who kept their lovely young mother busy and happy. "He used to call me 'Nusratam'—'My Nusrat'—and I called him Zulfi. We used to do crossword puzzles sometimes, mostly he won. He used to love playing with the children, but once they grew up, seven, eight, or nine, he thought the boys should be brought up strictly, they should stand on their own feet."[5]

" 'I ask only one thing of you, that you do well in your studies,' my father told us time and again," Benazir recalled in her autobiography. "As we grew older he hired tutors to instruct us in Maths and English in the afternoons after school, and he kept track of our school reports by phone from wherever he was in the world. Luckily I was a good student, for he had great plans for me to be the first woman in the Bhutto family to study abroad. . . . 'Pinkie will leave as a scruffy little kid and come back a beautiful young lady in a sari. Shah Nawaz will pack so many clothes his suitcase won't close,' "[6] Papa Zulfi predicted.

"He knew how to treat the children," Begum Nusrat said. "They could argue with their Father, and they used to disagree on certain subjects. [Benazir's] character is a lot like her father's; in her thought, her thinking, her character, she's a lot like her father—the exact shape of her hand. My husband had very delicate, nice, pretty hands. He had a lot of hair when he first married, very wavy."

Pakistan was quite dull to Zulfi after Geneva, and compared to New York. He kept busy, however, between his home and family in Larkana, and work he did there as president of the Abadgar Association, and his office and family in Karachi, where he served on the West Pakistan Social Welfare Council as well as the Pakistan General Cotton Committee. Yet none of those provincial boards truly challenged the restless talents of Zulfikar Ali Bhutto. They were small-town jobs with slow-minded men arguing over petty problems. He had supped with kings, and danced with queens, knew the highlife of London and the bright and seamy sides of the United States.

So that Summer, when President Mirza invited Zulfi Bhutto to join the "next cabinet" as minister of commerce, he accepted with alacrity. Dr. Khan Sahib had been shot dead that May, and the deputy speaker of East Pakistan's Assembly was fatally stabbed in September, after having been declared insane and dragged from the House. Every party had its own paramilitary force, and though Prime Minister Noon banned all private military forces in late September, it was too late for the ban to prove effective. "From a soldier's position it was quite clear to me that the general elections were going to be stand-up fights," Ayub Khan declared. "Whether the army liked it or not it would get embroiled, because in the final analysis it would become a question of maintaining some semblance of law and order."[7] "Law and order" is every army's excuse for a coup.

On 7 October 1958, President Mirza wrote to his prime minister: "After very careful searching of my heart I have come to the conclusion that the country cannot be sound unless I take full responsibility." The 1956 Constitution was thus unceremoniously added to Pakistan's growing constitutional scrap heap.

"By the time you get this letter Martial Law will come into operation and General Ayub, whom I have appointed as the Chief Martial Law Administrator, will be in position."[8]

"The hour had struck," Ayub reminisced. He had arrived in Karachi on 5 October 1958, and went directly to see President Mirza, who was "brooding, bitter and desperate," Ayub thought. " 'Have you made up your mind, Sir?' " the commander in chief asked. The president replied yes. Two days later, on 7 October, at 8:00 P.M., the curtain rang down on Pakistan's civil government. Mirza proclaimed martial law, or what Ayub, as chief martial law administrator liked rather perversely to call "the Revolution."[9] Very soon, however, the coconspirators, or rather founding fathers of that dramatic change in the course of Pakistan's ship of state, had a falling out.

Ayub did not trust Mirza, blaming first of all his lovely and talented Persian wife, Begum Nahid, for plotting to "finish off" his military partner in "revolution"—threatening supposedly to do to Ayub Khan what Ayub did instead to Mirza. Ayub commanded the army, after all; Mirza was only a lame duck president, with no government, not even a nominal cabinet, to turn to any longer for moral support. Each day that fateful October Ayub took more responsibility and initiatives into his own large hands, till Mirza tried in a last desperate effort to retain control to appoint Ayub prime minister over a twelve-man cabinet, whose youngest member was Zulfikar Ali Bhutto. On 27 October 1958, President Mirza swore in the new cabinet, with Generals Azam Khan, Burki, and Sheikh, directly under Ayub. Manzur Qadir was the foreign minister, Mohammad Shoaib in charge of finance, and Zulfi Bhutto given commerce.

Ayub considered the job of prime minister a demotion because he would be serving under Mirza. That very evening Ayub struck, sending his three toughest subordinates, Generals Azam Khan, Burki, and Sheikh, to Mirza's house with an offer: step down immediately but keep your pension and your life. The president wisely accepted, obviously understanding the alternative. The army's "legal experts" came up with the opinion, Ayub wrote in trying to shift the blame for his treachery, that the

> office of President was redundant. I said, "Now, don't you chaps start creating more problems for me. Why do you bother me?" . . . I had hoped that Iskander Mirza would settle down to the new situation. We had been good friends. I told myself that unless he did anything overtly wrong, it would be disloyal on my part to act against him. . . . But the clamour became louder from headquarters and from the army officers. They all said, "Whatever you may be doing, this man is going to nullify it all." They feared that the people would have no faith in our policies. . . . I said finally, "If you feel like that, I shall go and tell him so." They said, "No, you should not go. We will tell him on your behalf."[10]

It was Ayub's second "revolution" in three weeks.

Ayub assumed the title of president the next day, retaining the Mirza cabinet, including its youngest member, Minister of Commerce Zulfi Bhutto—soon to be entrusted with half a dozen other cabinet portfolios as well by Ayub, including that of foreign minister. Zulfi Bhutto's name, however, is not even mentioned in Ayub's autobiography, published in 1967, a year after he fired Zulfi from the cabinet. Small wonder Zulfi was overheard to say of Ayub's ghost-written work, "The book is full of lies! Black lies, white lies, all sorts of lies!"[11]

Mirza, flown to London with his begum, lived quietly there till his death on 13 November 1969, seventy years to the day from the day of his birth. His body was flown to Teheran, and buried there with state honors by order of the shahinshah. Zulfi Bhutto "stood by President Mirza throughout his long exile," Mirza's son, Humayun, wrote in the preface to his father's *Autobiography,*[12] sent to Prime Minister Bhutto during his last year in office. Zulfi, however, once characterized Mirza to his oldest friend Piloo Mody as "a gentleman who knew nothing about politics. He thought that by small gestures, pranks and tricks he could run the ship of State."[13]

President Ayub proved much more durable than his partner in "the revolution." Ayub was a Spin Taran Pathan, and had always remained in the army. Like many soldiers, he liked to pretend he was a "simple" man. For all that simplicity, however, he would hold supreme power over Pakistan longer than any other person had before his double coup. Only one other "simple Muslim" general, Zia ul-Haq (1925–88), would enjoy a longer tenure at the top of the Islamabad-Pindi pyramid of power.

Zulfi started his career on very cordial terms with Ayub, who needed a smart Sindhi adviser and preferred Zulfi Bhutto to M. A. Khuhro; the latter had been brought in as defense minister by Suhrawardy to keep Commander in Chief Ayub Khan in his place several years earlier. One of Ayub's first orders, in fact, was to arrest Khuhro for black-marketeering, not an uncommon sin in Karachi. Almost as important as Zulfi's youth and brains in helping to explain his rapid rise in Ayub's cabinet was his lack of prior political involvement. He had as yet relatively few enemies other than his inherited feudal rivals. His weaknesses were virtually unknown, but because of his father, everyone knew his name.

Zulfi's eight-year political apprenticeship under Ayub Khan gave him his first real taste of power and experience in handling public affairs. He had entered the military government by appointment, of course, so did not have to worry about election campaigns or pleasing constituents. Still, Zulfi knew more about politics than most of his cabinet elders, with the exception of Ayub. And Ayub appreciated how useful and clever Zulfi could be, taking him along for advice when he visited Quetta and Lahore early in 1959. General Gul Hasan, whose regiment was Proben's Horse, met Commerce Minister Bhutto at the Quetta Club soon after the new regime had taken over. "He came to the bar. We were having a drink," Hasan recalled, quoting Zulfi's first words to him, " 'I'm very sorry I'm late, but I just put the old man [Ayub] to bed, and I'm free now.' "[14] The general liked the new minister, who was a convivial drinking companion. "He was quite a pleasant chap, good mimic, good sense of humor, and presentable."

From Quetta, Ayub and Zulfi flew to Lahore in early February, where both addressed a number of public meetings, trying to justify the new regime and its tough policies. Ayub warned his business community audience against selling "their conscience for monetary gains,"[15] and urged them to look "forward instead of backward." "There is only one miracle that can change the destiny of our nation . . . hard work, honest work, clean work." Had he learned the word *destiny* from his bright young companion? Zulfi told students at Islamia College that "the unitary form of government was best for Pakistan"—a swift shift from the federalism he had been advocating just a few years earlier.

Zulfi helped Ayub in many ways. Because the president no longer had time to be commander in chief, he had to appoint a successor. Lieutenant General Mohammad Musa was promoted to general for that key post, which Ayub had held since 1951. Now Ayub worried about his equality of rank with Musa and other generals, however, and the potential dangers that posed. "I remember on one occasion we had a meeting of senior army officers and they submitted to me a paper which explained how everything was going wrong in the country," Ayub wrote. Ayub asked General Musa, "Does this paper represent your views too?" Musa replied, "Yes, all of us feel like that." Ayub exploded angrily: "So far as Pakistan is concerned it is there and it will stay and no one will be able to thwart its progress."[16] But such meetings with his officer "peers" left Ayub rather nervous.

"He asked me for my advice on how to place himself head and shoulders above their squabbles," Zulfi later recalled. "I told him that one way of doing it was to show complete impartiality, fairness and justice, and . . . since it was essential for him to be head and shoulders above the others, it would be better if he elevated his own rank from that of a General to that of a Field Marshal."[17] Ayub was delighted with that "brilliant idea," and on the first anniversary of his coup, proclaimed himself Field Marshal Ayub. Zulfi was then in New York with the UN delegation. Ayub's military secretary rang him from Pakistan "to thank me for making such a sound suggestion."

Similarly, Zulfi helped Ayub prepare the ground for writing a national constitution, and holding elections of some sort to muster at least some pretext of popular support for the "Revolution." Foreign Minister Manzur Qadir, a Punjabi jurist, suggested establishing a bureau of national reconstruction to find out what the people "wanted government to do" for them, and at the same time to "guide people" in "selecting honest men" to run the government. Ayub liked the idea, turned the job over to Zulfi, who was also put in charge of Information and Broadcasting, Village AID, Basic Democracies, and Tourism and Minorities. Zulfi's own ambition, of course, was to take over Qadir's ministry, but he had to wait a bit longer for that plum.

Ayub launched his Basic Democracies scheme in 1959. "It was basic in so far as the whole structure was to be built from the ground upwards," Ayub wrote, "and it was democratic in the sense that the affairs of the country were to be entrusted to the people within a constitutional framework."[18] Zulfi would later denounce it as "Basic Fascism," but in 1959 he still believed in Ayub Khan and was obviously supportive of the system over which he was given direct responsibility. One of his first acts as Minister for Basic Democracies was to tour Sind to help spread the word about this strange-sounding four-tiered system, elaborately designed to keep Ayub in control for as long as possible by reducing Pakistan's entire electorate to a mere 80,000 trustworthy voters. He went first of all to Larkana, then to Nawabshah to have dinner as the honored guest of Ghulam Mustafa Jatoi, one of Sind's other leading baronial landlords, whose father was a contemporary of Sir Shah Nawaz. The family owned more of Sind than all the Bhuttos combined.

"Everyone assumed that Jatoi and I knew each other, so I was not introduced to him and I did not want to enquire, expecting to get to meet him naturally," Zulfi recalled twenty years later in his death cell. "As I moved down the line shaking hands, I saw a young man with a smart movie camera filming me. He continued to do it until I got into the car. He did the same at the lunch." That night, when Zulfi drove to Jatoi House, it was "lit like a fire" and so many guests were there that it "looked as if the whole of Sind had been invited." Zulfi felt odd because he had not yet met his host, but when he stepped out of the car, the young man with the movie camera was waiting to receive him. Then "I knew." That was the first time Zulfi tasted Jatoi minced and stuffed chicken, which Ghulam Mustafa put on his chief guest's plate "by digging his fingers into it and simultaneously telling me that it was their speciality. We both kept the secret and gave everyone the impression that we had known each other for donkey's years."[19] It was the dawn of a powerful political alliance that would keep Zulfi Bhutto and G. M. Jatoi in joint control over Sind's public affairs for the next twenty "donkey's years."

Zulfi left Karachi in September 1959, flying first to Iran, then on to Ankara, Paris, Dundee, London, and Toronto en route to New York for the fall session of the UN Assembly. The chief foreign correspondent of *Dawn* (Karachi) at the time, Nasim Ahmad (1927–90), later Pakistan's ambassador to the UN from 1989 till his death in Tokyo, hosted a lunch for Minister Bhutto at the Dorchester. Zulfi "greatly impressed"[20] the crowd of overseas Pakistanis, mostly students, who heard him talk about energy and power as twin keys to Pakistan's industrial future. He was in top form whenever he addressed a large luncheon or dinner party. Nasim Ahmad recalled that Royal Salute was the Scotch Zulfi preferred, though he favored Chivas Regal, and Johnnie Walker, Black Label almost as much. Being in the Dorchester must have brought back memories of his honeymoon there, but Nusrat did not accompany him on this trip.

At the United Nations Zulfi addressed the Second Committee of the General Assembly on the price of primary commodities, arguing that the major importance of underdeveloped countries like Pakistan to the modern world was that they represented "power vacuums," and "one false step from any one of them and the world could be plunged into devastating conflict." One root cause of war was "economic disequilibrium," and as long as dangerous vacuums of power existed, with "yawning chasms of grinding poverty, ill-health and ignorance" all hope of lasting peace "must prove chimerical." Strong pressures kept building within underdeveloped countries, Bhutto noted, and of late a number of "weak and fumbling" Afro-Asian regimes "including that in my country, have been swept away."[21] Thanks to his government's new "far-reaching reforms," Pakistan had started to reshape itself; however, investment was inadequate, the rate of domestic savings too low, and population growth too high. The only way out was through international assistance.

Zulfi praised recent UN attempts to increase the flow of investments, and to raise the level of technical skills in underdeveloped countries, but argued that a new capital development fund was now needed. Pakistan had just removed a number of taxes and capital-participation requirements for foreign companies in order to stimulate investment, yet was desperate for more industrialization. Most important was the need to stabilize primary commodity prices at much higher levels. Pakistan's "total loss" from falling primary prices on world markets from mid-1957 to mid-1958 alone was more than twice the total financial aid received from different sources up till then.[22] Zulfi noted that the rapid rise in manufactured goods' prices had cut the purchasing power of all Pakistani exports to less than half of its 1948–9 value. The widening gap between underdeveloped and industrial countries urgently required immediate creation of a compensatory fund.

Zulfi completed his work at the UN in little more than a month, then visited Boston, Washington, Los Angeles, San Francisco, Tokyo, and Manila on his way home. He addressed Karachi's business community soon after his return to give an accounting of his mission. In Toronto, en route to New York, he had met "leading exporters, industrialists and businessmen, with a view to giving a boost to our exports to Canada." In New York, discussions with leading businessmen and industrialists had been profitable.[23] In Washington he had signed a treaty of friendship and commerce with the United States on behalf of Pakistan. On the West Coast of the United States he had looked into trade prospects with the region and became hopeful about improving trade with Japan and the Philippines. Production shortfalls in Pakistan would jeopardize such efforts; Pakistani producers must strive to meet growing foreign demand. The remarkable growth of the economies of Hong Kong, the Philippines, and Japan were in stark contrast to Pakistan's stagnation, but there was no sound reason for the disparity. The new government had introduced an expensive export bonus scheme in January, providing import licence vouchers for every export, but no real strides in that sector had yet been made. The talk was a blend of reportage and exhortation.

The Soviet Union's premier, Nikita Khrushchev, had come to the UN that September, and Zulfi raised with him the possibility of a Russo-Pakistani joint venture in oil exploration in Baluchistan. Hunt Oil had just abandoned its interest in the region. Khrushchev was sufficiently intrigued by so bold a proposition from a U.S. ally as to send a delegation of experts to Pakistan in September 1960 to explore oil potential along the Makran coast. The Soviet team recommended exploration of the potentially rich region, for which the Soviet government would float a long-term loan. In December 1960 Bhutto led a delegation to Moscow to negotiate the agreement.

Zulfi opened a second diplomatic front in 1959, when before the Political Committee of the UN. He delivered a short, strong statement in support of the "people of Algeria" against French Imperialism. It was Pakistan's first open stand on that important pan-Islamic issue, one that was to remain near the top of Zulfi's foreign policy agenda. He was not yet foreign minister, but he took bold initiatives that he hoped would meet with approval from his field marshal president in Pindi, Punjab's army general headquarters, to which Pakistan's capital had been moved on the first anniversary of Ayub's "Revolution." The Algerian question was still "under discussion," Zulfi told his Karachi business audience. "On this question and on Kashmir, I shall report my views to the President and Foreign Minister."[24] Thanks to his initiative on the matter, Pakistan recognized the Provisional Government of Algeria, building the first of several important pan-Islamic bridges across North Africa that Bhutto always kept in good repair.

From the birth of Pakistan Kashmir had been a primary source of conflict with India. The Princely State of Jammu and Kashmir, with three-quarters of its population Muslim, had more reason to expect to join the new Muslim nation of Pakistan in August 1947 than to be integrated into the Indian dominion. Kashmir's Maharaja Hari Singh was, however, Hindu, and at first tried to keep his state independent of both his larger neighbors. But in early October 1947 Pathan tribal "volunteers" had invaded Kashmir in British lorries and as they approached Srinagar, Maharaja Hari Singh fled and opted to join India. The first undeclared war had broken out and would continue to shroud Kashmir's lovely Vale in smoke and fighting for more than a year. A UN cease-fire agreement was effected at the start of 1949. Governor-General Jinnah had tried in vain to hurl Pakistan's entire army against the airborne Indian troops flown into Srinagar from Delhi on Governor-General Mountbatten's order, but British officers were still in command of both new dominions' armies, and Pakistan's commander in chief refused absolutely to launch his force against troops under Mountbatten's command. Jinnah felt treacherously betrayed, and authorized the expanded invasion by "volunteer" forces, but they proved less effective than India's army. Mountbatten and Nehru both promised Pakistan, however, that India would accept the choice of the Kashmir people as to their future in a fair and impartial plebiscite after all fighting had stopped and all invading forces were withdrawn.

No plebiscite has ever been taken in Kashmir as a whole. Pakistan refused to order unilateral withdrawal of its "volunteers" as long as Indian troops remained inside the Vale to "intimidate" Kashmiris. The UN cease-fire line became a de facto border between the western third of the former unitary state of Jammu and Kashmir, called Azad (Free) Kashmir by Pakistan, and the rest of the state, including the capital of Srinagar and its Vale, called Jammu and Kashmir, now a state within India. The UN subsequently sent several of its best mediators to try to resolve the preplebiscite deadlock but to no avail. A token UN force of monitors remains at checkpoints along the cease-fire line, yet sporadic firing and night crossings continue everywhere along that "border" wound, which has refused to heal.

In 1959 China was constructing a road through Ladakh's Aksai Chin that would link its western "New Province" of Sinkiang to Tibet. India insisted that Ladakh, Kashmir's northeast region, was entirely Indian territory, a position China disputed. Prime Minister Jawaharlal Nehru warned Chou En-lai not to underestimate India's firm resolve to defend its integrity against any invasion, including one from China, with whom Nehru had once hoped to forge a Third World pan-Asian alliance based on Five Principles *(Panch Shila)*, which began with "mutual respect" and ended in "self-determination." Zulfi's sharp political mind recognized this simmering conflict between India and China as a major source of potential diplomatic advantage for Pakistan if properly exploited. He was acutely conscious of how swiftly great powers shift allegiances, of how winds of diplomatic change move faster the higher their altitude, till jet speeds become possible at the summits. He had the foresight to envision possible advantages to Pakistan, therefore, in seeking ties not only with the Soviets and North Africa's Islamic nations but also with China. His "enemy's enemy" was, he knew, always a potential "friend."

Zulfi wrote Ayub shortly before leaving the UN General Assembly to say that for "the past several weeks" he had been "anxiously concerned" with the India-China situation in Ladakh and the impact it could have "on our position regarding Kashmir." He had noticed in the press that when Ayub was asked a question regarding Ladakh during an airport interview, he had replied to the effect that it was India's problem.[25] Zulfi was careful to write that though he "wasn't sure" of Ayub's exact wording, or whether it was "accurately" reported, "India and its friends" might "construe, and probably use" such a statement in future negotiations *against* Pakistan. "We can be taken to have tacitly recognized India's authority over that part of Kashmir which she controls at present. After all, it is by virtue of the present partition of Kashmir that India controls Ladakh and is in a position to declare that China's encroachment on Ladakh is an encroachment on India itself," Zulfi argued, adding that should Ayub's statement be left to stand, "we can be deemed to be stopped from saying in future that the responsibility for the preservation of the territory of Jammu and Kashmir is not that of India but of the Security Council." Pakistan had, after all, insisted that Jammu and Kashmir was *not* Indian territory, and that therefore its external defense was a matter for the Security Council alone to consider. Ayub's remark made an immediate, "authoritative pronouncement" necessary to safeguard Pakistan's consistent position on Kashmir. The pronouncement would not "embroil us with China" but could, on the contrary, be welcomed in Peking as a statement from Pakistan "questioning the very basis" of India's stand on Ladakh.

Neither Ayub nor the preoccupied Foreign Minister Qadir had been sharp enough to grasp what Zulfi appreciated the moment he read the news report, which had, in fact, been quite accurate. Ayub had other things uppermost in his mind and was never as adroit in handling reporters' questions as was his youngest minister. He preferred giving orders. Then too, Ayub had been focusing on his prolonged and vital Indus Canal Waters negotiations with Nehru, which were far more essential to Pakistan's survival than Kashmir because the headwaters of the Indus and Punjab rivers were under Indian control. Talks with the World Bank and Nehru were at a crucial stage; the almost-concluded treaty would insure Pakistan of 80 percent of the total flow of the Indus waters for its canals, and more than $740 million from the World Bank to build the huge Mangla, Tarbela, and Rohtas dams. Zulfi was aware of the importance of those negotiations, and assured Ayub that the effect any statement on Kashmir might have upon them would have to be carefully considered; still, "we cannot let India damage our entire position on Kashmir during the time these negotiations remain pending." Zulfi's solution was to make a statement, "unhostile in tone and confined only to a principle."

He enclosed a copy of his letter to Ayub in a letter to Foreign Minister Qadir, and explained his feeling that "we shall have to examine the whole question in depth and not let the India-China situation regarding Kashmir drift and develop to our detriment."[26] Zulfi thus prepared the ground for a meeting with Ayub and Qadir in early December, impressing them both in advance with his diplomatic sensitivities, global awareness, and shrewd strategic planning. In April of 1960 he was appointed minister for Kashmir affairs, Ayub turning over the portfolio held until then by himself. Zulfi retained control over the Ministries of National Reconstruction and Information, was also given Fuel, Power and Natural Resources and put in charge of Pakistan's Projects Division, but relinquished Commerce. Still the youngest member of the cabinet, he was no longer its most junior. Now he had Ayub's ear and full confidence, for the field marshal had learned at Sandhurst to recognize a first-rate mind when he encountered one—and Bhutto could be a charmer as well.

Zulfi later asserted that he had noticed certain "appalling weaknesses" in Ayub within a year of his assumption of the presidential office but had remained attached to him because, by Shakespearian logic, "the one who lives by honest dreams must choose the bad amongst the worst." Ayub's "insatiable greed" was what had "disillusioned" him about Ayub, though their "breaking point" did not come until after the 1965 War with India over Kashmir. Ayub's tall, dignified stature and militarily self-assured manner served him much the way that the same qualities had served President Dwight Eisenhower: each was widely admired and supported at home and abroad. Vice President Lyndon Johnson, for example, found Ayub "forceful and capable" and was "deeply impressed" with him and the progress made under his "dependable" leadership.[27] That assessment was conveyed to President Kennedy in 1961, and ensured more aid for Pakistan. Zulfi was willing to wait and work under the field marshal's wing, patiently accepting Ayub's leadership till Pakistan would be ready for the "best," which he alone—he believed—could offer.

Minister Bhutto again led the Pakistan delegation to the UN in 1960, when for the first time he broke ranks with the U.S. position on the People's Republic of China, abstaining rather than voting against Peking's membership in the world forum. Washington wired its dissatisfaction to Foreign Minister Qadir, who then wired Zulfi, retracting his "discretionary powers" on future UN votes. Zulfi never liked Qadir, and was hardly one to take orders from a colleague he considered inferior to himself, intellectually as well as socially. "I feel that the time has come for Pakistan to adopt an attitude in the United Nations more consistent with its recognition of the Peking regime than has been the case since 1954," Zulfi wired home, in October 1960, little more than two years before he would take charge of Pakistan's Foreign Ministry.[28] He bolstered his

China argument by noting how important it was to "strengthen our position" among Third World Asian-Africans. His hope was to forge a Third World alliance under Pakistan's leadership, the strategic dream of his larger global vision. He viewed Sino-Pakistani friendship not only as a counter to Indian hegemony in South Asia but as one part of his blueprint for an Afro-Asian "Third Force," a superpower equal to either the United States or the USSR.

The Foreign Ministry remained wedded to U.S. policy, however, with the CENTO and SEATO alliances, Qadir fearing that Afro-Asian "neutrality" would eventuate in Nehruvian Indian leadership. In Ayub's cabinet, Zulfi alone had the vision to recognize the global appeal of "neutralism" as a political philosophy. After returning from New York, Zulfi forcefully argued in the cabinet on 18 November 1960 that the time had come for Pakistan to shift toward the Soviet Union and China, voting for admission of the PRC at the next session of the UN and the opening of "direct negotiations" with the Soviets on a broad economic agenda. Qadir vetoed the suggested stance on China, but just a month after that meeting Bhutto led Pakistan's first official delegation to Moscow.

There was no direct flight to the Soviet Union from Pakistan, so Zulfi was obliged first to fly to New Delhi on 13 December 1960, where he and his delegation had to wait three nights at the Pakistan embassy before a plane took off for Tashkent on 16 December. Bad weather over that central Asian city, which was to host Indo-Pakistani peace talks half a decade later, forced a landing at Samarkand, birthplace of the founder of South Asia's great Mughal Empire, Babur. "The grandeur of Islamic architecture and culture so richly visible in this citadel of the great Timur and his descendants . . . made one feel proud to be a part of its history, race and religion," Zulfi wrote, feeling renewed pride in his Islamic "heritage."[29] They drove the next day from Samarkand to Tashkent, where the delegation was "warmly" welcomed and prayed at the famous Jamia Masjid. Zulfi flew into Moscow on 19 December, and entered the Kremlin the next day for final oil-contract negotiations. The sessions concluded New Year's Eve with a party hosted by Khrushchev, who "was in great form and proposed a number of toasts," all of which Zulfi drank. The first Soviet-Pakistani agreement released 120 million rubles of credit to Pakistan over a twelve-year period. A team of Soviet explorers, engineers, and scientists, soon arrived to search the sands of Baluchistan and Sind, where they found a rich field of gas, but no oil.

By the eve of his thirty-third birthday, Zulfi Bhutto had become Pakistan's second-most-experienced adviser to Ayub Khan on foreign affairs, crossing verbal swords with Foreign Minister Qadir in more than one cabinet meeting yet always remaining loyal to the team of cabinet government. He said just enough to make Ayub think that if ever a younger spokesman were needed, he was ready to take on the challenge. His UN experience had, moreover, helped Bhutto to focus his global vision more sharply, convincing him that in a world "dominated by the great powers and filled with the fear of a nuclear holocaust" the "umbrella of a World Organisation" was the best protector of "smaller non-nuclear states."[30] By 1961 Zulfi was hopeful that Algeria would soon take its place in the UN after seven years of war against colonial France, informing his friends in Karachi that when it did, "*Insha Allah,* a tragic chapter will close in history."[31]

Kashmir remained at the top of his South Asian foreign policy agenda, however. Only when that dispute was resolved, Zulfi argued, would peace and progress come to the subcontinent. The following year, speaking to the graduates of Sind University, Zulfi described "India's forcible occupation of Kashmir" as a "grave threat to peace." He condemned Indian "aggression" in attempting "to stop the march of the people of Kashmir towards their goal of self-determination," and asserted that Kashmir's "fraudulent accession" to India was the "greatest corrupting influence on the ethics of international relations." Self-determination is the "God-given right" of Kashmir's people, Zulfi argued, concluding that "There can be no peace without the solution of the Kashmir problem."[32] Neither Ayub nor Qadir had ever used such strong language in public pronouncements on Kashmir.

Ayub, of course, knew the precarious nature of his rule and appreciated better than Zulfi Bhutto that in "military thinking . . . the crucial factor was 'capability.'" Pakistan's moral and legal claims to Kashmir were certainly more persuasive than India's, but India's military capability remained far greater, which helps perhaps to explain Ayub's caution about using language as inflammatory as that of his young adviser. Ayub had come to enjoy his position as president, and rather liked the idea of continuing to rule a peaceful Pakistan, whose new capital, Islamabad, was daily growing, even as its "Basic Democracy" Constitution was finally being born in midsummer of 1962. A Constitution Commission, appointed by Ayub early in 1960, had finished its work in two years. That year the National Assembly met for the first time since Ayub's 1958 "Revolution."

Zulfi was one of Ayub's closest advisers on the Constitution and how best to manage Pakistan's transition from autocratic military rule to "democratic" government with at least some patina of popular political support. As minister for basic democracies, Zulfi helped prepare the ground for Ayub's first election. Eighty thousand martially loyal basic democrats had been chosen to represent some 90 million Pakistanis by an elaborate five-tier process of indirect screening, completed in January of 1960. More than seventy-five thousand of those basic democrats voted yes on the question of whether they had "confidence" in President Ayub Khan. Fewer than three thousand no votes were cast in that intimidating referendum, which Ayub hailed as a national "election."

Elections to provincial and national assemblies were held in the spring of 1962, with only the basic democrats enfranchised. All candidates ran as partyless independents for those high legislative offices. Zulfi ran for the National Assembly from Larkana, where one Abdul Fatah Memon had the temerity to file nomination papers contesting the constituency. After a week, Memon wisely "withdrew."[33] Having thus gotten himself "elected unopposed," Zulfi decided to campaign for some of his like-minded friends in Sind; primarily Yusif Chandio in Thatta, Ghulam Mehar in Sukkur, and Ghulam Mustafa Jatoi, who ran for the National Assembly from Nawabshah.

"He [Jatoi] was at the railway station to meet me with many of his friends,"

Zulfi remembered, flattered at finding Jatoi himself coming into his private railway saloon, followed by food enough for "an army," including the delicious Jatoi minced chicken. Zulfi had planned to "get down" at Nawabshah to make a public pitch for his friend, but Jatoi insisted that he "must be very tired" and was so "considerate" about his needs that Zulfi was brought around to think that simply by "stopping" at the station he had done "more than enough" to boost Jatoi's campaign. Years later Zulfi learned from Pir Pagaro, leader of Sind's populous and powerful tribe of Hurs, that the Pir had just then given Jatoi an "ultimatum": if "Bhutto-Sahib" so much as stepped down in Nawabshah to campaign for Jatoi, Jatoi would lose the Pir's support. Jatoi's clever strategy had kept both Bhutto and Pir Pagaro on his side, ensuring his unopposed election.

When the assembly convened, Jatoi and Zulfi became intimate political allies. Jatoi introduced Zulfi to another landlord member elected to that assembly from Punjab's Muzaffargarh district, then a singularly shy, handsome young man named Ghulam Mustafa Khar.

Some 150-odd members of the new National Assembly met on 8 June 1962, when martial law was declared to be "ended." But Pakistan's most popular old political leaders were not there, either having been jailed or disqualified by Ayub's martial-law Elective Bodies Disqualification Order (EBDO). Former prime minister Suhrawardy had been arrested on 30 January, and thereby prevented from winning his seat, triggering student riots that started in Dacca and soon spread all across East Pakistan. That 21 February was the tenth anniversary of Dacca's Martyrs Day, and before the next decade was over thousands of new Bengali martyrs would be born.

Minister Bhutto urged Ayub to reintroduce political parties, and the president finally agreed, taking the faded mantle of the old Muslim League as his party in support of his government's Political Parties Bill, which continued to exclude all those earlier disqualified from elections under martial law acts, such as the EBDO. Bhutto spoke eloquently in the assembly on 10 July 1962:

> The role of political parties is essential to every state. . . . If democracy is to flourish, you must have respect for the other man's point of view. . . . Sir, why was Pakistan created? Why did millions of people give their lives for the creation of our state? . . . The fountain head of our way of life was to be a democratic way of life. But unfortunately soon after the creation of Pakistan, we lost our great and beloved leader, the Quaid-i-Azam [Jinnah], and soon after his death we lost the Quaid-i-Millat [Liaquat]. Therefore, Sir, the political history of this country went through such . . . turmoil that in October, 1958, the Revolution had to come to try and put the state of affairs in order.[34]

His apologia for martial law went on to blame pre-Ayub politicians like Suhrawardy, "who played havoc and ran amuck with the destinies of the people," and therefore deserved to "be debarred from polluting the social and economic life of the nation." Zulfi insisted that during all pre-martial-law local elections "the whole machinery of the police and the bureaucracy were geared up . . . to stifle freedom of thought, freedom of expression. . . . To practise dictatorship in a democracy is the worst form of evil against society, and that is exactly how we were functioning before the present Constitution. . . ."

Zulfi repeated his warning against "the democracy of dictators":

> Sir, we suffered and continued to suffer from the petty-mindedness or feudal rivalry in our province [Sind]. I too am a part of that society. Perhaps one reason why I am here today as a minister is because I belong to this privileged class. Therefore, I do admit the advantages of the system. But . . . it has many inherent drawbacks. It leads to petty intrigues, it leads to victimisation of the people . . . to callousness towards poverty . . . to lethargy. So when feudal rivals clashed . . . there was no development, no factories, no roads . . . absolute darkness and miserable poverty prevailed. Only . . . the chosen few prospered. . . . But now . . . democracy is our creed. . . . Without democracy Pakistan cannot progress.

Then he cautioned: "By a process of democracy a coterie of dictatorship has been imposed time and again. . . . One must protect and safeguard, through institutions, through democratic measures, the indirect and back-door method of stifling the people's will. . . . People who believe in a totalitarian system impose a dictatorship by using democratic methods."

Then came a warning against what one day would become Zulfi's own means of retaining power:

> We know, Sir, how demagogues have taken issues to the people. . . . We know how, without ever believing in issues or in principles, people have tried to say that a certain issue is their religion. We have known how manifestoes have been drafted for political purposes with no intention of fulfillment. . . . It is ironical that elections were held for the first time when Martial Law was imposed in the country. No general elections were held before. . . . Under the old constitution . . . the poor individual who had the courage to file a nomination paper was huddled into a car and thrown into the desert. When after a day or two he returned, he enquired from the successful chief minister if there was no law and order in the land. The chief minister replied that in his realm there was no law but only order and that was his order.

In August of 1962, Zulfi flew with Ayub to Lahore, Karachi, Quetta, and Pindi, speaking from every platform and at every party meeting with his president. He was becoming Ayub's foremost adviser in political matters as well as foreign affairs. Zulfi's brilliant eloquence was a great asset to Ayub, who never liked long speeches or politicking of any kind. He felt fortunate to have a young man at his side, an Oxford man of good family one could trust, a man who could argue forcefully with the loudest and harshest of Pakistani students and other so-called intellectuals. The aging president began to feel positively paternal toward his youngest minister, and may well have considered him a possible heir to his political throne—not immediately but in another decade or so, if Bhutto mellowed with maturity. Ayub knew, of course, about Zulfi's passions and heady tastes. Every man's man enjoyed lovely women, as the field marshal himself did. Every soldier worth his salt liked strong drink. But Ayub had always been most discreet in his private affairs.

Zulfi had met a woman in 1961 for whom he cast all discretion to the winds, though he had, in truth, never been known before that to inhibit his passions. This "Black Beauty," as she was called by his closest friends, had a powerful impact on Zulfi. She was married when they met, to Dacca lawyer Abdul Ahad, a friend of Sheikh Mujibur Rahman, destined to be one of the first Dacca intellectuals slaughtered by West Pakistani troops when the Bangladesh struggle for independence began in 1971. Husna Sheikh was born and brought up in Calcutta during the 1930s, though she herself was not Bengali, her father having come there for work from Haripur on the North-West Frontier. Still a woman of charm and vivacity in her fifties, Husna must have been ravishing when Zulfi first met her in the full glow of voluptuous young motherhood. But it was not simply her physical charms that hypnotized him; she was the first woman Zulfi ever loved (except for Benazir) who could think about and talk politics much the way he did, never tiring of a subject most Pakistani women considered "men's business." The first night they met in Dacca at a large party, they found a place to sit quite alone, where they drank and talked politics, till "four o'clock in the morning," which surprised them both. "He was a brilliant man," Husna recalled. "He wanted to do many things," and "had an intuition that he didn't have enough time. . . . Mostly whenever we were together by ourselves we would discuss politics."[35]

Zulfi flew to Dacca as often as he possibly could, and before year's end Nusrat knew, of course, that something had possessed him. She challenged him. Zulfi not only confessed his new love but bragged of it, extolling the brilliance of his dusky new beauty. As mother of his four children and still no negligible beauty herself, Begum Nusrat refused to be shunted aside and remain silent as his village-cousin bride had done without cry or whimper. But all her tears and pleas proved unavailing because Husna's potent appeal was not simply physical but also cerebral. She pandered to Zulfi's not inconsiderable ego, whenever they discussed politics and world affairs, after the urgent flames of passion died down, for she understood power politics almost as well as he did. And she flattered him without seeming to flatter, even as she sated him without tiring herself, stimulating his mind, body, and spirit, rousing him to peaks of excitement he had never known, awakening powers left so long to rusticate back home he had thought them almost moribund.

Dacca with Husna beside him was now to Karachi what Karachi had been to Larkana when first he met Nusrat and danced close to her, what Berkeley had once briefly been with Carolyn. Nusrat then had been as clever as Husna was now in teasing him, catering to his ego, ensnaring his romantic spirit, exciting his passionate dreams and desires with every smile, wink, taunt, or defiant toss of her head. For Zulfi's proud, vain, arrogant, insecure, clever, scheming, easily bored, spoiled psyche, nothing was as comforting as a beautiful woman who devoted herself fully to his needs, desires, and dreams, rousing his hopes and calming his darkest fears, as his own mother once had.

"There was once he fell in love," Begum Nusrat recalled tearfully. "He had an affair with . . ."—but the name was too painful to utter aloud—"another person. . . . He used to come and tell me about it and apologize and say he didn't know what happened to him . . . and I said I didn't want to stay with him." Nusrat, whose Persian name means "victory," soon found that neither tears nor righteous indignation lured him back from the satisfaction—physical, mental, maternal, political—he had found in Dacca. "He had to be *best* in everything." Husna smilingly remembered, "the world's best lover, as well as leader! And he *was!*"[36] Husna's telling him that he "was" seems for a while to have sufficed to keep Zulfi so enamored and magnetized by his "Black Beauty" that in 1962 he actually threw out Begum Nusrat after one of their harsher fights over his constant infidelity.

"We were at Pindi, in Flashman's Hotel," Ardeshir Cowasjee, heir to Pakistan's first Parsi shipping empire, then still a friend of Nusrat and Zulfi's recalled. "Begum Nusrat had walked over from Mrs. Davis's Hotel to cry 'Zulfi has thrown me out,'" and asked Ardeshir and his wife, Nancy, for "help" in getting her to Ayub Khan or his begum. Nancy and Ayub's daughter were very close friends, so the President's House car was soon sent around to drive Nancy and Nusrat to Government House to see Begum Ayub. Nancy recalled how Nusrat tearfully recounted that she had been thrown out of her own home by her faithless husband, "bag and baggage." Before the day ended, Zulfi was called on the carpet by Ayub Khan and ordered "to take his good wife back" or "quit the cabinet." Zulfi acquiesced, but the next day he marched up to Ardeshir, whom he had once fondly addressed as "Tippu Sultan," hissing with rage, "I'll *never* forgive you for this!"[37] Nor did he ever forgive Ayub either. Zulfi's memory was as long as his temper was hot.

Ayub, however, quickly forgave Zulfi, as a good general must forgive those who accept his orders and surrender to his superior force. He might have appointed Zulfi foreign minister a bit earlier had it not been for the flap over Nusrat because Manzur Qadir, weak and tired, had stepped down before the end of 1962. Ayub chose, however, to appoint Mohammad Ali Bogra, who was still in Washington and could easily be deputed to New York as foreign minister, his last official assignment. In January of 1963 Bogra suffered a fatal heart attack, and Ayub gave Zulfi the chance he'd been waiting, working, itching for almost all his adult life. It was quite a birthday present, promotion to foreign minister at only thirty-five. Zulfi was then old enough to qualify for the presidency under the terms of Ayub's 1962 Constitution. He considered himself much better qualified for that job than its incumbent but would wait a few years longer before making his move to displace Ayub.

6

Foreign Minister to the Field Marshal

(1963–1965)

Zulfi went to New York again as head of Pakistan's UN delegation in late 1962, and was then engaged in a six-month series of talks on Kashmir with Sardar Swaran Singh, leader of India's UN delegation. India's Prime Minister Jawaharlal Nehru also went to New York for the UN session in late 1962, staying at the Hotel Carlyle on Fifth Avenue, just a few blocks from where Zulfi always stayed, the Pierre. Nehru invited young Bhutto to his suite for tea. They had first met in Bombay a quarter century earlier, when Zulfi was just a boy and the British still ruled over India. Pakistan was as yet unborn.

"We were standing on the top floor in the drawing room of his suite by a window," Zulfi recalled. "He had his leg on the railing. He was looking down at the traffic and talking about the past. He said 'Look how small we are,' as he pointed to the people and the traffic below. . . . I nodded. He then said this world has to be saved from destruction, from atomic weapons."[1] That was Nehru's last visit to New York. He died just a year and a half later. His immediate successor would be Lal Bahadur Shastri, who would be followed to the pinnacle of South Asian power, soon afterward by Nehru's politically astute daughter, Indira. In many ways Zulfi admired and sought to emulate Nehru, whose foreign policy of "nonalignment" and friendship with China had drawn heavy criticism from Washington yet paid handsome dividends in winning India support not only from the Soviet Union and China but throughout the Third World.

Zulfi's foreign policy was thus inspired in part by Nehru. Jawaharlal's Trinity, Cambridge training and Fabian Socialist philosophy were an education more radical than Zulfi's but both were heirs to substantial fortunes and paternal political wisdom. Motilal Nehru (1866–1931) had presided over India's National Congress, in 1919, before he managed his son's successful candidacy for that high office shortly before his own death. As chief architect of independent India's foreign policy, Jawaharlal had eschewed all Western pressures to join in Dulles's "pactitis," yet won support for Indian development from every major industrial power of the world, including the United States. Nehru's global vision and foresight gave India almost undisputed leadership of the underdeveloped Third World of Asia and Africa as well as eastern Europe's Yugoslavia. China's Foreign Minister Chou En-lai would challenge Nehru and India's primacy in that important uncommitted portion of the world, of course, as would Zulfi Bhutto later. Despite Bhutto's admiration for and emulation of Nehru's neutrality, Kashmir proved too explosive an issue for India or Pakistan to defuse short of war.

Zulfi's first major achievement as foreign minister was to conclude a Sino-Pakistan boundary agreement on 2 March 1963 that became the cornerstone of Pakistan's strongest, most important Asian alliance. The long-standing Sino-Indian cold conflict over the border turned into a hot war in November 1962, when Chinese forces poured over India's northern border and wiped out Indian troops, heading virtually unopposed toward Assam's capital of Gauhati. China's recently discovered road built across the Aksai Chin in Ladakh had triggered earlier border clashes with Indian police in that remote region, but by the end of 1962 it seemed as if Asia's two gigantic powers were headed for all-out war. President Kennedy, responding to the urgent call from his New Delhi ambassador, John Kenneth Galbraith, to whom Prime Minister Nehru had appealed for help in desperation, immediately ordered the airlift of enough U.S. military equipment to arm and outfit ten new northern-tier divisions of Indian troops. As the first U.S. military jumbo jets landed in West Bengal, China's forces swiftly opted for unilateral withdrawal, leaving Indian territory as suddenly and unexpectedly as they had invaded. Pakistan's conclusion of a boundary agreement with China in the wake of that Sino-Indian war, ceding the Aksai Chin region that it did not actually control at this time to its powerful northern neighbor, was viewed with anger and alarm by India. New Delhi charged Pakistan with having "unlawfully ceded" more than two thousand square miles of Indian territory to China, vehemently protesting in the UN Security Council that the government of Pakistan had "unilaterally altered" and "unlawfully" apportioned Indian Union territory to China in violation of UN resolutions of 17 January 1948 and 13 August 1949.

Foreign Minister Bhutto countered India's charges on 25 March 1963, calling them "part of a systematic and sustained campaign of propaganda against Pakistan." He insisted that "Pakistan has not ceded even one square inch of territory to China," but "has gained 750 square miles . . . which had till then been in China's occupation and control."[2] Bhutto used the opportunity of refuting India's charges to reiterate Pakistan's claims to all of Jammu and Kashmir, and its demand that a plebiscite to held throughout the militarily divided state. "The charge of aggression against Pakistan has been repeated *ad nauseam* by India over the last 15 years." Jammu and Kashmir was not, Zulfi argued, a part "integral or otherwise" of India but belonged to the people of Jammu and Kashmir, the future of whose territory must be decided with "an impartial plebiscite under the auspices of the United Nations." It was outrageous of India to claim sovereignty over it. He saw no substance, moreover, to India's charges that Pakistan had violated the Security Council Resolution of 17 January 1948 in reaching a border agreement with China. To the contrary: by agreeing to "delimit and demarcate" its boundary with China, Pakistan had helped to improve the region's prospects for peace. Foreign Minister Bhutto then listed no fewer than eleven violations of the same UN resolution that had been "committed by India in Kashmir" since 1948. Pakistan was ready to conclude a truce agreement with India, "here and now," so that both Pakistani forces and the bulk of the Indian army might be withdrawn in "a synchronised manner."[3] The withdrawal would enable a plebiscite under UN auspices.

That July, Bhutto spoke again on Kashmir, in Lahore, responding to what he called "vile and slanderous attacks" against Pakistan by Prime Minister Nehru. "Kashmir is to Pakistan what Berlin is to the West," Zulfi argued, echoing President Kennedy's cry at the Berlin Wall. Without "a fair and proper settlement of this issue the people of Pakistan will not consider the crusade for Pakistan" complete. The continued conflict in Kashmir "threatens the peace and security of the world," and was an issue hanging "heavily on the conscience of mankind." Pakistanis would not "forsake a righteous cause merely because more bayonets and bullets may be supplied to India from any source to consolidate her usurpation of Kashmir." How could "deception and fraud replace truth and virtue?"[4]

Zulfi's rhetoric in Lahore boiled close to the point of ignition. He had engaged in prolonged, fruitless negotiations with Swaran Singh for half a year—so long that he had learned to mimic each of his Sikh adversary's "traits and foibles," entertaining his "Ministry officials," as Salmaan Taseer reports. "How he scratched his ear when he was thinking; his nose when he needed time; or rubbed his forehead when being evasive."[5] Zulfi was almost as good at mimickry as he was at debate.

Jawaharlal Nehru, then in his last year of life, would not budge on Kashmir, fearing, he said, a "Hindu backlash" within India should a plebiscite be held. Nehru was unwilling to risk losing the "lovely" Vale, over which he continued to wax poetic, calling it his "ancestral home," though the Nehrus had migrated from Kashmir to Delhi and Allahabad several centuries earlier. He also compared Kashmir to a "beautiful woman," and like Zulfi, Nehru had a weakness for women.

A few days after his Kashmir speech in Lahore, Zulfi drove to Karachi, where he suffered such acute stomach pains at 3:00 A.M., shortly after going to bed, that his doctor rushed him at once to the hospital for an emergency gall bladder operation. Zulfi had suffered periodic bouts of malaria, compounded since childhood with influenza, but this episode was his most acute illness to date. His usually robust body remained so weak ten days later that when he spoke to conclude the National Assembly's foreign policy debate, he requested the chair's "indulgence" to remain seated.

"Unfortunately, India is the spoilt child of the world," Foreign Minister Bhutto, who well understood that syndrome, began. "India gets away with all its machinations by irrational explanations which the world only too readily swallows. The misfortune of this region is that the powers which are not familiar with India's mentality and do not understand India's approach to international problems are only too eager to accept India's policies at their face value. That makes it possible for India to continue to menace the peace of the region and the world."[6] Up until then India had committed aggression no fewer than five times: in Kashmir, Junagadh, Hyderabad, Goa, and China. In contrast to Pakistan's good relations with neighbors like Nepal, Ceylon (Sri Lanka), Indonesia, Burma, and Afghanistan, India's regional posture was one of "arrogance and intransigence." Yet India asserted it was "a peace-loving state."

Zulfi defended the wisdom of his ministry's negotiations with India on Kashmir and of his Kashmir policy against opposition criticism. To those who said that by entering into the talks with India "we compromise the Kashmiris' right to self-determination," Zulfi declared that in every round of his talks with Swaran Singh, "the right to self-determination of the people of Kashmir constituted our basic stand for a settlement." He went on to argue, on the other hand, that "India was a loser" in the negotiations because during the past fifteen years, its position had always been that "the problem of Kashmir had been settled and finished." By the mere act of reopening negotiations, therefore, realpolitical Bhutto insisted, "India admitted the existence of the Kashmir dispute," which it had come to settle on "an equitable and honourable basis." He took credit for having "made the Kashmir problem a live problem again." India also acknowledged that fact, Bhutto was proud to report. He assured his listeners that "India cannot keep Kashmir under subjugation much longer, [because] this state of affairs is bound to result in an explosion."[7]

In November 1962 Ambassador Galbraith had wisely suggested to President John Kennedy making reopened negotiations on the Kashmir plebiscite a prerequisite to the immense airlift of U.S. military aid to India, but Kennedy opted to send arms as swiftly as possible, abandoning any potential leverage Washington had to bring Nehru back to a plebiscite table. Nehru did, however, offer Pakistan a "No War Pact," but Zulfi rejected that as "sinister": "A No War Pact with India can have the effect only of lulling us into a false sense of security and making us feel that India would not resort to force against us. Then, we could become easy victims of Indian aggression."[8]

Foreign Minister Bhutto thus restored Pakistan's policy on Kashmir to its pre-Liaquat Ali Khan phase, proclaiming total mistrust of Indian promises as well as intentions in this area, vowing never to give up on the Kashmir "problem", no matter how long the struggle. "This problem, I declare, must be settled, and it will be settled, because no one can deny justice for all time to the people of Kashmir. Future history will show that the people of Kashmir will not forever be denied their inalienable right of self-determination." At the same time, he appealed to the West, to Pakistan's allies, to remember that "India has repeatedly said that Pakistan is India's Enemy Number One." "It is India that has committed aggression."

Before leaving the National Assembly, Zulfi bolstered the sagging spirits of his countrymen by informing them, "We have our friends. . . . We have assurances also from other countries that if India commits aggression against us, they will regard it as aggression against them . . . we shall never be alone in facing aggression." Here he was referring to Chinese "promises" he said he had received in Beijing during his recent visit there to sign the boundary agreement. He may also have had whispered words of reassurance from some of his closer friends in the Trucal sheikhdoms, possibly from North Africa, and from his good friend President Sukarno of Indonesia.

On his first visit to Washington as foreign minister, Zulfi met with President Kennedy in the White House in October 1963, a month before the president's assassination. "We talked of the Sino-Indian conflict, the decision to give India military assistance," Bhutto recalled, "the consequences of an India armed to the teeth for her neighbours, and the future prospects of US-China relations."[9] Zulfi urged the "bold and imaginative President" to recognize China, but Kennedy said that only "if the American people will endure me for a second term" would he possibly venture to "break new ground" in that part of U.S. policy. Bhutto liked Kennedy and the feeling seems to have been reciprocated, for he recalled that as the president shook his hand before leaving, he remarked, "If you were an American you would be in my Cabinet." Zulfi sharply retorted, "Be careful, Mr. President, if I were American, I would be in your place." At which they "both laughed heartily."

In New York, Zulfi attended the UN General Assembly session and enjoyed a number of elegant parties, private as well as diplomatic. He always found the high life associated with UN meetings stimulating, enjoying beautiful women, good food, and high spirits of every sort. One of the women he met at a cocktail party that fall was Rita Dar, daughter of Madame Vijaya Lakshmi Pandit (1900–90), the first woman president of the General Assembly. Mrs. Dar recalled how immediately after meeting her, Zulfi eyed her lasciviously, inviting her to "come up to his apartment . . . that sort of thing." To her irate refusal Zulfi replied, "Any woman in Karachi would be honored to be asked!" Mrs. Dar responded that she was not from Karachi but from Delhi, at which point he turned "abruptly" and stiffly "walked away."[10]

By late November 1963 Foreign Minister Bhutto was back in Karachi. In an address to the prestigious Lions Club there on the United Nations and world peace, he called for an early end to underground atomic testing and the spread of nuclear weapons. He advocated superpower disarmament, and urged cutbacks in conventional and nuclear forces by East and West. However, world peace was perhaps "least secure" in Asia: "This vast and ancient continent, inhabited by more than half the population of our planet, continues to be the scene of great convulsions which may well change the destiny of mankind." He noted that though Pakistan had friendly relations with China, Iran, and Burma, recently concluding agreements with each, border disputes with India remained "volatile," most of all in Kashmir. He hailed a recent pan-African meeting in Ethiopia and called for Asians to adopt greater "continental consciousness." Sukarno of Indonesia, one of Zulfi's role models, was planning to host a Pan-Asian Conference in Djakarta. Indonesia's Bandung Conference of 1954 had spelled out ten principles of "international conduct," among them was elimination of all forms of colonialism from Asia and Africa, and the famous "Five Principles" *(Panch Shila)* of peaceful coexistence that Nehru had so warmly endorsed. "The time has come," Bhutto urged, "to convene a second Asian-African conference to review the conclusions reached by the first and to revitalise and renew its pledges which still remain unfulfilled."[11] A few months later Sukarno announced that the second Indonesian Afro-Asian Conference would be held in April of 1964, but it was never convened.

In December 1963 a sacred hair of the Prophet Mohammad kept in the Hazratbal Shrine near Srinagar turned up missing, apparently stolen, triggering riots that swept across Kashmir and left death and fear in their wake. India flew in fresh troops and tightened its grip over Kashmir, whose special status under Article 370 of the Indian Constitution had been somewhat eroded by making Articles 356 and 357 applicable to Kashmir in October 1963. In January 1964 Foreign Minister Bhutto requested an urgent meeting of the Security Council to consider the "grave situation" in Kashmir, a situation he attributed to India's "unlawful steps" in "arrogant disregard of the resolutions of the Security Council."[12] He viewed recent Indian actions as part of a "sinister design" over the past fifteen years to "obliterate the special status of the State of Jammu and Kashmir" by refusing to honor its own earlier "commitment" to the holding of "a free and impartial plebiscite" under UN auspices.

En route to New York, Zulfi stopped in London, where he met with Foreign Minister Alec Douglas Home and Commonwealth Secretary of State Duncan Sandys to muster British support for his Security Council appeal. Interviewed by the BBC and asked, in light of Kashmir's having been discussed in the UN Security Council no fewer than 109 times over the past fifteen years, would the 110th discussion be "likely to bring the problem any nearer to solution?" Bhutto replied. "We are prepared to discuss it for a thousand times, to see that the problem of Kashmir is settled in an honourable manner." Pressed for the proposals he would present to the UN, he reaffirmed his faith that a "plebiscite is the solution . . . the basic factor involved is the right of self-determination for the people of Kashmir."[13]

As violence escalated in Kashmir, the Security Council, to avoid a further Soviet veto, appealed to India and Pakistan to settle their differences in Kashmir by direct negotiations. India then announced that Nehru was ready to talk with Sheikh Mohammad Abdullah (1905–89), the state's most popular political leader and a former chief minister who had been dismissed from that office in 1953 and held in detention for more than eleven years (with a three-month interval of freedom in 1958) for having publicly proclaimed in June 1953, "What is now wanted is that the people of the State be given an opportunity to decide their future freely and without fear."[14]

Sheikh Abdullah, hailed as the "Lion of Kashmir," held a press conference early in April, at which he cautioned that the only "alternative to a negotiated settlement" for Kashmir was "a clash of arms" between India and Pakistan, which would be "suicidal."[15] The sheikh opposed permanent partition and fearlessly declared from crowded platforms throughout the Vale, "They think of rearresting me, but if Abdullah is rearrested will the Kashmir issue be settled?"

The People's Republic of China had not as yet been admitted to the UN. Still, Chou En-lai flew to Pakistan at this time to underscore China's support for the Pakistan position on Kashmir: the problem should be "settled . . . in accordance with the wishes of the people." Nehru's health was fast fading and he hoped to avoid further conflict with China as well as Pakistan, hence, he had ordered Abdullah's release, wishing he could use that most popular Kashmiri to negotiate a peaceful settlement of this persistent South Asian problem. Zulfi Bhutto flew to Indonesia early in April to meet in Djakarta with Sukarno and Foreign Minister Subandrio, as well as with Chou En-lai. Zulfi was ready to sign an anti-imperialist resolution on Vietnam at that agenda summit for the still-scheduled second Afro-Asian Conference, but Ayub restrained him, fearing that such an announcement would jeopardize U.S. aid to Pakistan. Zulfi managed, nonetheless, to get his friend Sukarno to endorse a statement on Kashmir, similar to the one Chou had made in Pindi.

From Indonesia, Bhutto flew to Washington for a CENTO ministerial meeting in late April. There he addressed the National Press Club.

Most of what Bhutto said in Washington focused on the Kashmir dispute, "the bane of all troubles and problems, not only between India and Pakistan but in that whole region . . . even beyond the subcontinent."[16] The people of Kashmir were "in revolt, unmistakably in revolt." The problem "can be settled," "will be settled," "has to be settled"—only then would Pakistan be able to "live in peace with our great neighbour, India." He also touched on the need to end poverty and squalor throughout Africa and Asia. And on Pakistan's opposition to apartheid because "we believe in the equality of all men," adding that with India's "rigid caste system," its antiapartheid statements made him think aloud, "Physician, heal thyself." It was an effective, brief speech.

Soon after returning home, Zulfi flew from Pindi to New Delhi with Sheikh Abdullah to attend Nehru's funeral, just after his death on 27 May 1964. They had sufficient time in that hour's flight to discuss a fallback position on Jammu and Kashmir. The Sheikh advised Zulfi to hold tenaciously to the plebiscite demand for the entire state, but suggested that Jammu and Kashmir's partition below the river Chenab at a point called Pethlinot would be a "realistic" final position. Bhutto was "elated" by the Sheikh's flexibility; in earlier talks at Pindi with Ayub, Abdullah had firmly insisted that any partition was unacceptable. A few days later Zulfi met with Kashmir's other major political leader, Mirza Afzal Baig, who basically agreed with Abdullah's position. Baig, however, urged that Pakistan not abandon some forty thousand Muslims living in Doda District at that time, and Bhutto promised to keep them in mind.

Bhutto placed a wreath on Nehru's bier on behalf of Ayub Khan, and then spoke with Nehru's mourning daughter Indira Gandhi (1917–84) in Teen Murti House, where Nehru's aorta had burst. The USSR's Premier Aleksei Kosygin and Sri Lanka's Prime Minister, Mrs. Bandaranaike, were also present, as was Sir Alec Home and a host of Indian dignitaries. Bhutto joined the funeral procession despite Delhi's sizzling heat on that Thursday, the twenty-eighth of May.

The next morning, Zulfi called on India's philosopher president, Dr. S. Radhakrishnan, at Rashtrapati Bhavan. Radhakrishnan, who had lectured at Oxford, was a more polished orator than Bhutto. They talked about Kashmir, and agreed that its early "honourable and equitable" resolution would "open up" many possibilities for economic and cultural cooperation between India and Pakistan, to the "mutual benefit" of both nations. Having been born in South India, Radhakrishnan was hardly as emotionally committed to Kashmir as Nehru and other North Indian Brahmans, including Indira Gandhi.

From the president's office, Zulfi went to Lok Sahba, where he met with the leading candidate to be chosen as Nehru's successor, Lal Bahadur Shastri (1906–66). Small and frail though he was, Shastri would soon prove himself as strong a leader of India's polity as Nehru had been, equally determined to retain Kashmir's Vale as an integral part of India's union, even at the cost of war with Pakistan. Bhutto was to meet him again, in Tashkent, but only after a year of diplomatic sniping that had erupted in two rounds of war between their countries. Had Zulfi underestimated Lal Bahadur's resolve in their first brief meeting? Shastri's diminutive stature and lack of polish and sophistication may have misled Zulfi to misjudge his adversary's tenacity.

Acting Prime Minister Gulzarilal Nanda received Foreign Minister Bhutto coldly in the prime minister's handsome office in South Block that Friday afternoon. As Nehru's home minister and the senior member of his cabinet, Nanda was in charge during the transitional mourning interval. It soon became clear to observers that he hoped to remain in the office, which few Indians other than Nanda himself thought him fit to fill. Ascetic Hindu that he was, Nanda hardly found Bhutto's manner, appearance, or language congenial, never warming to this visitor. Nor did he say anything to give Zulfi reason to hope for an easy settlement of the Kashmir problem. Zulfi next went to see Congress President K. Kamaraj Nadar, the Tamil "King of Love" who would soon become India's "kingmaker," managing the succession to power first of Lal Bahadur Shastri and then of Indira Gandhi, outmaneuvering their most powerful competitor, Morarji Desai, who was obliged to remain alone in his Gujarati stronghold in western India. Kamaraj was willing to listen to Bhutto, smiling sweetly at his flawless English, little of which Kamaraj understood and virtually none of which he spoke.

Jaya Prakash (J. P.) Narayan (1902–79), though he held no office at this or any other time during the remaining fifteen years of his life, was one of India's leading Socialists and a disciple of the saintly Vinoba Bhave, with whose triconfederation solution for Indo-Pak differences in Kashmir J. P. agreed. Narayan was to be the spiritual leader of the Janata (People's) movement, destined to topple Indira Gandhi's Congress government in the mid-1970s. J. P., like Zulfi, had been educated in the West, had a cosmopolitan intellect open to winds of change and ideas from every continent. They talked cordially through the late afternoon and dined together in Zulfi's suite atop New Delhi's Ashok Hotel.

Bhutto met with Indira Gandhi again on the morning of 30 May, returning briefly to Teen Murti House shortly before leaving Delhi. Sheikh Abdullah came to the airport to see him off, but too many other ears were there for them to have further private conversation.

The early-May Security Council meeting on Kashmir in New York had simply urged both parties to "abstain from any act that might aggravate the situation" in Kashmir. It was less than Bhutto had hoped for, but Sheikh Abdullah's release, and Nehru's death before May ended gave Zulfi reason to hope for perhaps some movement toward a resolution of that premier South Asian problem, possibly along lines of Sheikh Abdullah's partition-fallback plan. Yet as June with its sweltering heat wore on in Karachi and Pindi, no change was seen in New Delhi's tough stand on Kashmir. Shastri was elected prime minister by his Congress Party, thanks to Kamaraj's support. Soon after taking charge in Delhi, Shastri had suffered a "minor" heart attack, which made him seem even weaker. India's high commissioner called on Foreign Minister Bhutto in Karachi near the end of June, using Shastri's illness as the reason for postponing talks or negotiations toward possible settlement of the Kashmir dispute during the July Commonwealth Conference, scheduled to be held in London.

Zulfi by now believed that no Indian leader was "serious" about reviving the plebiscite solution for Kashmir, and suspected that Shastri would use his physical frailty as an "excuse" for further procrastination. By mid-1964, then, Zulfi made up his mind that the only solution was a military one. Seventeen years, he insisted, was long enough for Kashmir to wait for an impartial polling of its people. He had no doubt what the result of a plebiscite under UN auspices would be. No Pakistani did. Nor did he believe that any shrewd leader in Delhi doubted the outcome. The crux of the matter was to arrange Kashmir's "liberation" without alarming Pakistan's Western allies enough that they would cut off the vital material required to achieve it.

In August of 1963 the United States had suspended a promised loan of more than $4 million for Dacca airport, when Pakistan signed its agreement on civil aviation with China. Throughout 1964 U.S. Secretary of State Dean Rusk reaffirmed the importance of "steadfast" U.S. support to "Indian economic development and defense efforts" as a counter to China's growing strength and potential danger. India was viewed by Pentagon as well as State Department policy planners as the major "counter" to China in Asia, despite what that might mean toward undermining the U.S. alliance with Pakistan.

Ayub was preoccupied by mid-1964 with the most formidable political challenge to his power. In December 1963 he had taken on another presidency, that of the Pakistan Muslim League as well as of the nation. His "constitutional" program of Basic Democracies, had promised national elections in 1964–65. East Pakistan continued to seethe with antipathy to West Pakistani "imperial" domination. Suhrawardy died suddenly, his heart failing in Beirut in September 1963, and Sheikh Mujibur Rahman (1922–75), his leading lieutenant, vowed to avenge his leader's suspicious death, but was left first to try holding the Awami League together. Mujib was too young and unknown in the West himself to run against Ayub for the presidency. A much older Bengali leader, former Governor-General Khwaja Nazimuddin, then in the last year of his life, brilliantly brought together all the leaders of a broad spectrum of political opposition to Ayub—from right-wing Jamaat-i-Islami mullahs of the West to left-wing Maulana Bhashani's National Awami Party of the East, from conservative Council Muslim Leaguers, under the North-West Frontier Province's (NWFP's) Khan Abdul Qayyam Khan (1901–89), and the Nizam-i-Islam Party old guard under Chaudhri Mohammad Ali to the radical Awami League of Mujib himself. The

Combined Opposition Party (COP), as they were called, found their ideal standard-bearer against Ayub Khan in Fatima Jinnah (1893–1967), the gaunt, gray-haired, ascetic sister of the Quaid-i-Azam, who at age 71 took up the challenging burden of campaigning nationwide against the field marshal.

"Mother-of-the-Nation" *(Mader-i-Millat)* Fatima Jinnah was still a formidable presence. Her name alone won her widespread support. She tried to lure Zulfi Bhutto to her side, offering him the same job he was then holding under Ayub, if he joined her crusade. Zulfi, however, knew Ayub and West Pakistan's tough Punjabi chief minister, the nawab of Kalambagh, too well to leave his secure perch at the Foreign Ministry to fly over to COP's mélange of wrangling outsiders. In the election roughly two-thirds of the eighty thousand enfranchised Basic Democrats did exactly as they were expected to do: reaffirmed their loyal faith in President Ayub Khan. That Miss Jinnah secured 36 percent of the votes (mostly from the East) was in itself a miraculous victory. She insisted ever after that she had, in fact, actually "won" that "rigged election," and should have become president of the Land of the Pure her Great Leader brother had founded. Ayub and his cohort "cheated" her, she believed, though few Pakistanis, in the West at any rate, were in 1964 ready to be ruled by a woman.

"Now Miss Jinnah had nothing in common with the various opposition parties yet she did not hesitate to come into the field," Ayub noted in his memoirs, typifying Pakistan's prevalent male chauvinism. "Since the entire Opposition campaign was to be based on emotionalism, her choice seemed logical . . . it was going to be emotionalism all the way . . . because she was after all an old lady and widely respected as sister of the Quaid-e-Azam."[17]

Zulfi helped Kalambagh stage the Lahore campaign at Mochi Gate, Pakistan's most crucial political testing ground. They had to be sure Ayub didn't "fall flat on his face," for he was scarcely an orator. Kalambagh took care of arranging for the transport and sustenance of the civilian "troops" to that well-staged scene, and suggested readings from the soul-stirring Urdu poetry of Allama Iqbal to keep the crowd in a good mood until the field marshal's arrival. Zulfi coached Ayub on how to speak, simplifying his political message to so huge an audience, teaching him to repeat the few most stirring phrases and important points, and how to avoid getting bogged down, as he usually did, with facts and figures, conceding no failure or fault, while slinging mud and dirt at the COP.

Zulfi was passing on to Ayub his favorite tricks of rhetoric, all of which he would soon unleash in most virulent form against him. He also launched his uncle-cousin Mumtaz's political career at this time. As foreign minister, Zulfi had no time to campaign in Larkana. He was, in fact, in New York for the special Security Council meeting in May when Mumtaz rang him up the day before Larkana nomination papers had to be filed to ask, "Who should be the candidate from here?" Mumtaz himself did not want to run. Zulfi replied, "You file the nomination papers. I'm coming back shortly and if you want to withdraw then, withdraw." Mumtaz filed and was elected unopposed, "So then he came back and said, 'Now, why do you want to withdraw? Why should we lose the seat?' So I stayed. He knew he could rely on me."[18]

Everything seemed to be going Zulfi Bhutto's way that year. Almost everything. Husna had left her husband and moved to Karachi. Zulfi found a furnished house for her in Clifton, just a few minutes from his own larger place at number 70, a newer walled building off Sunset Boulevard, where they enjoyed watching popular videos late into the night, whenever they were not talking politics, or making love. Begum Nusrat had the children at home for holidays, but both Benazir and Sanam, then aged 10 and 7, were at private English boarding school in Murree that year; Murtaza was 9, and at boarding school in Karachi. Only baby Shah Nawaz, then just 6, remained at home. Nusrat knew well what her husband was doing, and decided that she wanted no more of him. Ayub's warnings and avuncular advice had wrought no change in Zulfi's heart or behavior. He was a bit more circumspect in his nocturnal meanderings for a while, but eventually, Nusrat recalled, "I said, 'I don't want to stay with him. I told him I'll take two children and you keep two children and I'll just go away.'" That was when she finally left him, flying home to her father and sisters in Iran, hoping to find peace there. She could no longer pretend to be happily married to a man who was never at home or, worse still, who yelled at her and lied whenever he came within range. She tried to take Shah Nawaz with her, but Zulfi adamantly refused to let any of his children leave the country. He attempted forcefully to stop her from going as well, but soon recognized that she was determined to leave him. He knew that in many ways she was much stronger, "the man in their family," as her father had once told him. So he had let her fly, but not one of *his* children was permitted to leave *his* country with their mother! Nor would he let any of them write to her as long as she was outside Pakistan.

For six months in Persian exile Nusrat heard nothing from any of her precious children. It was too high a price for her to bear. Begum Bhutto returned for her children, and tried her best thereafter to "pretend I don't see" Zulfi's nocturnal comings and goings. She was always the perfect hostess, of course, at official parties, which helped her somewhat to keep the pain and sorrow that weighed so heavily on her mind at bay. Some eight years later, however, Nancy Cowasjee recalled that Begum Nusrat "attempted suicide by swallowing an overdose of barbiturates."[19] Her husband's physician, Dr. Nasir Sheikh, reportedly averted the tragedy at Karachi's Parsi Hospital.

By the fall of 1964 Pakistan and India had moved closer to war. On 24 September Foreign Minister Bhutto endorsed a warning letter just received from the high commissioner in New Delhi, alerting him to India's acquisition of powerful new arms. "India is going way ahead," Bhutto told Ayub's Secretariat, "Her temporary difficulties should not deceive us. She is taking from both Camps and ignoring protests in the wake."[20] Shastri's defense minister, Y. B. Chavan, had proposed building up India's army to a well-equipped standard force of 825,000 men, with forty-five air squadrons and a navy that included submarines. Chavan had flown to Washington in May, requesting F-104 fighter-bombers of his Pentagon counterpart, Robert McNamara. Hindustan Aeronautics of Bangalore had just built and successfully tested two HF-24s,[21] and Vice-Premier Kosygin of the Soviet Union had offered to produce MiGs in

India during his visit to Delhi for Nehru's funeral, but Pakistan's air force was armed with F-104s, giving Pindi an edge over India that Chavan was eager to remove. He was scheduled to meet President Lyndon Johnson on 28 May, but flew home for Nehru's funeral instead that morning in Secretary of State Dean Rusk's plane. At the end of August Chavan flew to Moscow for a two-week tour of Soviet air and naval centers; he traveled in a submarine in the Gulf of Finland, and met with General Malinovsky, then minister of defense, and Nikita Khrushchev, just weeks before the latter was ousted by his Kremlin comrades.

At the urging of the U.S. ambassador to India, Chester Bowles, President Johnson invited Prime Minister Shastri to Washington that fall. When Ayub learned of the invitation, he voiced so strong a personal complaint that Johnson backed down, pleading "indisposition." Shastri was bitterly offended by that diplomatic slight but went to Canada as planned in October, refusing even to speak to Johnson when the White House called him in Ottowa,[22] then accepting with alacrity an invitation from Kosygin to visit Moscow on his way home. Shastri stopped in Karachi on his return flight, and met with Ayub on 12 October 1964 for a private and informal summit, their only meeting prior to Tashkent.

By that time there were daily "incidents" along the cease-fire line that separated Pakistan's Azad Kashmir from India's Jammu and Kashmir state. Shastri warned Ayub that there was "danger" that they might "escalate into something bigger." India's diminutive prime minister told Pakistan's martial president that he would like to see the incidents in Kashmir stopped, but Ayub countered that after seventeen years of India's "broken promises" and "hostile attitude" toward Pakistan there was an "almost undeclared war" going on. He complained about recent murders and violent evictions of large numbers of Muslims from India's Assam and other eastern states sparsely populated by the Naga and Mizo tribes into which landless Bengali Muslims from East Pakistan had migrated. Those peasants had cleared and cultivated jungle land in India, only to be terrorized now, butchered in their villages, or forced to flee for their lives. Shastri promised to "look into" that, expressing his willingness to meet with Ayub to discuss all such matters and "iron them out." Then Ayub reverted to Kashmir, an "even bigger problem." Shastri declared himself not "strong enough" to carry "his people" with him toward renegotiation of the Kashmir dispute "at present."

Ayub sent a note on the talk with Shastri to Foreign Minister Bhutto as well as to Foreign Secretary Aziz Ahmad.[23] The Indians were obviously "preparing for War," and soon would have their own MiG factory assembling the USSR's most modern weapons. They might even get F-104s if Washington felt alarmed enough about losing India's friendship and nonaligned "support" to its rival superpower. Time was therefore seen to be on the side of early action by Pakistan, as Zulfi Bhutto had been urging all that year. Ayub, moreover, would soon have a fresh electoral mandate, his position as president secure for another five years. Ayub personally was convinced after seeing frail little Shastri that his Indian counterpart was too enfeebled to respond with more than a whimper to any frontal assault or outright war. Shastri had turned over his Ministry of

External Affairs portfolio to Swaran Singh by this time, and Bhutto's contemptuous mimickry of the white-bearded Sikh with whom he had long argued over Kashmir was well known to Ayub.

On Zulfi's recent visit to Djakarta, President Sukarno promised him Indonesian "naval support" for Pakistan in any war against India, even as China had promised "military support." Chavan appealed to Great Britain for military aid, but the best the British would do for India was to provide naval assistance under the cover of a joint U.S.-UK naval presence in the Indian Ocean in support of Malaysia. India's policy of nonalignment still paid off, however, helping New Delhi secure military muscle from the West as well as from the Soviets. Washington was alarmed to find Foreign Minister Bhutto flirting so ardently with Indonesia and China, and no longer felt fully confident of Pakistan's "most-allied" status.

All eyes in South Asia were focused on Jammu and Kashmir, but the first round of fighting erupted at much lower altitudes early in April 1965. The Rann (salt marsh) of Kutch, the often submerged lower borderland between Sind and Gujarat hardly seemed worth fighting over. Pakistan's long-standing claim to the northern half of the Rann, covering over 3,500 square miles, had continued to be disputed by Indian policymakers from Nehru to Swaran Singh. In March 1964 Pakistan had established two police posts in the Rann of Kutch, which India soon asserted were well inside India's border, and hence countered with an armed post of its own.

A warning was issued by Indian Home Minister Nanda in Lok Sabha on 7 April that New Delhi would take "effective measures to remove intrusions."[24] The next day an Indian note to Pakistan's Foreign Office protested a "massive concentration" of Pakistani arms on the border closest to the new outposts. U.S. Patton tanks had been moved into position; the Indians had nothing better than distant Centurions with which to confront them. At 3:00 A.M. on 9 April 1965 Pakistan's infantry advanced toward India's Sardar Post in the Rann. Firing ensued. India accused Pakistan of "aggressively" initiating the shooting; Pakistan insisted its patrols were moving "peacefully" between Ding and Surai when "entrenched" Indian forces "opened fire" against them. Each side thus accused the other of starting the conflict that immediately drew blood. Skirmishes continued between the border patrols for two weeks, and on 25 April Deputy High Commissioner P. N. Kaul was called to the Foreign Office in Karachi, and warned that aggressive Indian intrusions into Pakistani territory would be met by force. In New Delhi the conflict was described as "almost a state of war." Pakistan now had ninety-six Patton tanks at the front, and India claimed (Pakistan denied) that thirty of them advanced on the morning of 26 April to take the Indian post of Biar Bet. Each side declared it had inflicted heavy casualties on the other. One Pakistani general exulted, "We gave them hell!"

Zulfi Bhutto was in Sind at this time. Though he made no loud or boastful proclamations, he agreed with the consensus of foreign observers that Pakistan had soundly "beaten" India in the mini-Rann "war," a softening-up prelude to the real thing.

On 1 May 1965 Ayub informed Pakistan in a broadcast to the nation that India's "attack was foiled" in Kutch and "our forces" were in the "disputed Rann" to stay—at least until monsoon floods would drive all patrols from the region. Indian naval units moved into range of the Rann, however, and by late July some twenty thousand Indian troops were marshaled just beyond the Rann's waterline, poised to move into position inside the salt marsh as soon as the monsoon ended.

Immediately after the Kutch incident, General Gul Hasan, then director of Pakistan's military operations, met with President Ayub in his office. Ayub "took me along to brief Bhutto and Aziz Ahmad, Foreign Secretary at that time. So Ayub asked me, 'Do you think India will attack?' I said, 'Frankly no. They had chosen Kutch, way out on a limb, thinking they could get a bit of territory, but we responded, and now both armies were in place. At the same time we were planning to send these Mujahids across the border in Kashmir. A cell was constituted on Ayub's orders, and Aziz Ahmad was a member of that cell."[25] Pakistan's Kashmir infiltration was code-named Operation Gibraltar. Armed commandoes were trained in the hills to cross the cease-fire line from Pakistan and Azad Kashmir to "stir . . . uprisings" in Kashmir's Vale. The uprisings, it was assumed, would trigger Indian responses inside Kashmir, which would in turn stimulate local Kashmiri Muslims to fight for "freedom." Once the monsoon ended and the pot was briskly boiling round Srinagar, Pakistani troops could move in swiftly to "liberate" the Vale in the next operation, code-named Grand Slam. It all seemed perfectly simple that May, though General Gul Hasan had reservations, cautioning his more gung-ho colleagues in Pindi's GHQ, "Look here, we're sending in all these guerrillas. We'll escalate!"

Sheikh Abdullah had been touring abroad, openly stating in London that if a plebiscite were fairly held, the result would be Kashmir's accession to Pakistan. Upon his return to New Delhi in May 1965, the "Lion of Kashmir" was arrested and flown to South India for detention, rather than being permitted to fly north to Srinagar, where New Delhi's leaders feared he would have been received as the popular hero he remained, stirring demands for Kashmiri freedom.

As tension escalated in Kutch, Patton tanks were being moved to the border, despite Washington's insistance that such equipment was never to be used against India. President Johnson abruptly canceled a scheduled meeting with Ayub in April 1965. Bhutto immediately informed the ambassador of the Soviet Union that Ayub would be happy to visit Moscow instead of Washington in early April. Premier Kosygin jumped at the diplomatic opportunity, arranging the first Soviet-Pakistani summit on remarkably short notice. Bhutto, during the Kremlin launching of the joint oil and gas exploration venture had established cordial personal relations with Nikita Khrushchev, but Ayub's visit was a major turning point in Pakistan's international position, making it clear that India would no longer be the sole South Asian power to take arms from both the East and West. Within three years of Ayub's visit Soviet military equipment started to flow into Pakistani cantonments. Of more immediate importance, however, was the joint statement issued at the end of that brief summit, in which Soviet "support" of those fighting for their "self-determination" was affirmed, a clear reference to Kashmir.

The following month, Shastri flew to a "cool" reception in Moscow, where he soon learned that Kosygin and Leonid Brezhnev were both eager to "befriend Ayub" and hoped thereby to "wean Pakistan away from the CENTO and SEATO military pacts."[26] Shastri was, of course, assured that the Soviet Union and India were still "brothers," yet neither of the two top Kremlin leaders was willing to commit support for India's claim to the entire Rann of Kutch, or to "condemn" Pakistan's actions there, as Shastri had hoped they would.

On 27 May 1965 Zulfi reported Shastri's "monumental failure" in the USSR to Ayub, stressing in sharp contrast "the significance and the success of the President's visit."[27] There was no reference to Kashmir in the Soviet-Indian communiqué, no support for India in the Rann. Zulfi hailed Ayub's visit as "a milestone in the changing realities of Indo-Pakistan relations with the Soviet Union." His global strategy for Pakistan's emerging leadership of the Third World was, he felt, at last paying superpower dividends. Kosygin and Brezhnev, of course, were playing their own game. They sought not only to wean allies from the West but to secure a more important role with their southern neighbors, greater diplomatic influence over both India and Pakistan. The summit meetings with Ayub and Shastri were early moves in that direction. Hosting the postwar peace conference in Tashkent would be a more important later coup.

British Prime Minister Harold Wilson and Lord Louis Mountbatten initiated proposals for the settlement of the Kutch dispute, and by mid-May negotiations toward that end had begun in New Delhi as well as in Karachi and Pindi. Pakistan tried to get India to agree to reconsideration of *all* borders, thus bringing Kashmir into the negotiations, but Shastri was adamant in refusing to open up the Kashmir debate at a Kutch conference. New British proposals for resolving the border dispute were mooted in June, and when Ayub reached London from Cairo for the Commonwealth Conference on the morning of 16 June 1965, he told a cheering crowd of his own countrymen that "we do not want war. But if conflict is forced on us, we shall give a crushing reply to the enemy."[28] *Dawn's* special correspondent in London, young Nasim Ahmed, soon to become Bhutto's personal secretary, was most impressed by Zulfi Bhutto's eloquence in London, where he insisted that Kashmir was "at the heart" and "the root" of Indo-Pakistan problems.

Ayub discussed Kutch with Wilson at Marlborough House on 16 June, and Shastri met Ayub next day for an "impromptu" summit on the lawns of that Commonwealth Conference venue. Wilson met separately with Shastri and Ayub at his own home, Chequers, a few days later. Rumors of imminent "agreement" were rife, but the next day, 21 June, Zulfi informed the press that "we must go to the genesis of the crisis if we want genuine and lasting peace in the sub-continent. . . . The Rann of Kutch . . . forms part of a much bigger issue. The heart of the Indo-Pakistan dispute lies in the Srinagar Valley."[29]

The British government drafted a peace proposal on Kutch that was to be signed in New Delhi and Karachi on 30 June 1965. In Delhi the night before, six influential opposition members of the Lok Sabha, led by future Foreign Minister A. B. Vajpayee, wired President Radhakrishnan, denouncing the proposed cease-fire as a "gross betrayal" of Indian "interests," and urging the president to "intervene immediately" to derail the "dishonourable surrender." The final agreement had recognized Pakistan's right to patrol the northern portion of the Rann, and promised Indian withdrawal from Sardar Post, but because monsoon-driven water was moving deep into the Rann by this time there was more urgent pressure for signing the cease-fire that stipulated complete withdrawal in seven days of all troops in Kutch to their 1 January 1965 positions. Pakistan was particularly pleased with the final stipulation in the agreement that called for a tribunal of three members, none of whom would be Indian or Pakistani, one each selected by the two sides, the chair to be jointly chosen by both India and Pakistan or the UN secretary-general. The tribunal was to meet as long as necessary to resolve all further disputes, and both governments agreed to accept its findings without question and to implement them as quickly as possible.

Zulfi considered acceptance by India of the principle of international arbitration a major diplomatic victory, one that should lead to the same principle's being applied to the Kashmir dispute. Pakistan had, therefore, won the battle not only in Kutch but also at the London conference. Ayub was delighted at finding no messages from angry opposition leaders waiting for him in Pindi. "Never before have India and Pakistan been closer to war than during recent weeks," Ayub told his nation after reaching home, hailing the two agreements that "clearly demonstrate that India and Pakistan can and should resolve their disputes through peaceful means."[30]

Shastri made only a low-key reference to the agreement in his 1 July broadcast to the nation. The *Times of India* sardonically labeled the agreement, "Peace with Honour." For Indian readers reared on Western history, that hollow phrase used by Disraeli to characterize what he had "accomplished" at the Congress of Berlin in 1878 seemed most appropriate to Shastri's "victory" in London. J. P. Narayan, however, welcomed the agreement as a "diplomatic triumph." Most Hindu fundamentalist leaders nonetheless viewed it as "abject surrender" to Pakistan's bullying. July 4th was proclaimed Anti–Kutch Agreement Day in Delhi by those leaders, and thousands of angry Hindus marched while shouting anti-Pakistani slogans.

India's Parliament was scheduled to reconvene on 16 August, but before that date the opposition concluded plans to censure Prime Minister Shastri for his Kutch "surrender." High Commissioner G. Parthasarathi was recalled to New Delhi's Foreign Ministry and Sardar Kewal Singh was sent as his successor to Pakistan. The new Indian high commissioner called on Foreign Minister Bhutto on 10 August to convey Foreign Minister Swaran Singh's greetings to his "old friend." A meeting of both ministers had been arranged for 20 August in New Delhi to set in motion the tribunal part of the Kutch agreement. It was Kewal Singh's unhappy duty, however, to inform Bhutto that India had captured a large number of Pakistani "infiltrators" crossing into Kashmir during the past week. India's army and border patrols along the cease-fire line in Kashmir were strong enough to "deal" with the situation, Kewal Singh explained, but if left unchecked by Pakistan that sort of infiltration would "spoil" the atmosphere created by the agreement. Bhutto replied that "goodwill" could not be created in a "vacuum." He had told Swaran Singh the same thing about Kashmir many times. India's "unilateral and arbitrary . . . so-called integration" of Jammu and Kashmir State into the Indian Union through Delhi's application in January of Articles 356 and 357 of the Indian Constitution to Kashmir was, in Pakistan's view, the cause of current trouble in that region. Kewal Singh agreed that "good-will" required more than a vacuum to survive, but reported that Indian troops had caught many infiltrators armed with plans to "sabotage" Kashmir and with military equipment designed to do just that. Operation Gibraltar was in full swing by early August, but was not going as well as Ayub and Zulfi had hoped it would.

On 10 August 1965 a proclamation was issued by the Revolutionary Council of Kashmir to all "Brave Kashmiris," calling upon them to "arise, for now is the time."[31] It reviewed the history of India's "oppressive and treacherous rule" over Kashmir's "sacred soil," noting India's "utter contempt for world opinion." It stated that none of many dreadful acts listed had broken the "will of our people." The enemy was now "on the run," and people must solemnly pledge to take up arms and fight till all "usurpers" were expelled, all jailed leaders freed, and the "people's will" was allowed to determine the "future of our land." The Revolutionary Council declared all treaties and agreements between India and Kashmir annulled; the new National Government of Jammu and Kashmir run by the Revolutionary Council was "the sole lawful authority in our land" and would collect all taxes. Any Kashmiri, who refused to pay or preferred to cooperate with the Indian government or its "puppet administration" in occupied Kashmir would be deemed a traitor and be dealt with accordingly. All "sane and freedom-loving elements in India," particularly "brave Sikhs," were invited to lend "active assistance"—an invitation that to the government of India was a declaration of war. The Revolutionary Council appealed "to the world to support this freedom movement," and to the people of Pakistan to join in the struggle. It echoed Zulfi Bhutto's earliest speeches in the United Nations, "Remember that if we go down the light of freedom will be extinguished forever." That proclamation ended with all caps: "ARISE NOW, OR THERE WILL BE NO TOMORROW."

Few Kashmiris in Srinagar's Vale read or heard that proclamation, and of those who did most merely shrugged. There was no mass uprising. Virtually no young men in Srinagar were ready as yet to barter their lives for "freedom." For New Delhi, however, the UN-enforced cease-fire begun on 1 January 1949 was now dead. Five days later Indian troops reoccupied three strategic Pakistani posts in high-altitude Kargail that had been taken earlier but then vacated in the aftermath of the Kutch agreement.

Foreign Minister Bhutto had given Ayub his assessment of the potential international implications of a war in Kashmir in light of the recently concluded Rann of Kutch conflict. The roles of the Great Powers were so pervasive that all international disputes were "invariably" of "concern to them," Zulfi admitted, for Ayub was primarily worried about how the United States would view a move into Kashmir. Bhutto agreed that of all the powers the United States had "most to lose" from a general Indo-Pakistan war because it would "make nonsense" of recent U.S. arms shipments to both sides. And from "the ashes of destruction" China was bound to emerge stronger, as would communism in general. Because the Chinese could take "military action" in support of Pakistan in "*disputed*" Ladakh, the United States would "not find adequate justification for intervention," as it had in Vietnam. Given the extent of its military commitment in Vietnam and the Dominican Republic, moreover, the United States was already "fully occupied." Then, too, he argued, India's own weakened internal political and economic condition left it "in no position" to "risk a general war of unlimited duration" against Pakistan. Zulfi's own "authoritative sources," moreover, had convinced him of Pakistan's relative military superiority," an assessment that appeared to be validated by the Kutch conflict. To try to restore its army's morale and to redeem its position with its people, however, India might attempt "a general war of limited duration" against Pakistan, though *not* "against the Punjab Frontier where Pakistan forces are well poised." The latter assessment proved totally and disastrously incorrect. Zulfi expected India to move against East Pakistan in the hope, perhaps, of stirring up Bengali support there. That strategy could be countered by Pakistan forces striking "north to join up with Nepal and completely isolate Indian forces in Assam." He hoped, of course, that China would also move in that eventuality, forcing Indian troops in Assam "to fight on two fronts." None of that happened. Zulfi trusted too much in political-power "gamesmanship."

Bhutto bolstered his prowar arguments by reminding Ayub that the "morale of our nameless soldier on the front line is high," and the "justice of our cause is not in doubt." Time was on India's side: with arms pouring in from both superpowers, within two or three years India's military capability would be such that "Pakistan would be in no position to resist her." India's "ultimate objective" was nothing less than the "destruction" of Pakistan. Thus, the time to "hit back hard" was "now," to make it virtually impossible for India to embark on a total war against Pakistan for the next decade."[32]

By mid-August some 7,000 Pakistan-trained "freedom fighters" had been launched over the line from Azad Kashmir to trigger the "revolt." The Indian army then moved in strength to secure the Uri-Poonch sector of Azad Kashmir, crossing the cease-fire line that for over a decade and a half had served as the de facto border. The Indian army at the time had more than 650,000 troops, well over thrice the number of regular Pakistani soldiers, and with some 1,600 planes and 1,500 tanks, India had from 30 to 50 percent more heavy armor than its smaller neighbor. Under the command of General J. N. Chaudhuri, it had drawn up plans to attack Punjab if Pakistan launched an invasion in Kashmir.

Before the end of August 1965, Foreign Minister Bhutto briefed President Ayub again, insisting that "the success of the current movement in Kashmir is not only vital and crucial but it will be a decisive factor in the history of Pakistan." The freedom fighters had not fared well in the Vale, however, and experienced officers at GHQ, like General Gul Hasan, advised Ayub against "escalating." Bhutto countered: "I will not . . . even consider allowing this movement to die out because from the point of view of foreign policy as well as the requirements of internal politics . . . such a course would amount to a debacle which could threaten the existence of Pakistan." He counseled Ayub to resist mounting Western and UN "diplomatic and other pressures." "If we allow the tempo [of the infiltrators' attack] to die down even temporarily under such pressure we would not only have compromised ourselves but . . . give the Indians the opportunity to ruthlessly stamp out the rebellion." He wanted to call for more aid from China and the Soviet Union, and to invite Indonesia to send a military mission to Pakistan. He advised that the Azad Kashmir government appeal for Muslim volunteers from Algeria, the U.A.R., and Indonesia. "We must go all out to incite the Nagas and the Lushais in Assam and the Sikhs in the Punjab," he added.[33]

Zulfi also drafted a brief for Ayub on Pakistan's "political objective" in Kashmir. It was first of all to "de-freeze and activate the Kashmir problem so as to compel the United Nations to take positive and effective action to bring about a settlement based on the UNCIP [U.N. Commission on India and Pakistan] Resolutions." If Pakistan demonstrated "sufficient effectiveness in Kashmir," Zulfi felt certain that the UN would soon call for a cease-fire with a "Kutch-type" method proposed for settling the dispute. "Time is therefore of the essence in striking a few hard and decisive blows against India in Kashmir."[34]

On 29 August 1965 Ayub Khan sent the following top secret order, "Political Aim for Struggle In Kashmir," to his army commander in chief, General Mohammad Musa:

> To take such action that will defreeze Kashmir problem, weaken India's resolve and bring her to a conference table without provoking a general war. However, the element of escalation is always present in such struggles. So, whilst confining our action to the Kashmir area we must not be unmindful that India may in desperation involve us in a general war or violate Pakistan territory where we are weak. We must therefore be prepared for such contingency.
>
> 2. To expect quick results in this struggle, when India has much larger forces than us, would be unrealistic. Therefore our action should be such that can be sustained over a long period.
>
> 3. As a general rule Hindu morale would not stand more than a couple of hard blows delivered at the right time and place. Such opportunities should therefore be sought and exploited.[35]

A few days later, before dawn on 1 September 1965, Pakistan artillery opened fire across the frontier in the Bhimbar-Chhamb area. At daybreak heavy Patton tanks started rolling east in what was the start of Operation Grand Slam, designed swiftly to cut Kashmir off from India at its narrow southern neck. India's thousand troops in the sector fell back, but before nightfall the Indian air force had flown up and destroyed thirteen tanks and many other vehicles. Full-scale war had begun.

On 4 September, China's Foreign Minister Chen Yi flew into Karachi for talks with Bhutto, expressing China's support for Pakistan's "just action." That same evening the UN Security Council passed a "sense" resolution calling for a cease-fire and return of both sides to their respective positions behind the cease-fire line in Kashmir. Pakistan's advance tanks were by then more than twenty miles inside Jammu, and the next day, General Musa told his men: "You have got your teeth into him. Bite deeper and deeper until he is destroyed. And destroy him you will, God willing."[36]

India struck back on 6 September with a massive three-pronged tank advance toward Lahore from Amritsar and Ferozpore. That imminent threat to Punjab's capital forced withdrawal of Pakistan tanks from Jammu farther north, and their hasty retreat on 7 September when an Indian tank force headed for Sialkot, the rail junction north of Lahore used for supplying the Chhamb sector force. India's strategically brilliant pincers attack caught Pakistani military intelligence fast asleep; the director general of Inter-Service Intelligence (ISI), Brigadier Riaz Hussain was unable to inform Ayub Khan of the actual whereabouts of India's only armored division. "Ayub was furious," Zulfi recalled. "Ayub Khan gave hell to Riaz Hussain. . . . The Armoured Division of India was not a needle in a haystack. . . . With a quivering voice, Brigadier Riaz Hussain replied, 'Sir, from June 1964, Military Intelligence has been given political assignments on elections and post-election repercussions.' "[37]

Foreign Minister Bhutto had assured his president that India would not attack in Punjab, and now faced the embarrassing job of trying to explain that fatal error. Zulfi's quick mind and endless fund of ingenuity soon rose to that tough challenge in his "final analysis of the situation," sent to Ayub even as India's army was rolling to within the outer canal perimeter and suburbs of Lahore. The action "cannot be explained except in the light of positive United States complicity." It had become "increasingly clear" to him, he suddenly revealed, that since 1959 the United States had been trying to "secure a foothold in India"; hence, the advance by India's army toward Lahore convinced him of a "positively malicious" U.S. attitude toward Pakistan. "It is imperative in the present circumstances to issue a denunciation of the United States' complicity with India." Zulfi had convinced himself, of course, of the truth of this explanation of his failure to predict India's countermoves to Operation Grand Slam. "We must recognize the fact that the United States requires nothing short of capitulation from Pakistan." It was an explanation so convoluted that it must have sounded quite mad to the field marshal's ears.

Further: "If the United States should succeed in its designs against Pakistan it would only mean the isolation of our leadership not only from the mainstream of African-Asian trends, but it would also serve the purpose of isolating them from their own people. The next inevitable step will be in the direction of liquidating our national leadership which would then find itself in no position to offer effective resistance." He had crossed the line of rational analysis now, as Ayub was quick to perceive.

"If we face this situation resolutely there is no doubt that the hundred million people of Pakistan will triumph. The only language which the United States and its henchmen will understand at this juncture is the language of determination. . . . we cannot do better than to denounce its complicity in the present crisis."[38]

There is no record of Ayub's response to the memo but from that point at least the president of Pakistan better understood the mind of his foreign minister. "Your cousin is a madman," Ayub Khan later warned Mumtaz Bhutto. "Don't follow him! He'll lead you astray, get you into trouble!"[39] But the darkest fears and suspicions that Zulfi conjured up during this national crisis in his desperate need to prove himself right (did he still think himself the child obliged to resort to bravado to gain admittance to Cathedral Boys School? Or the foreigner required to prove to his tutor that he was capable of swiftly mastering Latin?) were soon to sound plausible to millions of Pakistanis. Zulfi's self-justifying rationalization to Ayub Khan was precisely the sort of "explanation" of Pakistan's inability to win the 1965 War in Kashmir that would reflect the suspicions, prejudices, and fears deep in countless Pakistani hearts and minds. Especially in those of the younger generation of students and soldiers, half-employed workers, and unemployed degree holders who viewed Hindu India as too cowardly, too weak, too "nonmartial" to beat any Pakistani army without total great power support! Hence, in 1965 it was the United States; in 1971 it would be the other great power, the Soviet Union. Otherwise, after all, who would there be to blame but oneself? How could it possibly be one's own mistakes, one's own failure to anticipate the obvious or the less than obvious?

When India opened the floodgates of its dams, inundating many square miles of Punjab, trapping close to a hundred Patton tanks in deep mud in the Khem Karan sector, halting a Pakistani attempt to outflank India's smaller force there and thereby stopping them from rolling on to Delhi, that was viewed as "Hindu treachery." How else could Pakistani Muslims sustain their myths of martial superiority? Less than a week earlier, after all, Indian armor had reached the outskirts of Lahore. Pakistani soldiers and generals fought as bravely, as courageously as they had ever done, of course, but they were fewer in number than the Indians they faced. India's high command had outmaneuvered its enemy. For Zulfi Bhutto that was *impossible,* unacceptable, intolerable. And most young Pakistanis felt exactly the way he did, believing that nothing less than complete victory was required in fighting India. The alternative could only be "our destruction as a self-respecting nation."

Just as Quaid-i-Azam Jinnah had won Pakistan against all odds after a Muslim League struggle many had deemed impossible, so from the ashes of this national humiliation and coming diplomatic "betrayal" a new fearless young Quaid-i-Awam (Leader of the People) would soon emerge, launching his own People's Party aimed at restoring, ressurecting, and rebuilding Pakistan as a truly "self-respecting nation." The 1965 War became Zulfi's political booster rocket, launching his mind into space to salvage Pakistan's depressed psyche, to snatch retrospective victory from the jaws of an internationally manipulated defeat. He would be ready to face both superpowers fearlessly. Little Zulfi, sling in hand, courageous enough to take on the twin Goliaths of treachery with nothing more than naked pride in himself and deathless faith in his nation and its people.

In mid-September Indian forces moved from Rajasthan onto the sands of Bahawalpur and Sind, retracing the exodus of Bhutto's paternal ancestors, impelling Zulfi to send another urgent "*Top Secret*" memo to Ayub:

> The next target will be East Pakistan. . . . The defence of East Pakistan would need to be closely co-ordinated with Chinese actions both in NEFA and also possibly in the region of Nepal and Sikkim. It would be necessary . . . to provide the Chinese with a link-up with our forces in that sector. . . . I envisage a lightning thrust across the narrow strip of Indian territory that separates Pakistan from Nepal.
>
> From our point of view this would be highly desirable. It would be to the advantage of Nepal to secure its freedom from isolation by India. It would solve the problems of Sikkim and Tibet and for us it would give us a stranglehold over Assam, whose disposition we could then determine.[40]

Pakistan's Air Marshal Asghar Khan and General Gul Hasan had flown to Beijing with Zulfi Bhutto on 9 September, Pakistan's last desperate appeal to the Chinese for military aid in the war. The Chinese could deliver nothing but words of strong moral support. Indonesia at least sent a few naval vessels to Karachi; Iran, Turkey, and Jordan sent messages of pan-Islamic support with token military hardware. Most Afro-Asian nations were distressed to find two South Asian countries locked in war, as was the West. For the first time since the Suez War, the United States and the Soviet Union were in agreement: it was urgent to stop the rapidly escalating conflict as quickly as possible. UN Secretary-General U Thant flew to Karachi, Pindi, and New Delhi, trying to persuade the leaders of India and Pakistan to call back their armies. The United States stopped all shipments of arms, munitions, and spare-part replacements to both nations. The U.S. freeze was viewed as total betrayal in Pakistan, seen by Zulfi Bhutto as "proof positive" of Washington's treachery.

Ayub, of course, knew how perilously low his army's supply of bombs and bullets was by the third week of September, and knew painfully well that more than a thousand of Pakistan's bravest soldiers were dead, fourteen of his best planes destroyed, and almost two hundred of his best tanks out of commission, many bogged down in mud. His military experience led him to conclude that accepting U Thant's proposal of an immediate cease-fire would be wise for Pakistan's armed forces and for the nation's future survival. His age no doubt contributed to his decision, though at any age he would have been less passionately impulsive than Zulfi Bhutto in arguing that a cease-fire meant surrender, and surrender was nothing less than the total destruction of Pakistan as a "self-respecting nation." The field marshal understood enough about war to know that the first trick of his Grand Slam had been trumped. Stubbornly prolonging the battle now would bring no victory, only heavier casualties and the loss of all future support from Washington, Moscow, London, and the UN Secretariat. Ayub tried at first to link a cease-fire to settlement of the Kashmir "dispute," through a self-executing agreement similar to that devised to resolve the Kutch conflict, but Shastri refused to consider discussing Kashmir as part of the "simple cease-fire" the Indians were willing to accept. On 20 September the Security Council "demanded" that both sides cease firing in two days, and promised to consider "what steps could be taken to assist towards a settlement of the political problems underlying the present conflict."[41]

Foreign Minister Bhutto immediately flew to New York, and on the night of 22–23 September 1965 addressed the Security Council, convened past midnight. He thanked the secretary-general for his "endeavours" and all "peace-loving countries" for taking "interest" in a war "imposed on us by a predatory aggressor." He enumerated India's various acts of "aggression" against "small" Pakistan since 1947, insisting "we have always known that India is determined to annihilate Pakistan." Jammu and Kashmir was not now and had "never been an integral part of India. . . . The people of Jammu and Kashmir are part of the people of Pakistan in blood, in flesh, in life, in culture, in geography, in history . . . in every form." Then he swore, "We will wage a war for a thousand years, a war of defence."

Bhutto's rhetoric thrilled every Pakistani who heard it, especially the men back home, who knew they had lost the war but whose dream of victory was being kept alive by the words carried by wireless radio to Karachi, Pindi, and Lahore. "The world must know that the 100 million people of Pakistan will never abandon their pledges and promises. . . . Indians may abandon their pledges and promises; we shall never abandon ours. . . . We shall fight for honour; we are not aggressors; we are the victims of aggression." When he cried out, "We believe more than ever before that justice is bound to prevail for the people of Jammu and Kashmir," Zulfi Bhutto's voice became the favorite of millions, if not tens of millions, of inarticulate West Pakistanis who felt exactly as portrayed.

"The whole world believes in the right of self-determination," Bhutto declaimed. "Must it be denied to the people of Jammu and Kashmir merely because power must prevail over principles? Power shall never prevail over principles. . . . The will and spirit of our people can never be destroyed."[42]

He then transmitted Ayub's message ordering Pakistan's armed forces to "stop firing" on 23 September at 1205. "But a cessation of hostilities is not enough. The Security Council . . . must now address itself to the heart of the problem . . . the future of Kashmir. It can no longer make a plaything or a toy out of 5 million people." Zulfi then went much beyond the marching orders received from Ayub, for he was back in New York not as a lonely young man contemplating "soaring" giants scraping the sky. He himself was the giant now, addressing the world's mightiest council and being heard in all four corners of the earth.

"This is the last chance for the Security Council to put all its force, all its energy, all its moral responsibility behind a fair and equitable and honourable solution of the Jammu and Kashmir dispute," he notified the World. "History does not wait for councils, organisations or institutions, just as it does not wait for individuals." Was that aimed at Ayub or Shastri? "Ultimately we shall have to be the final determiners of our own course. . . . Let me tell the Security Council, on behalf of my Government"—this was quickly to be denied and refuted—"that if now, after this last chance that we are giving the Security Council, it does not put its full force . . . behind an honourable settlement of the Jammu and Kashmir dispute, Pakistan will have to leave the United Nations. . . . Within a certain period of time, if the Security Council is not able to act in accordance with the responsibility placed on it, in accordance with its honour under the Charter—which believes in self-determination—Pakistan will have to withdraw from the United Nations."[43]

The more outrageous his rhetoric became, especially when his stage was wired to the world, the more heroic Zulfi Bhutto appeared to Pakistani audiences. But not to Ayub or Pakistan's establishment elite, for they recognized how dangerous so loose a cannon could be to the country's ultimate survival and economic development as well as to regional security. Even so, Zulfi Bhutto alone, of all the coterie who ruled Pakistan under Ayub Khan, was to emerge in the aftermath of the 1965 War a national hero. He alone turned military defeat, or at best stalemate into personal victory. He journeyed to Washington to try to convince the State Department and White House of the need to pressure India into accepting a self-operating agreement on Kashmir similar to the one that had followed the Rann of Kutch dispute.

In speaking with all four permanent members of the Security Council, Zulfi found only France receptive to appeals on Kashmir. Yet he feared that France might ultimately "give in to massive Indian pressure," as the British had in 1964, when Zulfi was hoping it would move the Security Council to pass a new resolution calling for a settlement based on a plebiscite. He thought to use the "Chinese factor" to insure that France remained in Pakistan's corner, however, believing economic self-interest now weighed heavily in French diplomacy and would favor admission of China to the UN. Zulfi met with French Foreign Minister Couve de Murville in New York on the eve of his 28 September speech to the UN General Assembly. In "Self-determination and Kashmir" Bhutto hailed "the emergence of the peoples of Africa and Asia from an era of colonial domination" as the "outstanding development of the present century."[44] He bemoaned the "lack of Chinese participation" in the UN, insisting that without the People's Republic of China, Assembly deliberations had "a distinct air of unreality." China and "our brothers" in Iran, Turkey, and the "great bloc of Arab countries . . . to which Pakistan is linked . . . not merely by ties of religion and of common culture but by common adherence to the idea of justice and peace" were thanked for their moral support, as were the president, government, and people of Indonesia for their "brave and unstinting support."

Zulfi noted that Indian leaders argued that self-determination was a "disruptive principle" and that a plebiscite in Kashmir would jeopardize the very survival of India "as a democracy, as a secular state." "Forcible annexation of Jammu and Kashmir by India is not a guarantee of Indian secularism, democracy or territorial integrity. On the contrary, it keeps alive those very fears and suspicions which made it impossible for the Muslim minority to accept a united Indian State."

The force of that argument appealed not only to Pakistan's populace but to liberal opinion the world over, for in defending the right of free choice for the people of Kashmir, Bhutto was only articulating Woodrow Wilson's Fourteenth Point of self-determination. He went on then to develop the legal position relative to earlier UN resolutions on Kashmir, dating back to 13 August 1948 and 5 January 1949, concluding that the Security Council remained "bound" to enforcement of the impartial plebiscite as the mode of determining the "ultimate fate" of Jammu and Kashmir State.

What Pakistan wanted, Zulfi concluded, was "peace and justice." Justice at times "demands a change" in the status quo, Kashmir being a prime example. "A cease-fire and its observation do not amount to peace. What is needed is firm action to eradicate the incentives to violence . . . to remove the seeds of war." "Pakistan has accepted the call for the cease-fire with the confidence that it would lead . . . to a self-executing machinery for a final settlement of the Kashmir dispute." He proposed that to initiate such a settlement, India and Pakistan "withdraw their forces from the State of Jammu and Kashmir," and the UN "send a force consisting of contingents from African, Asian and Latin American countries" to keep the peace. A plebiscite should be held "within a reasonable period." This eminently rational suggestion, Zulfi knew, had no chance of meeting with Indian approval.

The passion of his closing remarks, however, reflected the deepest feelings, fears, hatreds, and hopes, and highest aspirations of millions of Pakistanis.

> When Pakistan, a country much smaller than India, was invaded by India, the sufferings of both Pakistan and Jammu and Kashmir were fused. These sufferings formed a single resolve to fight against India's aggression. . . . These passions may be disregarded in the calculations of power politics, but history deals far more justly with them. When we say that we are giving the United Nations a last chance to settle the Jammu and Kashmir dispute, we are saying that we are determined not to let a righteous cause be abandoned. It is not the will of Allah that the victims of injustice and aggression should have no higher court of appeal.

In his fervent outcry and prayer for justice for the people of Kashmir, Zulfi Bhutto "fused" not only the sufferings of Pakistan with those of Kashmir but his own battered, defeated, and unappreciated feelings, dreams, ambitions, and desires with those of the people of Pakistan. Like them, he felt small and oppressed, bullied by Ayub and India, coldly "disregarded" by the cruel "calculations" of heartless UN councils, and great-power politicians. Yet "history" would vindicate them both, Bhutto and the people, for theirs was a "righteous cause," and Allah would save them all in His Court of Last Judgement. From the ashes of defeat, as from the ashes of death, Zulfi believed, Sufi mystic that he was, reputations, like devout souls, could rise again, winning the final victory of righteous "surrender"—Islam—to Allah over satanic injustice.

Feeling beleaguered at the UN, surrounded by hostile larger powers, most of whom were "conspiring" against him, against Pakistan, Zulfi looked to Allah for comfort and to his Islamic brothers—Arab, Persian, Indonesian, Malayasian—for support. He felt cheated and rejected by most of the great powers, finding a friend in France alone. He would meet in Paris with France's foreign minister, hoping to secure the modern arms and nuclear technology Pakistan would require to win the next war.

In early October 1965 he met with Morocco's Foreign Minister M. Benhima at the UN, thanking him for his strong support in the debate. Benhima divulged that India was making "frantic" efforts to win North African diplomatic support by issuing very "strong statements" against South Africa and Rhodesia. Zulfi also warned his foreign secretary that he had been told that India was moving "closer to Israel" because the Arab states, who had met in Casablanca in September, firmly supported Pakistan's cause. "India is seeking to enlist Jewish influence in USA," Zulfi wired the ministry, and "Israel is actively working on behalf of India and Washington."[45] On 14 October, Zulfi met with Arthur Goldberg, U.S. ambassador to the UN, who wanted to know about his talks in Paris, as well as those in New York with Soviet Foreign Minister Andrei Gromyko. Bhutto offered very little about Paris but mentioned Gromyko's offer to host an Indo-Pakistan peace conference in Tashkent. Goldberg assured him that the United States "would not be hostile" to Pakistan's acceptance. Bhutto charged that Indian troops in Kashmir had launched "an unbelievable campaign of genocide," which Goldberg said he would "look into."[46]

Foreign Minister Bhutto kept busy that October. He met with Romania's foreign minister, who also said he was willing to help reach "settlement favourable to Pakistan." He met repeatedly with Gromyko, India's major "defender" on the Security Council. Bhutto wanted the secretary general to visit India again to press New Delhi for a diplomatic solution to Kashmir, but because "India refuses to agree," Gromyko refused to act as Pakistan's spokesman. Bhutto met with France's Ambassador Seydoux, with Britain's Lord Caradon, and with Dean Rusk of the United States, in talking to whom he always felt frustrated, much the way he had begun to feel whenever he met recently with Ayub. Because China had exploded its first atomic bomb in late 1964, Zulfi feared that the United States was ready to "save" India from any possible future Chinese move with its "nuclear-armed" Seventh Fleet or by placing its "much more devastating hydrogen bombs at her disposal."[47] By now he firmly believed that India would not have "embarked on her bold adventure" of invading Pakistan in early September had the United States not "changed its policies," permitting India to "turn" her "six mountain divisions which were formed and equipped by the United States for the purpose of facing the Chinese," repositioning them "against Pakistan in Kashmir."[48]

At the end of October Zulfi addressed the Security Council on Kashmir one last time before flying home to report to Ayub and his cabinet. The council had been convened, over strong opposition from India's Foreign Minister Swaran Singh, to consider cease-fire violations by Indian troops. When Bhutto started to expatiate upon "the extreme measures employed by India to wreck vengeance on the people of Jammu and Kashmir," the council president asked him not to "dwell upon the internal situation in the State," triggering a heated debate inside that chamber. Swaran Singh walked out. Zulfi was permitted to finish his lengthy speech, and the peroration sent waves of pride and painful pleasure throughout Pakistan, where his words were heard time and again.

> We are fortified by the faith that, despite India's arrogance and obduracy, despite its flouting all canons of civilized conduct, despite the armed might which it deploys against Kashmir's helpless people, this long-drawn-out tragedy can end only in the victory of the people of Jammu and Kashmir and in the vindication of the honourable position which our country and our people have taken. . . . And we tell you, . . . we shall face complete extermination; we shall face destruction; we shall never dishonour our pledge. We shall fight by the people of Jammu and Kashmir, and we shall honour that pledge irrespective of what the Security Council does, irrespective of what the great powers do. This is a part of our faith; it is ingrained and enshrined in our very civilisation. . . . That is why we are able to face aggression from a country six times our size. . . . We stand for a righteous cause; that is why we are brave. We fight for justice; that is why we are brave. And, finally and ultimately, whatever you do, we must triumph; we must succeed because justice is with us.[49]

It was Zulfi Bhutto's fiercest anti-Indian speech at the UN, perhaps because he knew it would be his last before the world body that year, his final statement prior to the Tashkent meeting, which was set to start in early January 1966. Zulfi flew back to Pakistan a popular hero, the only truly popular hero of the 1965 War. Not just in Sind but in Punjab as well he was embraced and hailed everywhere for his eloquent defiance of the giants that surrounded Pakistan. East, North, and West welcomed this son of Sind with tears of pride and passionate kisses. Sindhi landlord-educator G. M. Shah, who had known Zulfi's father well, and was an intimate of his first father-in-law, Wadero Ahmad Khan Bhutto, welcomed this hero-prince home to Karachi by arranging his purchase—for Rs. 40,000 (c. $2,000 today)—of 96-year-old Khan Bahadur Ullamah Ikanah's unique library and collection devoted to Napoleon: thousands of leather-bound first-edition volumes and hundreds of memorabilia, including many beautifully framed cameo photos and paintings of Zulfi's foremost role model. "Every painting, every picture, every book, every statuette," Education Minister (in Benazir Bhutto's cabinet) G. M. Shah recalled a quarter century later on his Islamabad veranda, still amazed at how he had "managed" that transfer of treasure from one Napoleon lover to another. The books remain the treasured heart of Zulfi Bhutto's library of some ten thousand volumes at 70, Clifton, where pictures of Napoleon's bust in painted porcelain continue to guard the elaborately carved doorframe of the entrance.

Zulfi briefed an all-day meeting of the Pakistan cabinet on Monday, 8 November 1965 on the Security Council resolutions and prospects for negotiating a peaceful settlement with India over Kashmir. He was pleased to report some change in the Soviet Union's traditional pro-Indian attitude; it appeared that there was "considerable anguish" within the USSR over "the unsettled conditions" in South Asia, and the Kashmir dispute now "posed a dilemma" to Moscow. Worried about Pakistan's closer ties to China, the Soviets did not want to "write off" Pakistan by giving its "complete support" to India. Nor did they wish to abandon India to greater "inroads" of "American influence." Zulfi excelled at such analyses, and even Ayub was sufficiently impressed with the sophistication and sobriety of his foreign minister's global report to take Zulfi with him to Washington in mid-December; he'd been invited by President Johnson for a pre-Tashkent talk. Shrewd as he was, the field marshal may have taken Bhutto along more to keep him in sight than for his invaluable assistance, knowing how popular a hero he had become throughout West Pakistan.

Zulfi accompanied Ayub to the White House on 14 December but met President Johnson only after Ayub had spent half an hour alone in the Oval Office with him. Bhutto was left to cool his heels in an anteroom. After what no doubt felt much longer than it actually was to the impatient minister, the heads of state emerged. Johnson strode right over to the dejected Bhutto, and as Zulfi later reported to Mumtaz, "He held my arm and he took me aside and he said, 'Young man, you're doing *extremely* well! I've got my eye on you. You're going to go very far!' When he told me *that,* I at once knew that I'd got the *boot* [the Sindhi word he'd used, in fact, was *gaddi,* which means "finger up"] and that was what they had discussed!"[50] On the eve of the South Asian summit in Tashkent, then, Zulfi knew he would soon be out of Ayub's cabinet, and that the next stage of his political career had better start with the new year.

7

Winters of His Discontent

(1965–1969)

Zulfi flew to Tashkent on the eve of his thirty-eighth birthday to attend the peace conference, to which Premier Aleksei Kosygin had invited Prime Minister Shastri of India and President Ayub Khan of Pakistan. That first South Asian summit, in the USSR's Central Asian capital, was convened on 4 January 1966. Sardar Swaran Singh and Defense Minister Y. B. Chavan accompanied Shastri, and Minister of Information Habibullah Shahabuddin and his secretary, Altaf Gauhar, accompanied Ayub and Bhutto. Zulfi's intense dislike of Altaf Gauhar dated from Tashkent because Ayub confided more in the secretary, who was ghostwriting his autobiography than he did in his by-now-mistrusted foreign minister. Ayub would not be ready to sack Bhutto for another six months, yet both men knew that their days of wine and hunting together were over. Much as had happened in the White House, Ayub went off to chat with Shastri immediately after the first day's formalities ended. Bhutto tried to join them, but the field marshal shook his head and extended a stern forefinger, making Zulfi back off.

If, after Washington, Zulfi retained any residual doubts as to his position in Ayub's cabinet and in his heart, the last would disappear in Tashkent's frigid environment. Zulfi was hypersensitive to how others felt about him, and picked up every nuance of voice and body language, especially of those above him on the power ladder he had resolved to climb to the top. Swaran Singh's walkout from the Security Council when Zulfi last spoke there was bad enough; that sort of rudeness was to be expected from the enemy, as Zulfi now thought of India and its highest officials. Ayub's snubs cut deeper, for while the president met daily alone with Shastri, as he did with Altaf, Zulfi was left to sit in another room and listen to Swaran Singh and watch his too-familiar gestures. They both went through the stale arguments on everything from why they could not discuss the future of Kashmir to prisoners of war and confiscated cargo. Zulfi was frustrated and bored.

Kosygin worked hard that week in Tashkent, shuttling between Ayub and Shastri when they were not locked together, arguing in Hindustani, trying to reach some agreement that could dispel the hovering clouds of war from Himalayan heights in Kashmir. Altaf typed a two-page draft of an agreement Ayub was ready to sign, but Shastri, as Kuldip Nayar (who flew with Shastri to Tashkent as a member of the Indian press corps) put it, was "a tough negotiator," raising a last-minute request for a "no-war" clause. So Ayub "wrote in his own hand on that typed draft that Pakistan would renounce the use of force in settling disputes with India." Then Shastri agreed to the draft.[1]

Bhutto hit the ceiling when he read the insert in the pages Ayub later showed him, threatening to fly home to "expose" Ayub's treacherous "surrender" to all of Pakistan. Ayub felt constrained to delete the insert, though Altaf and Minister Shahabuddin, agreed to it, as did Pakistan Commerce Minister Ghulam Farooq. They all knew, however, that Zulfi was the most popular figure in Pindi and Lahore as well as in Karachi and Larkana, and to risk antagonizing him enough that he would fly home alone would be tantamount to igniting a new "revolution." So Ayub resiled.

Bhutto immediately called Andrei Gromyko, who was with Shastri when the telephone rang. Zulfi tried "diplomatically" to explain that Ayub had agreed to "renounce force" only because India had "agreed" to a "plebiscite" in Kashmir. Gromyko's angry reply: "It is a lie."

Here again it was Zulfi Bhutto's hard line, his tougher stand than Ayub's on negotiating with India, his absolute refusal to be cowed or to "give up" Kashmir, that more accurately reflected the feelings of most Pakistanis—at least in the West—than did the position of the more moderate Ayub, whose days of robust health, as of power, were by now much diminished.

Kosygin and Foreign Minister Gromyko were obliged to use their most persuasive powers on Lal Bahadur Shastri in the final days of Tashkent's summit, for Zulfi had stalemated Ayub by his rage and threats. India's army was poised and waiting in Punjab to move against Lahore if the talks broke down. Shastri knew the dangers of withdrawal without a no-war pact or clause in the Tashkent declaration but was sick of war, having spent most of his days in high office fighting on one front or another. He was weary as well of the tough negotiations in cold Central Asia. On 10 January 1966 Shastri and Ayub signed a declaration (Bhutto refused to allow it to be called an agreement) stating their "firm resolve to restore normal and peaceful relations between their countries and to promote understanding and friendly relations between their peoples." By the end of February all troops were to be "withdrawn" by both countries to positions they had held prior to 5 August 1965, and both sides agreed to "observe the cease-fire terms on the cease-fire line." It was hardly a diplomatic victory for either side, yet perhaps it accurately reflected the battlefield realities the conference had been called to articulate.

Lal Bahadur Shastri retired to his dacha at about ten o'clock after attending a Soviet banquet to celebrate the declaration. He was to stop in Kabul next day on his flight home to Delhi. Ayub had invited him to stop in Pindi as well, for those two leaders had come to respect one another after a week of often-acerbic meetings. Ayub considered Shastri "a man of peace who gave his life for amity between India and Pakistan."[2] Shortly after 1:00 A.M. Shastri was awakened by the sharp pain of his third, fatal heart attack. His physician, R. N. Chug, was asleep in another room, and reached Shastri too late to resuscitate his patient. Death occurred at 2:00 A.M. on 11 January 1966. Ayub Khan, who looked and sounded "genuinely grieved," served as one of Shastri's pallbearers, a personal act of Indo-Pakistan friendship. Bhutto was also awakened early, and his foreign secretary reportedly told him, "The bastard's dead!" Zulfi asked, "Which one?"[3]

Zulfi never forgave Ayub for having humiliated him in front of Nusrat. He was too proud a wadero to allow anyone to chastise him, as Ayub Khan had done, for what he considered solely his own business. He was, after all, Zulfikar Ali Bhutto! The upstart general, who was a field marshal thanks to Bhutto's ingenuity, was no more than a Pathan commoner, a brash usurper of presidential powers, Washington's "stooge" and "messenger-boy." Ayub's bungling weakness in the war with India, just concluded in such abject failure, was the final straw. "I had submitted my resignation in Tashkent," Zulfi later recalled, but "Ayub Khan implored me to withdraw it" in the "national interest."[4]

On the return flight, Zulfi knew that his days of living in his military dictator's long shadow were almost over. Eight years was enough. Not only his birth, but his brains, his guts, his Pakistani spirit and courage were all better, stronger, than Ayub's. Zulfi had long sensed a mystic identity with Pakistan, with his people, and that was enough to sustain and support him, no matter what a stupid old general or weak dictator might say or try to order him to do. Ayub would soon learn that no one was powerful enough to humiliate, or order Zulfikar Ali Bhutto around. He nominally remained foreign minister till June, but the day after reaching Rawalpindi, Zulfi flew home to Larkana, where he stayed at Al-Murtaza for more than a month, hunting and planning his next moves.

"He was pushed out," Mumtaz said, talking of what Ayub had done to Zulfi, "but then I think he was not too sorry. The timing was perfect. He was the hero of Tashkent, and everything worked for him." At Larkana, Zulfi had, of course, taken Mumtaz, his "brother" (*Ada,* as they called each other in Sindhi), into his confidence, discussing with him his plans for leaving the unpopular government and starting a new party, a people's party, one that would appeal to Pakistan's masses, to its workers and peasants as well as waderos.

Zulfi consulted many other intimates at this time as well, including Sheikh Kaiser Rasheed, who had served under him as director of the Foreign Ministry and would remain his confidant, a companion in the darkest days of his depression, one of the few men he ever trusted. Another of his closest friends was Ghulam Mustafa Khar, who had become a "regular visitor to my home," Zulfi recalled after his first election to the National Assembly three years earlier.[5] Ghulam Mustafa Jatoi and Khar were inseparable in those days, and Jatoi always brought Khar along to Zulfi's place. The younger man, who was soon to be known as "Sher-i-Punjab" ("Lion of the Punjab") became Bhutto's favorite nocturnal party buddy, and one of those early committed to Zulfi's embryonic People's Party. Not Jatoi, however, for he "kept on good terms" with Ayub as well as Zulfi, after their split, which on later reflection Zulfi considered "typical" Jatoi behavior, finding his old wadero friend "deep as the deepest well in

Sind's desert."[6] After Tashkent, Ayub had gone off to hunt in Jatoi's Pye Forests at Nawabshah, confiding to Zulfi's old friend how he was going to "fix" Zulfi, which Jatoi, of course, relayed to Bhutto the next time they met. They continued to meet socially because Jatoi's passionate first wife, Maria, had been one of Zulfi's special favorites from the very first time he laid eyes on her.

Bengali-born J. A. Rahim, intellectual theoretician of Zulfi's new party, was also recruited by Bhutto from among his Foreign Ministry friends. Rahim had been a member of the Indian civil service prior to Partition, joining the Pakistan Foreign Service at its inception, rising through its ranks to become the ambassador in Paris. Zulfi got to know Rahim more intimately during his flying visit to France in 1965, for Rahim also enjoyed late drinking parties. Paris was his last assignment before his mandatory retirement. Rahim's father, Sir Abdur Rahim, had been a judge on the Pakistan High Court and speaker of the Central Assembly, and a close friend of Suhrawardy's, whom J. A. met in his Kutchery Road house in Karachi. Zulfi needed some one with J. A.'s knowledge and wealth of experience, and would for almost a decade keep Rahim's brains on tap—first in founding the People's Party and then in his cabinet, until the night Rahim offended him in 1975 and almost paid with his life for having left Zulfi's dinner party before the "Raja of Larkana" (as J. A. dared to call Bhutto) made his inordinately late entrance. Zulfi was never one to take an insult to his *Izzat* (honor) lightly. The only jokes he enjoyed were those he sarcastically aimed at the egos of others.

Ayub stopped at Larkana, after finishing his hunt in Nawabshah, in mid-February, to urge his lame-duck foreign minister to speak out in favor of the Tashkent Declaration in the National Assembly. Opposition party pressures kept mounting to hold public debate on the War of 1965, which had "recklessly endangered" the lives of 50 million East Pakistanis in the West's "adventure" to "liberate" Kashmir. Ayub Khan had been attacked from all sides since his failure to win Operation Grand Slam, and the opposition tide kept rising in the aftermath of Tashkent, from the East as well as in the West. He had tried his best to save Pakistan, Ayub felt, and thought he had done a decent job of averting greater disaster in signing the peace agreement with India. Yet now he was viewed as the weak villain, while Zulfi Bhutto's popularity kept rising. So Ayub broke his journey in Larkana, and stayed at Al-Murtaza persuading, or as Zulfi preferred to put it, "begging me to return to Rawalpindi . . . pathetically."[7] Ayub wanted Zulfi to come to the defense of his government in the assembly debate. "He was determined to keep me associated with his government at all costs until he had re-established his authority," Zulfi reflected later on this murky episode of his political career.

Ayub still hoped to advance the prospects of Indo-Pakistan friendship in the spirit of Tashkent, which his long private talks with Shastri had embodied. He replied to Prime Minister Indira Gandhi, Shastri's successor, from Larkana, the same day he received her message, handed to him there by High Commissioner Kewal Singh, that India was ready to resume flights to Pakistan: "I am glad to learn of your constructive decision in a matter which is of mutual benefit to India and Pakistan. I am also issuing immediate instructions to our civil and military authorities to permit the resumption of air flights . . . on the same basis as that prior to 1st of August, 1965."[8] He went on to express his hope that "lasting peace" would soon "return to the Sub-Continent," and concluded by stating his "personal admiration" for the positive manner in which Mrs. Gandhi had responded to the Tashkent Declaration. Zulfi's response to his appeal for support in the upcoming assembly debate was also "positive," but hardly what Ayub had hoped for, or anticipated.

Bhutto rose on the National Assembly floor on 16 March 1966 after prolonged, acerbic attacks against the government for its mishandling of the war, most of which came from the opposition ranks of members from the East led by the venerable Shah Azizur Rahman. "What is our state and what is our status?" he began in what Ayub had thought would be *his* government's major rebuttal speech, but which Zulfi wrote and delivered as the opening salvo of his own still-inchoate Pakistan People's Party.

> What are our objectives? What are our motivations? Pakistan is a great ideal. A member of this House has said that Pakistan is a man-made country. Pakistan is not just a man-made country. It is a God-made country. . . . It is a beautiful thought. It is a creation of excellence. That is what Pakistan is . . . not just the sandy desert of Sind or the rugged nobility of Baluchistan and the enchanting lushness of Bengal or the inspiring plains of the Punjab or the raw courage of the land of the Pathans. . . . Indeed all these things . . . go to make Pakistan. . . . [But] there is something much more to Pakistan. It is the blessing of Allah. Pakistan is the creation of the surge of Islamic nationhood. Pakistan is the product of an earth-shaking idea. It is a revolution cut out of the heart of history. . . . Pakistan is a live revolution.[9]

The modifier *live* was, of course, understood by all who heard him as Zulfi's implicit judgment that Ayub's "revolution" of 1958 was now dead.

Zulfi waxed more romantic, almost poetic, as his imagination soared. In his speeches at the UN the year before, he had touched the hearts of millions of Pakistanis. Here he was speaking directly to Pakistan's elite in the National Assembly.

"Pakistan is a mystical idea . . . Pakistan is the heart-throb of the people. Pakistan is the culmination of the aspirations of the Islamic Order," exulted this youthful foreign minister of Ayub's dying regime, a year and a half before the birth of his Pakistan People's Party, but with that party already born in his mind. "I now come to the war. . . . History knows of two wars: one is the immoral war of avarice and exploitation and the other is a war of resistance against domination and exploitation . . . a war of national self-assertion. Small powers have stood against mighty empires, little people have had to fight for their freedom and they have succeeded because theirs have been . . . just wars. . . . A just war cannot fail . . . no matter how great is the empire that is pitted against it. . . . In what category does the struggle of the people of Jammu and Kashmir fall?" Could any among those hundreds of elected representatives of the people of Pakistan disagree with his answer? Was there an opposition leader so old or so cold?

> But is it fair for the foreign minister of Pakistan to pronounce whether we fought a war of exploitation or whether we fought a just war? . . . The whole world supported the people of Jammu and Kashmir during the September war. Have you ever asked yourselves, . . . why is it that the international community supported Pakistan in the September war against India? . . . They supported Pakistan because the struggle of the people of Jammu and Kashmir was a just struggle. . . . The world of Islam, perhaps for the first time in its history of 1300 years, was united . . . right from Algeria to Indonesia in support of Pakistan . . . because ours was a struggle for the cause not only of the people of Jammu and Kashmir but because our fight and our struggle was for a better world order, a finer society, for greater justice and articulation of right against wrong, and that is why 5 million people of Jammu and Kashmir and the people of Pakistan were supported by the Afro-Asian countries, by the world of Islam.

"India, in size and territory, in resources and in diplomatic ingenuity," continued the most ingenious, or at least the most imaginative, diplomat of his day, "is a great country. India is like Europe, the whole of Europe without Russia . . . yet it stood alone." For rhetorical reasons Zulfi now conveniently scuttled his earlier, equally inaccurate argument that both the United States and the Soviet Union solidly backed India in the war. "So they stood absolutely alone, forsaken and naked in this struggle and the late Lal Bahadur Shastri, at the height of the war, had to say that India is all alone. . . . India—Europe without the Soviet Union—standing against the Denmark of Asia. Why? Because justice was with us."

Zulfi then argued that India "was the aggressor." "It must be clearly understood that Pakistan did not start this war," but "we had every moral and legitimate reason and justification to support the people of Jammu and Kashmir in their legitimate right for self-determination. We are ourselves the product of self-determination. . . . We support *bellum justum,* a just war of the people for their liberation. In spite of that, many patriotic friends of ours have said our defences were bad and we took a terrible risk, that it was possibly an adventurism and that there was an element of immaturity in it." Ayub's words of reprimand, spoken in darkest moments of martial despair, under the stress of learning that Indian armor had rolled within range of Lahore, were words Zulfi would never forgive: "adventurism," "immaturity." The truth often hurts.

Just as Zulfi could never admit he had been wrong in predicting an Indian strike at Bengal rather than into Punjab, he now adamantly refused to accept Ayub's reproach. "This is not so. It must be appreciated and understood that this was a heroic struggle. This was a heroic support for a great and noble cause. This is one of the factors which makes Pakistan great and which will make Pakistan a pioneer and a pillar of strength and morality for the whole of Asia and Africa." Out of the days and sleepless nights of dark depression he'd known as the tide of war turned against Pakistan, out of the personal humiliation and rejection he had felt in Washington and Tashkent, Zulfi had formulated his plan. He would not only lead Pakistan at the helm of a new people's party but forge a Third World alliance of all Muslim, Asian, and African peoples, most of the world being kept down by the great powers. It was to become the ultimate goal of his life: to lead Pakistan, to energize and transform his people into "a pillar of strength and morality for the whole of Asia and Africa" to follow—soaring like the skyscrapers he had once personalized in New York.

"Let us look at history," Zulfi continued, for he liked to think himself a student of history.

> If the whole world can be plunged into the war of 1914 for the assassination of an Archduke . . . should we not be committed to the five million people of Jammu and Kashmir and support them in their struggle for independence? In 1939 Britain and France declared war against Germany, because there was a commitment . . . to support the international frontiers of Poland. . . . A commitment had to be honored . . . and that is why Britain and France are great powers. . . . What happened in Korea? Again nations fought . . . a war of liberation and a just war. . . . Can Kashmir be an exception? . . . The argument, that the future of fifty million people of East Pakistan was jeopardised for five million people of Jammu and Kashmir, is a pathetic argument. . . . It is a bankrupt and an immoral argument. . . . [T]hen you will say, let Baluchistan go, let Sind go, and parts of Pakistan will be slowly and gradually swallowed up by India. . . . It must not be forgotten that it is not just Jammu and Kashmir which is at stake. . . . India cannot tolerate the existence of Pakistan and that is why on the pretext of Jammu and Kashmir war was unleashed on Pakistan. . . . [T]hat is why twelve or more of India's finest divisions, whose guns and wheels were greased by great powers, made an onslaught on Sialkot and Lahore to destroy Pakistan, because in the destruction of Pakistan lay India's most sublime and finest dreams. But the aggressor was brought to a halt.

Then Zulfi rewrote the "epic struggle," which he dubbed "a most glorious chapter in the glory-studded history of Pakistan," which "stood" like an "impregnable wall. This nation stood to a man against the terrible onslaught. Ours was a mighty victory. It was a victory of the people of Pakistan, a glittering crown was worn by the armed forces of Pakistan." Few things are more important to Pakistanis than military prowess, and Zulfi Bhutto's tributes to the armed forces would soon help him to wean them from their top generals, first Ayub Khan, and later his successor, Yahya Khan.

Zulfi insisted that for

> 200 years we have not fought a war. . . . Our people have fought in Tripoli; they have fought in Italy. They have died in foreign lands, fighting for their foreign powers. But what a magnificent and beautiful difference there is in fighting for your own country. . . . We are talking in terms of

> losing five thousand men or so. One hundred million of the people of Pakistan have fought a war. . . . We never lost anyone. Each one is a martyr to the greatness of Pakistan. Each one of them has contributed to the glory of Pakistan. . . . These lives have never been lost; they can never be forgotten. . . . We were the smaller country pitted against a powerful one; and we could not only hold that country at bay, but ours was the victory.

Pakistanis who heard such oratory could not help but feel the thrill of national pride. Was it any wonder they would soon hail him as Quaid-i-Awam (Leader of the People)?

"Now, I come to the famous Tashkent Declaration," Zulfi announced, after speaking for more than an hour. Then slowly and painfully he read out each word of each article of the declaration, after which he reiterated for each article, "The Prime Minister of India and the President of Pakistan agree." That initially innocent-sounding phrase was repeated so often, in so lugubrious, so sarcastic a tone that it soon acquired the chilling character of Mark Antony calling Brutus "an honorable man" in his funeral oration for Caesar. About the fourth article he explained:

> [The] Prime Minister of India and the President of Pakistan have agreed that both sides will discourage any propaganda against each other. Now, . . . it is a different thing to pursue one's legitimate right to support the right cause and indulge in propaganda. Propaganda means vilification, propaganda means slander, propaganda means abuse. No respectable and self-respecting country would like to indulge in propaganda. This is not our policy. We do not indulge in propaganda. . . . But as far as Jammu and Kashmir is concerned, as far as the question of liberation is concerned, as far as the question of eviction of Muslims is concerned, as far as the question of justice is concerned we are not precluded from espousing and propagating these causes.

"I tell you, we are true," Zulfi declared in closure. "There can be no force, no strength greater than truth. Truth is on our side. . . . [H]istory has shown that truth is on our side. . . . India will have to abandon its colony in Jammu and Kashmir. . . . [O]ur cause can only succeed if we pursue our struggle because ours is an honourable struggle sanctified by law and protected by Allah."

It was hardly surprising that Ayub Khan should start suffering from high blood pressure and heart problems soon after Zulfi's address to the National Assembly. He had, after all, specially stopped in Larkana to request, to urge his foreign minister to speak out in support of his government and its diplomatic commitments.

Three days after this performance, Zulfi was ordered to resign as secretary-general of Ayub's Muslim League by the league's executive committee. It was clear to all who had heard him in the assembly that Zulfi Bhutto was ready to launch his own party. He remained in the cabinet till midsummer, for there were several trips to which he was committed: to Turkey, to Iran, and to Indonesia, where Sukarno had asked for him because they were close friends. Ayub had no intention of venturing on any of those missions himself, finding it hard even to keep up his diminished domestic schedule. No matter what Zulfi said to the contrary, Ayub knew the war had been lost, not won, or was at best a draw, which might suffice to launch the meteoric rise of a young populist politician but was hardly the stuff a field marshal could thrive on. He fell into so deep a depression that his hitherto robust, vigorous body soon broke down.

In mid-1964 Zulfi helped convince Ayub of the wisdom of establishing closer economic and diplomatic links with Turkey and Iran, Pakistan's regional Islamic allies (all three were already in CENTO). The trio formed an alliance called Regional Cooperation for Development (RCD), which initially focused on increasing trade and the exchange of industrial information among themselves. It also had the exchange of arms and military cooperation with the West as an even more important element because all three were close to the Soviet Union's southern border.

Zulfi visited Ankara for a CENTO meeting in mid-April 1966, writing to Ayub on the eve of his departure to suggest a more radical Pakistani policy toward the Soviet Union: "It is essential for us to improve our relations with the Soviet Union . . . independent of any important initiative taken by the Soviet Union."[10] In planning his new party and future, of course, Zulfi had decided that his appeal to Pakistan's educated youth and the more radical masses would be enhanced by improved relations with the USSR, as well as with China. He felt betrayed, moreover, by the United States, and knew that Washington would side with Ayub once their split became irrevocable. In Ankara U.S. Secretary of State Dean Rusk approached Bhutto on 20 April requesting ("almost pathetically" Zulfi wrote in his memo of that meeting) Pakistan's assistance in trying to arrange a meeting on Vietnam with China's Foreign Minister Chen Yi, in either Geneva or Monaco. "I said to him that so far the United States had not taken Pakistan into confidence. Many preliminaries were involved and we were not prepared for such a mission." Rusk dropped the idea next day ("He resiled from his specific request and did not manifest the anxiety of the previous evening").[11] But after turning down Rusk's request, Zulfi saw an opening and made a suggestion to Ayub: "As the question is so important to peace in Asia and to the world at large, now is the time for Pakistan to play its part. We should not hesitate only because there is risk of failure . . . the situation is ripe for a meaningful initiative by an Asian State having good relations both with Peking and Hanoi and with Washington. In the present vortex of international affairs, it seems that it is Pakistan that happens to be that State. The gauntlet should be thrown . . . it would establish our bona fides with both sides."[12]

But Zulfi's bona fides with Ayub had by now disappeared. The ax fell in mid-June. Ayub informed Zulfi that it was best for both of them, and for Pakistan's international credibility, that he take "sick leave" abroad. It was now clear to all Pakistani leaders, of course, that "Bhutto-Sahib" was out, never to return to any cabinet over which the field marshal would preside. Travel in Europe had always appealed to Zulfi, and a trip would give him time to plan his political strategy more carefully. The one promise Ayub demanded of him in return for no public sacking was that he would make no political speeches during his "long leave for health reasons," as the Pakistani press reported the joint "declaration" of the two men.

"I informed Ayub Khan that I would oblige him," Zulfi later recalled, "and that my departure would be honourable." Minister Roedad Khan, who had first met Zulfi in 1959, when Khan was district magistrate in Peshawar, immediately heard that "Mr. Bhutto had received his marching orders" from Ayub Khan. "Though it was very late, I told my friend Mr. Ghulam Ishaq Khan [now president of Pakistan], 'We must go meet him. . . .' He had a house then at Civil Lines in Pindi and he was sitting all alone with a glass of whiskey in his hand, and . . . I extended my hand, but he didn't take it. He threw his arms around me, and said, 'The way Ayub Khan treated me today, you will not treat your orderly!' . . . I remember both of us telling him, 'Better days will come.' . . . Next day he left for Lahore by train, and I saw him off late at night at the Cantonement . . . Station. . . ."[13]

A posh government parlor car had been put at Bhutto's disposal, and he left Rawalpindi in the dead of night, 20 June 1966, headed for Lahore, where he had been invited to lunch by his old feudal competitor-colleague, the Khan of Kalabagh, then still governor of West Pakistan. Word raced down the tracks ahead of the train, and was "whispered" over the telegraph wires strung above, so that by the time his train pulled into central Lahore station there was not an inch of empty space on its platforms. "Almost the whole of Lahore gave me an emotional welcome and kept requesting me to make a speech," Zulfi reminisced from his prison cell over a decade later, noting that had he broken his word to Ayub and delivered a political tirade then, "the whole of Lahore would have been set on fire, but I refused to utter a single word."[14] He stood on the steps of his car, tearfully waving to the wildly cheering crowd. He was garlanded, his hands were kissed, and he was lifted onto the shoulders of some of the "thousands of students and well-wishers" who flocked there just to catch a glimpse of him, shouting, as Salmaan Taseer, a Lahore student himself at the time, reported, "Bhutto *zindabad*" (Long live Bhutto), "United States *murdabad*" (Down with the United States), and . . ." anti-Ayub slogans as well. . . . [T]ears poured down his face as he was carried out of the station . . . the handkerchief which he used to wipe his eyes was sold later for Rs. 10,000. Pakistan's redeemer seemed to be at hand."[15]

The fire and brimstone Zulfi had spouted at the UN, and the cries of victory and hope he so eloquently articulated in the National Assembly proved to every young Pakistani that here at last was a leader who had the country at heart. Not since the early death of Quaid-i-Azam Jinnah had any one as potent in his political appeal as Zulfikar Ali Bhutto, the new Quaid-i-Awam, appeared to redeem the "Land of the Pure."

But first Zulfi would take full advantage of his promised holiday abroad, for Ayub retained firm control over the army, and without the army, Zulfi knew, no redeemer, no matter how popular or how great his mystic powers might be, stood a chance of storming Pakistan's many barricades to the twin citadels in Rawalpindi and Islamabad. Zulfi spent a month in Sind, dividing his time between Larkana and Karachi, before sailing for Cairo and Lebanon, where he stopped en route to Europe that late summer of 1966. In mid-July he wrote from the grand Hotel Phoenicia in Beirut to his old Sindhi political opponent Kazi Fazlullah, back in Larkana, with whom he had parted political company in 1962. He was eager now to mend broken fences.

> You may blame me. Such things do unfortunately happen in politics but . . . with goodwill, misunderstandings can be drained out. . . . Generally speaking, it is true that people out of office tend to be ultrasensitive and those in office are inclined to be oversuspicious. We have no cause now to be either suspicious of each other or . . . ultrasensitive. . . . [W]e can start fresh, on a new slate. . . . To use the farmer's raw sense we both come from the same village. . . . I am again reminded of what you told me in Larkana soon after you retired from politics when you quoted Warren Hastings' speech at the time of his impeachment. . . . History alone can determine whether the qualities and services of a man out weigh human failings and weaknesses, and is there a soul without error?[16]

From Beirut Zulfi flew to Geneva and thence to Paris, where he met J. A. Rahim, and they spoke at much greater length, in more detail, about the new People's Party that Zulfi would unveil in little more than a year. Zulfi admired, as he later wrote of Rahim, "his brilliance, intelligence, knowledge, grasp of Foreign Affairs, his gift of writing, his political appreciation and other qualities. . . . He believes in socialism and in modernism. He has no time for obscurantists. He has a secular mind. Above all, he more than anyone else made me decide to launch a new Party. Actually he and I founded it in Paris in 1966. He worked hard on the Foundation Papers."[17] Rahim was the party's foremost theoretician, a Bengali Nietzschean philosopher, whose Manifesto noted that the "general will" of the Muslims of South Asia "founded the State of Pakistan," as "a monument to their unfulfilled hopes and aspirations. They wanted its citizens to live in freedom, a nation progressive and prosperous, powerful and pledged to shield from oppression Muslims" in India as well.

> The new State so resplendent with noble purpose, as it seemed in the beginning, has fallen prey to internal weaknesses, grown forgetful of its own people's welfare. . . . No people in their right senses can desire the aim of the state's policy to be the increase of poverty, general misery of the masses, rampant corruption, demoralization of all classes. The people . . . desired the opposite of the condition to which they have been reduced. . . . Pakistan is . . . one of the poorest among nations . . . ignorance, intellectual sterility, ill-health, dishonesty, crime, corruption, superstitions. All the forms of oppression by authority and by those who exercise the power on account of their riches are to be found here . . . every government of this country has followed the policy of concentrating expenditure in the domains that benefit the privileged classes. . . . To make matters worse, there are men, some of whom hostile to the very conception of Pakistan, who are now condemning all Pakistani Muslims,

> except themselves . . . as unbelievers, if they will not subscribe to the sanctity of economic exploitation and social injustice. This appeal to ignorant fanaticism is dangerous not only to the State but to the unity of Muslims as Muslims.[18]

Rahim drafted the Manifesto and had it printed in October of 1966, writing to Bhutto then from Paris, "In the words of the Chinese oracle I Ching, 'The ablution has been made, but not yet the sacrifice.' The manifesto has been printed and herewith the Movement has been launched. I will go on with the work according to the plans. . . . We expect you to act on your part. There is no point in keeping silent. The people must be taught that there is an alternative to the present regime of corruption. The manifesto is the guide." Rahim believed that he and Zulfi were, indeed, about to inaugurate an era of idealism and selflessness for Pakistan. "We expect you to keep firmly to its sixteen doctrines. When dealing with politicians, who will come to you to bargain for themselves, please do not compromise on the doctrines. Without the Doctrine the Movement is nothing."[19]

The Manifesto began with the fourfold motto of the Pakistan People's Party (PPP): "Islam is our Faith, Democracy is our Polity, Socialism is our Economy, All Power to the People." Rahim and socialist Dr. Mubashir Hasan (b. 1922), whose huge old Gulberg house in Lahore would become the venue of the Pakistan People's Party's actual birth in December 1967, were to serve as the ideological parents of the PPP's doctrine. They were both theoreticians, the only doctrinaire intellectuals among the new party's activist founding fathers, each in his own way a world remote from Zulfi Bhutto, yet both wise enough to appreciate that Bhutto alone had the wealth as well as the charisma and popular vote-getting appeal to give their radical doctrines any chance of winning mass support in the conservative polity.

From its conception, then, the People's Party was a unique and unstable blend of the right and left wings of West Pakistan's leadership spectrum, for Zulfi, Mumtaz, Khar, and Jatoi could hardly be accused of harboring any radical-socialist leanings, nor could barristers Rafi Raza, or Hafeez Pirzada, who early joined them. Mubashir, on the other hand, could never be accused of feudal or reactionary proclivities. Rahim was more difficult to label, sometimes as radical as Mubashir, at other times more reactionary, or authoritarian, than Zulfi. "I am now working on an elaboration of the programme that can make a reality of 'Mass Mobilization,'" wrote Rahim to his leader. "If I can manage it, I shall have it printed as a brochure." Though J. A. had retired from the Foreign Service, his pension was not as yet "in sight," so he lived penuriously in Paris, closing his letter to Bhutto with "You must take the lead now."[20]

Zulfi had, in fact, gone to London from Paris and spoken in the City's old Conway Hall on 13 August 1966 to a large gathering of Pakistani students living in the U.K. "I am not supposed to be in good health," Zulfi began. "But I can assure you no matter how poor my health, it is sufficient for India." Still, he had promised Ayub *not* to speak on politics, as the one condition of his paid vacation, so he informed that well-informed audience, "It is not necessary for me to call a spade a spade and to talk on matters, to which, I am sure, you have given very deep and profound thought. . . . I hope you will forgive me, if you are expecting me to talk on the Tashkent spirit or any other intoxication."[21] What he did talk of was how Pakistan was "the voice of a hundred million people articulated on the purity of an ideal," and how, though India is "threatening us with the atom bomb . . . science and technology are everyone's right. . . . Progress and scientific technology cannot be restricted. If India has the bomb, that does not mean that we are going to be subjected to nuclear blackmail." He reiterated his firm intention of "standing by" the people of Kashmir, upholding their "right of self-determination." Here again Zulfi struck the perfect note of fearless idealism that most inspired young people, especially his listeners, who felt guilty for not having been nearer to the "front" during the 1965 War. Pakistan without Kashmir, Zulfi declared, was "a body without a head and it is a very beautiful head." He called upon all Pakistanis to espouse every Afro-Asian cause and struggle, insisting in language he had picked up from J. A. Rahim and Mubashir that "we are the proletariat of the world . . . therefore, we have to cooperate, collaborate, get together, assist one another . . . and finally the right cause and justice must prevail. . . . Our people deserve it. For centuries they have lived in misery, squalor, filth and poverty."[22]

Zulfi was aware by now that Ayub and his closest colleagues considered him most "dangerous," a "Maoist" as well as a "madman." "Must we be labeled as communists and various other things merely because we fight against domination, because we rise against exploitation?" he asked the cheering students. "Must we be blackmailed and must we be slandered in this fashion merely because we say it is not the law of God that people should not have a better life . . . just because we want to defend our sovereignty and integrity and our national independence and give our people a more egalitarian future?" By appealing to the masses, his People's Party would, Zulfi hoped, win the next general election. By focusing Pakistan's foreign policy not only on Kashmir's liberation but on justice and equity for all of Asia and Africa, Zulfi dreamed of forging an alliance of most of the world, one capable of complete independence of both superpowers, which he as the most popular Leader of Pakistan might one day "chair" and guide. Even as General de Gaulle spoke of a "European Europe," Zulfi now spoke of an "Asian Asia." He included China, of course, in such an entity and would welcome India, once India abandoned its "great fraud," by liberating Kashmir. "We do not have to fight for a thousand years," Zulfi said about India, "but please do not forget there is a thousand years' conflict behind us. Must our quarrel with India be eternal? It can come to an end on the basis of justice and equity . . . which means not only the right of self-determination of the people of Jammu and Kashmir, but proper treatment of minorities."[23]

Preoccupied as he was with starting his party, Zulfi concluded this London speech by waxing poetic about "the true nature of politics" and the "great role" of a "politician," who has to be a "mathematician . . . to calculate, take into account everything," and "a musician and romanticist. . . . [He] must know the tempo of the time, the rhythms of revolution."

Soon after he returned to 70, Clifton in Karachi that October, Zulfi received a message from his venerable neighbor across the road, Fatima Jinnah, who was eager to speak to him. "I walked to her residence [Mehota Palace]. She was sitting in the hall with her white poodle beside her," Zulfi recounted. "She got up and with a smile waved her thin finger at me, and said, 'I told you to leave him. It is all your fault.'"[24] Her hair was as white as her sari, and at 73, "Fati," as her Great Leader brother always called her, was in the last year of her life. She reminisced about Jinnah's friendship with Sir Shah Nawaz, the old days in Bombay with Dr. Patel and his "tea parties," and Liaquat Ali Khan and his begum. But mostly she talked about how she'd been "robbed" of the presidency in the last elections, which she had felt so confident of winning. "She told me that the rigging was done very shamelessly in the Punjab," Zulfi remembered. "She said with some rancour, 'Being old and alone, I had to reconcile myself with the situation . . . [but] I trounced Ayub Khan.'" That was the last time he spoke to her, but Zulfi felt no regrets at having refused to join her East Pakistani-dominated combined opposition. He could never have subordinated his ego to a woman's in politics, nor, for all his fondness of Husna's friends, would he have felt comfortable working in harness with so many Bengalis. Nor did he want merely to serve as foreign minister. He would be his own foreign minister soon, and until then would feel free to attack Ayub's foreign policy with impunity.

Well before Zulfi returned from abroad Ayub had appointed Syed Sharifuddin Pirzada to serve as foreign minister. Pirzada led Pakistan's delegation to the UN that year and, among other business, strongly supported a resolution that called for a world conference against nuclear weapons proliferation. Zulfi was in Larkana, for his annual shooting party on the eve of his thirty-ninth birthday, when he read Pirzada's statement in Karachi's *Dawn*. Taking time out of his holiday to criticize his successor, he warned that "with India remaining outside the net of the resolution" Pakistan could hardly risk supporting it.

> Pakistan will always find it difficult to quantitatively keep pace with India; but qualitatively we have maintained a balance in the past and will have to continue to maintain it in the future for our survival. It is for this reason that as Foreign Minister and Minister-in-charge of Atomic Energy, I warned the nation sometime back if India acquires nuclear status, Pakistan will have to follow suit even if it entails eating grass. . . . My criticism of the resolution is not opposed to national interest and security. Quite the opposite; it has been made in the interest of the nation and should be welcomed. It is dangerous to take aim with a gun loaded with blank cartridges.[25]

West Pakistan's military establishment appreciated Zulfi's knowledgeable reference to guns, and how best to use them against any enemy, especially India. East Pakistan at this time was exploding in mass protests against its "colonial exploitation" by West Pakistan, and Sheikh Mujibur Rahman's six-point platform of the popular Awami League he now led became a veritable blueprint for charting Bengali independence from Punjabi and Pathan rule.

Mujib's Six Points called for the establishment of a "federation" of "parliamentary" government, run by a legislature "directly elected on the basis of adult franchise." Each wing would be virtually autonomous, the central federal government empowered to deal *only* with matters of defense and foreign affairs. The wings alone would have the "right of taxation," and send "a fair share" of their "separate or freely convertible currency" to the center for supporting defense and foreign affairs. Because East Pakistan jute earned most of Pakistan's vitally needed foreign exchange, point five demanded that "separate accounts" be maintained for the foreign exchange "earned by each wing," and that East Pakistan—called Bangladesh (Land of Bengal) by Awami Leaguers—raise its own militia or paramilitary force, thus allowing its people to escape further oppressive "protection" by West Pakistan troops.[26]

Soon after adoption of the famous Six Points by Mujib's Awami League in February of 1966, Ayub called Mujib a "secessionist." Police arrested him and raided the headquarters of his party, triggering riots and strikes in Dacca, Chittagong, and other major centers. By year's end East Pakistan trembled at the brink of civil war. Compared to Mujib, of course, Zulfi sounded positively tame and loyal. Hundreds of Bengalis were killed and thousands wounded by the end of 1966, raising the darkest prospects of military escalation in the minds of the leaders in Pindi's GHQ, and in the bright offices of Pakistan's grand new capital, Islamabad, built as a monument to Ayub's "revolution."

Zulfi hoped to compete against Mujib on his own turf, writing a column called "My Debut in Journalism" for Dacca's *Pakistan Observer* on 12 January 1967, testing those Bengali waters in English, to see what sort of political prospects he would have in the East.

> Although by birth I am an agriculturist and by education a lawyer . . . it is politics above all that inspires me and kindles in me the flame of a lasting romance. Politics is a superior science and a fine art. . . . Politics must be clean. It must not be based on negative considerations and rest on sloganism and on the lure of banditry. Too many people are understandably suspicious of politics. Tragically, we have played politics with politics. We have soiled a great art. No longer will people repose confidence in politicians who exploit their sentiments and abandon principles at the wink of favours.[27]

Zulfi, of course, meant Mujib, but as General Gul Hasan, who knew them both, later reflected, "The tragedy of Pakistan is that we had only two popular politicians, neither of whom could ever tell the truth."[28] "The politicians of Pakistan are facing a crucial new test as destiny stands at the dawn of a New Year," Zulfi wrote. "A new look and a new style will have to emerge. The old ways will no longer appeal to the people. . . . Drawing room intrigue has reached a nauseating limit which the people will not tolerate in the future"—an explanation of why old parties like the Muslim League were dead. As for Mujib's Awami League and Ayub's heavy-handed repression of it, he wisely commented, "Terror and repression have never been the final answers to political problems. . . . The administration should not be intimidated nor should the administration intimidate. Criticism must be genuine and solutions should be just and realistic. In a free and independent society, there is no room for bitter personal animosity. The line between government and opposition should not be based on vendetta and abuse, but on a sincere difference in principles."[29] Zulfi was openly appealing to Ayub to forget the "bitter personal animosity" that kept them apart. Compared to Mujib, after all, and his "secessionist" army of Bengalis, Bhutto was a politician with whom any Sandhurst graduate could surely enjoy "a sincere difference in principles."

But Ayub would never forget his last year in harness with Zulfi, and could stomach no more of him. The field marshal's newly appointed army commander in chief, Shi'i General Agha Mohammed Yahya Khan, however, and even more perhaps, Yahya's leading adjutant, Major General S. G. M. Peerzada, would both find Zulfi Bhutto just the sort of tough-minded political leader with whom the Army *could* work and play.

"Politics is a many splendoured thing," Zulfi now wrote, almost singing it, waxing romantic. "A politician must be patient and he must also dare. At times it is not patience but risk and daring which are required. But the risk and daring must not smack of adventurism because it is fatal to play with the future of a whole people who repose confidence in their leaders." The fine political line between "risk and daring" and "adventurism" that he drew and believed in his own mind was, of course, the line dividing *his* sort of patient but popular politics from Mujib's "fatal" variety. Zulfi had met Mujib many times at Husna's Dacca parties and personally they got along quite well for both were convivial men who enjoyed all the good things in life, yet they never trusted each other. Perhaps each recognized too much of himself in the other's raucous laughter, loud and racy humor, and intoxicated passion for power that shone from moist and slightly bulbous eyes. It was more than the innate mistrust most West Pakistanis felt for Bengalis, and the other way round. Each must have seen in the other's inflated ego the mirror of his own singular ambition, and each understood that in any nation, or jungle, only one supreme leader, one king, could emerge and survive at a time.

Ayub was almost kidnapped, nearly assassinated, during his presidential tour of East Pakistan in December. Zulfi wrote to "warn" the foreign secretary at that time that India was determined to "dismember" Pakistan. Ayub now feared that Mujibur Rahman was India's leading agent in Dacca, and that the Awami League was but an East Bengali branch of Indira Gandhi's Congress Party. More of Mujib's supporters were arrested in December 1967, and thirty-two Bengali members of his party were soon to be charged with helping him to "treacherously conspire" with India against Pakistan.

That same month, Zulfi's Pakistan People's Party was born in Lahore, on 1 December 1967, at Dr. Mubashir Hasan's sprawling home. Zulfi spoke and showed the assembled delegates a "sketch" of their new party's flag, a white crescent moon and a star against a three-colored background of Islamic red, black, and green. The party's motto was repeated with passionate approval by all, the most radical and the more devout of delegates: "Islam is our Faith, Democracy *(Jumhuriet)* is our Polity, Socialism *(Musawat)* is our Economy, All

Power to the People." Mujib's Awami League in the East, much like Wali Khan's National Awami Party of the North-West Frontier Province, Zulfi argued, was more attuned to Indian "political philosophy" than to ideas and ideals unique to Pakistan.

"We have to seek a Pakistani interpretation," Zulfi told Muzaffargarh's Bar Association in January 1968, explaining why he "embarked on the formation of a new party." His was a "progressive party," yet purely Pakistani in inspiration and direction.

> Our task is to bring about a general understanding between all the parties of the left and of the right. The Pakistan People's Party can form a great bridge, because we have come with a clean slate. . . . We have not come with any past prejudices; we have no personality clashes; we have not done anyone down; nobody has done us down. We are all wedded to certain basic principles. We have said that Islam is our faith and for Islam we will give our lives. . . . It is the basis of Pakistan. There is no controversy on that and if any party were not to make Islam as the main pillar of its ideology, then that party would not be a Pakistani party. It would be an alien party. That goes without saying. . . . At the same time, we believe . . . that Islam and socialism are compatible. . . . We believe that the country must have a socialist economy because for 20 years we have seen only loot and plunder.[30]

The People's Party Manifesto adopted by the convention in Lahore on 1 December first gave as the "ultimate objective" of the party "the attainment of a classless society," which was deemed "possible only through socialism," meaning "true equality of the citizens, fraternity under the rule of democracy" within an "order based on economic and social justice." Orthodox critics of the People's Party would insist that socialism was contrary to the ideals of Islam, but the Manifesto argued that the party's "aims follow from the political and social ethics of Islam. The Party thus strives to put in practice the noble ideals of the Muslim Faith."[31]

Idealistic West Pakistanis flocked to the People's Party because it promised real changes in the nation's "economic structure," leading peacefully to a "juster socio-economic structure," putting an end to the "unjust order" that had kept its stranglehold over progress for the past twenty years, ever since the "Land of the Pure" was born. Young lawyers like Khurshid Hasan Meer (b. 1925), who had served as Rawalpindi secretary for the Combined Opposition Party (COP), which had run Fatima Jinnah against Ayub, eagerly volunteered their time, talents, and energy in the service of the party. The best-educated young women, such as Dr. (Miss) K. Yusuf, a dedicated teacher and later principal of a woman's college in Lahore—and like her friend Khurshid Meer, Kashmiri by birth—also volunteered to work for the party because its Manifesto and leader seemed to offer Pakistan great hope for a better future for its undernourished, ailing children.

Orthodox Muslims as well as secular Communists rallied to the party, magnetized by the force and fearlessness of Zulfikar Ali Bhutto's fiery speech. He met every criticism head-on, thundering in one instance: "The present system is not Islamic. . . . This highway robbery that is taking place in the name of capitalism . . . does not even exist in . . . America. . . . When the living conditions of the masses are deteriorating, how can this system have the sanction of Islam? Go into the interior and see how human beings live. They have no future, their children have no future. . . . We have to tackle basic anomalies. . . . Change this system and put an end to exploitation. This can only be done by socialism. That is why our party stands for socialism."[32]

Zulfi by now felt little more than contempt for Ayub and his mediocre cabinet and government by "half measures." He lashed out at what he called a system

> half democratic, half dictatorial. It has half a war with India, half a friendship with China, and it is resisting America by half. . . . We need a vigorous, dynamic, dedicated, sincere leadership and we require an ideological programme. . . . For 20 years Pakistan has just dawdled along. . . . Basic Democrats are the Brahmins of this country . . . corruption in Pakistan today has become a major industry. . . . Where is the security? You are from a backward area. I am from a backward area. The administration has virtually broken down. There is no such thing as an automatic process of law and order. . . . The cost of living is going up. . . . The economic situation is deteriorating. . . . We were supposed to be a second Japan. I do not see where this second Japan is.[33]

Even as Zulfi spoke, Ayub, who had suffered a pulmonary embolism, could not leave his sickbed. Colonel Mohiuddin, the field marshal's personal physician, was at his side day and night, but all who then saw Ayub thought him a "lost person," as several of them informed G. W. Choudhury.[34] Other than his family, only Ayub's closest advisers were permitted near the president for a month and a half: his principal secretary, S. Fida Hassan; Foreign Secretary S. M. Yousuf; Commander in Chief Yahya Khan; former Commander in Chief Mohammad Musa; Altaf Gauhar of Information and Broadcasting; and Inter-Services Intelligence Director General Akbar. Near the end of January 1968, the president's illness was reported to be influenza; in early February, "viral pneumonia." Few people realized that Pakistan's first field marshal and longest-surviving head of state lay near death's door, unable to breathe without pain.

While Ayub fought for his life in Rawalpindi's presidential palace, Mujib, who had languished for months in Dacca's central prison, where he was being kept in preventive detention, was charged on 2 January 1968 with treacherously conspiring to lure East Pakistan from its capital in the West with arms from India, secured through First Secretary P. N. Ojha, of Dacca's Indian embassy. The Agartala Conspiracy trial as it was called, named for the Indian border town where Ojha purportedly met regularly with Mujib's lieutenants, who then reported back to their leader in his Dhanmandi home in Dacca, dragged on for almost a year. It proved better as a platform for Mujib's Six Points than as a way of legally removing him from the pinnacle of Bengali popularity. The trial took so long and the "evidence" proved so tenuous that the case was finally thrown out of court before the end of 1968.

With Ayub confined to bed, and Mujib behind bars, Zulfi's was the most powerful political voice heard in Pakistan during most of 1968. "Politics is an objective science. It is like mathematics," rapt audiences heard from Zulfi—at last ready to play the political power game he had studied at USC, Berkeley, and Oxford for Pakistan's highest stakes. "We must always have the initiative in our hands. We must not fall back, because the moment we do this we shall have to keep falling back. Pressure is both a worm and a monster. If you stand up to it, it is a worm; if you fall back, it becomes a monster. That is why in politics it is so important to keep the initiative. . . . We need internal strength. . . . You cannot beat and exploit the people and then ask them to unite, because they must have a stake in uniting."[35] Zulfi urged his listeners to join him in building a cadre of "sincere, honest, able, dedicated, and hard working" public leaders. Was it any wonder that he touched the hearts as well as the heads of Pakistani youth?

"You may have heard about the Government's attempts to disrupt our meetings," Zulfi told a crowd in Mirpur Khas that February. "They thought we would be frightened by 'goondaism'. But we will never get frightened. . . . We . . . will never retreat. . . . They have locked up the poor in jails. Well, let them put me in jail if they have the courage to do so."[36] Zulfi knew that Ayub's government was investigating his personal use of government tractors and other land-clearing equipment in Larkana, considering filing charges of corruption against him at this time, but he also knew that Ayub and other ministers were at least as guilty of similar private misuse of public funds—and as a barrister, he understood how hard it was to prove conspiracy by direct evidence. As Ayub's law secretary would note in Zulfi's file, "It may not be possible to succeed in obtaining a conviction in an ordinary court."[37]

To Sind's Nawabshah Bar Association, Zulfi shouted: "I say that if the Government is corrupt, then people will be corrupted. . . . If your hands are clean and if you think only of the common good and of the progress of society . . . modern society provides . . . means, to deal with this problem. . . . A tendency has grown to say that the majority province of this country is showing signs of secession. I would refuse to believe that the majority of this country would want to secede from itself. What the people of East Pakistan want is 'friends, not masters.'"[38] The quotation was the title of Ayub's just-published autobiography, which made no reference to either Zulfi Bhutto or Tashkent. As Zulfi and his new party followers were more openly subjected to arrest for violation of the military Defense of Pakistan Rules, he became an advocate of freedom for Mujib and the dropping of all charges of "secessionist" treachery against him and his Awami League. He was young Pakistan's clarion of freedom, spokesman for every oppressed group in the country. "If you deny the people their rights, their fundamental and inalienable rights, if you do not make them partners in power, . . . if you suspect them, if you maltreat them, . . . you are bound to get alien agents among the people."

Ayub's regime stood "completely exposed and isolated," barrister Bhutto told fellow members of Sind's bar. There was nothing to fear, he assured them, because the regime's stability was artificial. "This Government moves like a maniac, throwing people into jails. How long can you fill the jails? Ideas cannot be imprisoned. Principles cannot be imprisoned. . . . [A] political party cannot be imprisoned. That is why we believe it is essential that every one of us should work together in the service of Pakistan, because this is our own country."[39] Bhutto at his best was not only a Pakistani patriot but a man of liberal and democratic ideas and ideals who won widespread support by virtue of his sincere belief in the ideals as he articulated them in public. One part of him did so believe, and that was the youthful Zulfi, who had deeply admired Hans Kelsen; whom Piloo loved as "Zulfi, My Friend"; and who inspired millions with his soul-stirring rhetoric and fearless ringing phrases. But there was the *wadero* Zulfikar Ali Bhutto as well, and *his* mind lived in an earlier century, *his* heart beat to a tougher feudal tempo, one rarely revealed in public.

Zulfi spoke of Pakistan's internal "basic contradictions" as "massive and complete," reflecting perhaps his unconscious awareness of his own basically contradictory personality.

> External policy is also a reflection of these contradictions. On the one hand . . . we were told there should be joint defence with India. That was said to be essential and inevitable. We ended up with a war against India. Then we were told that America was the only country which was Pakistan's natural friend and ally. . . . Then, all of a sudden, we discovered . . . the Soviet Union and . . . the People's Republic of China and it was essential for us to have good relations with these. . . . We were also told that we will do nothing for one great power which should be against the interests of other great powers, and yet we remained members of CENTO and SEATO . . . which China and the Soviet Union regard as unfriendly.[40]

Zulfi believed his timing in launching his People's Party perfect: "We cannot live in the present mess . . . we should be a model, self-respecting nation. . . . That . . . can only come if there is . . . a sense of trust, because there have been many betrayals." His feudal past kept roiling up from the inner darkness of his *Siasat*-consciousness.

Two weeks later he was in Khairpur, telling the bar there that Pakistan's economic and political crisis was "getting bigger and bigger."[41] The major reason was that government had "lost its sense of direction. . . . When you lose your sense of direction and you lose your sense of purpose and rely on brute force, then it is not really possible to bring satisfaction to the people." What Pakistanis wanted, Zulfi insisted, was "security, justice and satisfaction." But with a system so corrupt as Pakistan's, security remained "an illusion." For Zulfi, satisfaction meant "sharing power," making decisions that "determine your destiny." Even with Husna now, as with the others, there could be no "satisfaction" for Zulfi, without power. "If you are excluded from taking part in the decisions which affect you . . . then you are not only dissatisfied but you are disillusioned; you get frustrated and your sense of despair increases."[42]

"When you have something to conceal, and this Government has a great deal to conceal and a great deal to hide," Bhutto told his listeners—knowing well whereof he spoke—"it will naturally be sensitive to public meetings and to free expression. . . . You have a situation of tyranny in Pakistan. . . . Increase in tyranny leads to greater contradictions." As to the question of East Pakistanis wanting "to go their own way": "I cannot believe that. They are the majority. The seeds of Pakistan were sown in Bengal. The Muslim League was born in Bengal. . . . How can a majority leave? . . . How can it . . . become a part of India? The people are not secessionists, but what they want are friends. . . . if we treat them like servants and think that we have the legacy of the British to rule them, then they may want to go. . . . The Bengalis are being denied their rights . . . the time has come when all of us should actively work for the future of this country." Those who later accused Zulfi Bhutto of wanting to lose Bengal misread his earlier resolve. He wanted a united Pakistan, one nation under him, led by him and *his* People's Party. Believing in his own "destiny," just as Napoleon had done, he believed that would be best for Pakistan.

"Mistakes can be made by anyone. I have made mistakes," Zulfi admitted in Larkana a few days later.

> I am not here to argue that I was always right or to rationalise my mistakes. That I leave for those who think . . . they can commit no errors. . . . I am asked very often by these gentlemen, who commit no mistakes, how it is possible that I was eight years in the Government and why I am now against the Government. . . . But I would like to ask what right have these individuals to ask this question? . . . I was not insulted or humiliated like them. I was not kicked in the pants. My nose was not rubbed in the dust . . . I left in honour.[43]

Here spoke Larkana's wadero, for whom nothing was more important than honor *(Izzat)*, and few things sweeter than to say of his opponents how they had been "humiliated," "kicked in the pants," their noses "rubbed in the dust." *Siasat!*

Zulfi continued:

> I do not believe in rancour and bitterness or in a negative approach to politics. I believe in being broadminded and not letting personal considerations influence political decisions. And it is for this reason that I have not so far revealed so many things that have happened. But if lies are spoken about me, I should, at least, have the right to speak the truth. . . . Whoever says that I have sought a compromise is a liar. . . . On many occasions close associates and relatives of the President have come to me for a compromise; I have not taken a single step to have any compromise with the Government. . . . But my opponents want to drag people to their level of politics. . . . I would like to say something about the Head of the Provincial Government [General Musa]. . . . I know that he holds high office and that I am an ordinary citizen but I also know that it was I who approved his selection as Ambassador to Iran. It is not dignified for a Head of a Provincial Government to disapprove of a political leader sim-

> ply because he is small in size. . . . Is height the only thing that matters? The remark was in bad taste. The Governor may have power and authority but these don't last forever. And, as for people who are not tall, the Vietnamese have shown what they can do. I say to this former Army Chief, 'Learn from the lessons of Vietnam, from what little men have done in fighting the giant Marines, from the performance of little men without shoes . . . against the well-equipped Americans. What made the difference was the strength of their conviction and their belief in the righteousness of their cause.' Such are the people who are ruling this country—clowns and charlatans.[44]

Zulfi was firing his slingshot at Musa, who like Ayub, stood well over six feet, a muscular martial giant. But Larkana was Wadero Bhutto's turf, and no man in that crowd was not laughing with, or applauding, Pakistan's most popular hero.

> Please search your hearts and tell me: are members of the legal profession satisfied with the prevailing conditions; are the farmers satisfied; are the labourers satisfied? Who is satisfied—the refugee from India, the Punjabi, the Baluchi, the Sindhi, the Pathan, the Bengali? The Government today is isolated and it has alienated each and every segment of society and each and every part of the country. How long can this go on? The President has been talking about the stability of the country. . . . For two months he is ill, and who takes the decisions? Where are the institutions which provide for genuine stability? This is not the way stability is given to the people. Corruption has increased. Lawlessness has increased. Crime has increased. Prices of commodities have increased. Blackmarketing has increased. Interference in the judiciary has increased. Everywhere there is contamination. Everywhere there is chaos and trouble.[45]

Zulfi was at his best for he was, after all, at home and hence felt completely confident in his powers and his political destiny. He called Ayub's administration a "tissue-paper tiger," urging his friends not to be intimidated by authorities who might ask them why they had met with him. His courage was contagious:

> Remove the clouds of fear from your minds. Fight for the right cause. Do you think that I am not made to suffer? You do not know . . . the mean and despicable things that are being done. . . . But still I am there. Soon these will be things of the past . . . this kind of intimidation is only a balloon. Your resistance will cause it to burst. . . . This Government . . . has sown the seeds of its own destruction and is helping them grow into a bumper harvest by adding doses of fertiliser with each reckless act of tyranny and corruption. . . . The evils of the present system are like a cancer in our body politic. it cannot be cured by a few doses of medicine. It . . . has to be removed by surgery.

Zulfi hoped that Ayub's illness and the rapidly deteriorating economic situation that was doing so much to contribute to popular unrest would convince Ayub that he had better step down. Elections could then be held that would bring his People's Party to power, at least in the West. His party's manifesto and program were designed to appeal to Pakistanis of every region, Bengali peasants and workers as much as Sindhis. The latter, however, spoke Bhutto's language, and understood that his "All Power to the People!" meant first of all the people of Larkana and Sind.

Zulfi addressed the first Sind provincial convention of his new party on 21 September 1968 in Hyderabad, telling the delegates of Ayub's "shameless lack of faith" when his government "fell at the feet of the great powers" by accepting the 1965 cease-fire.

> My brothers, I wanted to resist Pakistan's enemies, but my opponents dubbed my patriotic feelings as emotionalism. . . . After coming back from Tashkent, wherever I went the people received me with affection. They asked me scores of questions on seeing tears in my eyes. But I could not tell them what befell my dear motherland. My silence was exploited as cowardice. As a matter of fact I am neither a coward nor emotional. I had kept quiet only because the enemy forces were looking for a suitable opportunity. I knew that a single sentence from my mouth could spark off a civil war in the country. I remained silent to avoid a civil war. . . . But now that years have passed I cannot remain silent. . . . Brothers, I am proud of having been trained politically by the ever-conscious patriotic people of Pakistan. They are my real teachers. This is why the politics of the people and humility are in my character.[46]

Humility hardly seemed the appropriate word for someone who had just said that "a single sentence from my mouth could spark off a civil war."

Zulfi went on. "I am not greedy for wealth . . . I never unduly drew a single penny from the national treasury for my own person. In the land reforms, my family surrendered its 40,000 acres of fertile land for the people. . . . Today, the Government accuses me of having taken an undue advantage of my official position. . . . I challenge all individuals who have been associated with the Government to declare their assets before they entered Government and as these assets now stand. I will do the same." He called out to Ayub in particular: "Come on, Mr. President, let us both take the initiative and account for our past and present assets. Let us tell the nation what you had before you became President and about all that I had before becoming a Minister. Let the people know what you gained and what I lost while in office. . . . Are you willing to do that? . . . I could have amassed wealth like you have done." Pakistanis love courage and admire heroics. Most of those who heard Zulfi Bhutto speak in 1968 loved and admired him for his fearless challenges to the giants of Pindi, Lahore, London, and New York.

A charge of criminal conspiracy had been filed that year against Zulfikar Ali Bhutto for having hired government tractors at low concessional rates to clear more than one hundred acres of land he owned in Larkana, some of which he supposedly held "illegally," "conspiring" to get his servants and distant or junior family members to say they were the owners. "Friends and elders, I am a human being," Zulfi admitted to the delegates. Then,

> I can make mistakes. But I am not guilty of any impropriety . . . I never took undue advantage of my office as Minister. However, I have committed one sin; and that is that I have been associated with this Government for eight years, although I served the country to the best of my capacity. . . . If like other Ministers I wished to have my own factories, and bank balance abroad, I could have done all that so easily. To hire tractors at half the fixed rate would be like picking up pennies after passing up pounds. . . . Once a representative of a Big Power took me by my arm and said: "Mr. Bhutto, you will get anything you wish provided you make some adjustments." . . . I jerked my arm away and said: "Never try to do this again. If you have purchased a few mean individuals, you should not think that everyone in Pakistan is a Mir Jafar [traitor]. I am not like those who have acted against national interests and stacked away funds. . . ." When I narrated this incident to the President, . . . he paid tributes to me. But I would like to ask him now whether this is the reward for those tributes?

The Wadero Khan Bhutto part of him felt grossly betrayed by Ayub, and angry that such petty charges should be filed against *him*. A few wretched half-rate tractors! How small-minded of Ayub, who so often had dined at his table, slept in his finest guest house, hunted his wild boar and birds. Bhutto's generous feudal heart, his noble spirit, could not understand such betrayal, so petty an attack—except as a desperate ploy to tarnish his name and thereby undermine his party's potent challenge.

> My brothers: The Pakistan People's Party is a principled, ideological party. . . . But the Government has resorted to personal vindictiveness. . . . We speak of principles, . . . the Government orders . . . "the tractors case" against me. . . . We mention socialism, the Government orders that we should be detained under D.P.R. [Defence of Pakistan Rules]. . . . We demand justice for the people of Baluchistan, instructions are given to arrest Akbar Bugti. . . . We say the people want democracy, it is said the Basic Democracies are good enough. . . . We demand food and clothing for the people, they give us bullets and beatings. . . . Tell us, Khan sahib, which of us has trampled on democracy, you or me? . . . If this situation continues . . . the people will rise in rebellion, and there will be bloodshed and civil war in the country. I am not prophesying. It is logic. I might be accused of spreading rebellion. Well, I will do that, if needed. I fear no one.

Prison for preaching "rebellion" would be preferable to the ignominy of being locked up for conspiring to misuse tractors. So he hurled another stone, this time at "Ayub Khan Sahib," shouting his name and, "I am not a coward. I cannot be browbeaten . . . I am not scared of your guns. . . . Come on, take up your guns; I have the power of the people with me. It is more powerful than an atom bomb. I have burnt my boats for democracy and socialism. . . . I will not let you sit in peace."

Zulfi had taken off his jacket and rolled up his sleeves. He was clenching his fists, showing the stuff he was made of, ready to stand his ground and take on any giant. As the fearless Quaid-i-Awam, forty years young, he was ready to take on the entire establishment. "Dear brothers . . . As justice is hard to get in a country where there is no rule of law, you remain at the mercy of bureaucrats . . . a class holding the reins of government in its cruel and blood-stained hands. . . . The Government keeps on harping about my isolation . . . claims I have no friends . . . [but] the poor people are in fact with me. God is with me."

Faced though he was with criminal prosecution, Zulfi stood ready to lead the march toward "economic justice," insisting that "socialism" must be adopted to make Pakistan a "real democracy." His party's program included demands for "abolition of feudalism, protection of the rights of peasants, and labourers; their right to strike; nationalisation of all key and basic industries, nationalisation of transport; and nationalisation of education."

> My dear friends, it is said that I am a wealthy man and a feudal lord. It is said that I have no right to struggle for socialism without distributing my wealth among the people. . . . Socialism can be introduced only when all means of production are brought under state control. But even so I hereby announce that if my wealth can be of any good to the nation I will not hesitate to give it away. But I cannot be so foolish as to hand my wealth over to capitalists and feudalists under the capitalistic system, so as to enable the rulers to make more money and spend more on their luxuries. Well, if the Government is courageous enough to introduce socialism, it is welcome to do so. I will be the first man to place my wealth under national control.

He paused, as if only just hearing what he had said, reflecting on the implications of his promise as it echoed in his mind. Then, more soberly,

> But. . . . you cannot fool the people by such useless arguments. I believe in socialism; that is why I have left my class and joined the labourers, peasants and poor students. I love them. And what can I get from them except affection and respect? No power on earth can stop socialism—the symbol of justice, equality and the supremacy of man—from being introduced in Pakistan. It is the demand of time and history. And you can see me raising this revolutionary banner among the masses. I am a socialist, and an honest socialist, who will continue to fight for the poor till the last moment of his life. Some ridicule me for being a socialist. I don't care.

Such was the dialogue Zulfi had aloud with himself. He had read enough French and Russian history to know something of the revolutions that had racked both lands, changing the polity of Europe. And he had read enough Marxist tracts to know something about socialism, but Zulfi compared himself most with "those Ahl-i-Quraish who rejected worldly riches, and lined up behind the Prophet of

Islam in order to spread the message of liberty, equality and peace. I only want to bring about a revolution in Pakistan."

Presidential elections were to be held in the winter of 1969–70, but Ayub had recovered strength enough to fly to London for medical treatment and had decided not to run. Zulfi's campaign trail took him from Hyderabad through the North-West Frontier, Ayub's home province, where he continued to launch fearless attacks against the ailing dictator.

> I am not afraid of you. . . . Why don't you put me in jail? . . . If you put me in jail the people will turn you out of the Government. . . . You are running your Government with force and suppression. . . . We are struggling for democracy and we shall continue to struggle. . . . 22 families . . . have usurped the economic resources of the whole country. . . . It was said before Martial Law that there were 600 zamindars and out of them only 200 were ruling the country. Now a score of families wield power. . . . Even in America, the centre of capitalism, such a wretched system does not exist. . . . We demand justice and fair play.[47]

That October Ayub was celebrating his "Decade of Revolution." Zulfi insisted it was time to start a true "People's Revolution."

"They accuse me of lust for power and say that is why I am opposing this Government," Zulfi told a public meeting in Abbottabad on 29 October 1968. "It is this Government which is drunk with power. . . . Brothers! I have committed mistakes in my life. I am a human being and to err is human . . . but for my sins and blunders I shall repent before my Creator and beg forgiveness. I shall not go to the President. . . . My greatest mistake has been that I was associated with this Government. . . . They have sucked the blood of the entire country. They have usurped the wealth of the whole nation. It does not lie in their mouth to accuse us."

By early November the People's Party had gained supporters and momentum wherever its popular leader had spoken. First in Sherpao, then in Peshawar, Zulfi kept jumping onto platforms, ripping off his jacket or vest, exciting West Pakistanis to cheer as he pugnaciously attacked the tottering Ayub and his ancient regime. Zulfi told his mass audience in Peshawar, "The people have no place in the government. The military rules here. . . . It is nevertheless a weak Government. . . . How can it be a powerful Government when all its policies are anti-people? . . . It is a Government of cannibals. It ate up Khwaja Nazimuddin and Suhrawardy—I do not want to name Madar-i-Millat [Fatima Jinnah] . . . and still it feels hungry. But it will not be able to eat me."[48] His facial expressions and body language showed even more than his words the contempt he felt for the fading dictator he'd long advised and assisted. "Prices are sky-high. None of the basic necessities of life are available. . . . Yet, it is argued that the country has made tremendous progress, and that there is stability. What stability? The stability of the graveyard?" He continued to attack government for having filed criminal charges against him.

Zulfi felt his messianic destiny was no less than to awaken "the masses" and "eliminate poverty and misery." He challenged the government to "arrest me," indeed, even to "kill me, if they wish," fearlessly insisting that he was "determined to expose them." Was he not, after all, *Zulfikar,* the righteous "Sword" of martyred Caliph Hazrat Ali, "friend of Allah"?

Ayub saw by now that there was nothing he could do to silence Bhutto short of arresting or shooting him. Five days later, when the president came to Peshawar to address a government rally he was shouted down by angry young Wali Khan and Bhutto supporters, and actually shot at. Ayub beat an undignified and rapid retreat. Zulfi by then had rolled on to Rawalpindi, where Mustafa Khar and Mumtaz Bhutto were waiting for him at the Intercontinental Hotel. Thousands of students, many from neighboring Gordon College, had just been chased from the Intercontinental lawn by police. Pindi was the site of the army's general headquarters, so crowds of raucous students were not welcome there, especially on the anniversary of the Russian Revolution!

"When I arrived at the Hotel Intercontinental I found the whole Mall area thick with teargas smoke," Zulfi later recalled. "About one-and-a-half hours after my arrival . . . I received a telephone call from the Polytechnic informing me that the police had opened fire there resulting in the death of a student, Abdul Hamid. I was told that the students were insisting on taking the body . . . to the President's House and that they wanted me to lead the procession."[49] Zulfi advised restraint, sending some People's Party lieutenants to the college several miles up the road to explain why he could not risk coming himself. Advocate Khurshid Hasan Meer drove with him the next day to offer condolences to Abdul Hamid's parents, and on the following morning Meer was arrested. Zulfi entrained that day with Mumtaz and Khar for Lahore, where an army of People's Party devotees, including his old comrade Dr. Mubashir, awaited him at the packed station.

In an interval of less than a year, Zulfi had roused most of West Pakistan, bringing its young men from a state of apathy or despair to the brink of righteous revolt against the military dictator and his major pillars of support. He had set Sind on fire, stirred the Frontier, and taken Pindi by storm, yet it was Lahore, Zulfi knew, that was the true testing ground for any man who hoped to lead Pakistan. Since the dawn of Mughal imperial rule in the early sixteenth century, Lahore had been the heartbeat of Punjab, and Punjab was not only the initial capital letter of Pakistan but the premier province of its West wing. Quaid-i-Azam Jinnah had known that, which was why he had insisted on holding his Muslim League meeting in Lahore in March 1940, when the prophetic Pakistan Resolution was adopted as the battle-cry single platform of Muslim India's premier party.

Now Quaid-i-Awam Bhutto was ready to bring his own political revolt to its boil, to launch his last push in Lahore. He took the jacket from his back and threw it into the cheering, almost worshipful crowd of frenzied young men, who looked to him as more than a political leader. He was the stuff true martyrs and mahdis were made of. There was fire not only in his speech but in his bloodshot bulbous eyes that rarely closed, hardly ever rested any more—sleep had abandoned Zulfi by now. The colors he had chosen for his People's Party flag, red, black, and green, were the colors of the Prophet and of Ali's martyred sons

Hassan and Hussain, the colors of Islam itself the world over. Zulfikar Ali Bhutto had in this year of frenetic campaigning and ceaseless oratory transformed himself into a living symbol of Islamic sainthood, a fiery *Shaheed* (martyr), ready to sacrifice his wealth, his life, and every drop of his blood for his people and their Land of the Pure, and for Islam, in the sacred service of God Almighty. He had come back to Lahore, more than two years after his eloquent silence and tear-filled eyes alone had spoken of Ayub's "betrayal" in Kashmir and Tashkent, to a hundred thousand or more young men who gathered at the railway station—hanging from precarious perches like bats in midday, filling every platform, tightly locked together on every step—just to catch a glimpse of this *Shaheed*. Yes, here at last was a political saint born in Pakistan! Here was the flesh that Sufi Saint Lal Shahbaz was made of, the spirit that gave Shi'is the world over strength to flagellate themselves mercilessly with chains of steel yet feel no pain, only the ultimate soul-seering pleasure of surrender (*Islam*) to God. There were, Zulfi told them in the forty minutes of Urdu fire that poured that day from his drooping martyr's lips in Lahore, "only 12 months and 19 days left" to Ayub's dying dictatorship. The lion-crowd sent its approval in so deep-throated a roar that it reached the President's House in Pindi 160 miles away. Ayub, awakened by the sound, called his giant Governor Musa, ordering him to arrest that "crazy young troublemaker."

Musa signed the order to detain Zulfikar Ali Bhutto under Rule 32 of the Defense of Pakistan Rules on 12 November 1968. Bhutto was charged with creating "disaffection to bring into hatred and contempt" the government of Ayub Khan. In the early hours of 13 November, Zulfi was arrested at the home of Dr. Mubashir Hasan, who was also taken into custody, together with Mumtaz Ali Bhutto. They were flown to Mianwali Jail, a prison fortress on the bank of the Indus.

"I was confined in an old cell full of rats and mosquitoes," Zulfi reported in his affidavit to the High Court of Lahore, "the charpoy [bed] was tied to a chain. There was an adjoining little room meant for toilet purposes. But it was so dirty it was repulsive to enter. . . . The food consisted of two chappaties made of red wheat with dal which had stones in it or two tiny pieces of meat. A strong light shone for 24 hours throughout my stay there making sleep at night extremely difficult."[50] He was kept in solitary confinement and given no writing paper until 18 November, although he had made that immediate request. Nor were letters or telegrams delivered to his cell, the first, but not the last or the worst, prison cell Zulfi Bhutto would be forced to inhabit before he died.

"This morning some ringleaders of this movement of lawlessness have been arrested," General Musa announced the day the Bhuttos were jailed. He would "brook" no further "disorder." There is a limit to everything."[51] As word spread through Lahore of the Quaid-i-Awam's arrest, students and young workers rioted through the streets and alleyways of the old city. Down the rail line at Multan, thousands had gathered on tracks and station platforms to greet their hero but instead heard the news of his arrest. Trains were soon being stoned and cars ransacked as the frustrated faithful searched in vain for their Quaid-i-Awam.

Barrister Bhutto rejected all charges filed against him as "a tissue of lies, malicious in intent and dishonest in purpose. My utterances and remarks, made extemporaneously in the course of long speeches, made in many places, have been deliberately torn out of context and even fabricated."[52] He denied having disclosed any state "secrets"; having "incited the masses" or students to break any law or take violent action; having conspired to "plan" or "plot" the "overthrow" of Ayub's government by "force"—that "figment" of the government's "imagination" was but a "symptom of its ailing condition." The "voices raised in the streets" were a "spontaneous verdict" by the people against the government's excesses, corruption, and selfish purposes, proof positive of how "derelict" the system was, how widespread the resentment against Ayub.

"Our people . . . feel the pain of privation and yearn for the happiness of their children," Bhutto stated in the affidavit he filed in Lahore's High Court 5 February 1969, one month after his forty-first birthday. "Their poverty is unimaginable but yet they hope for a better future. . . . Starvation has dried the milk in the mother's breast and suffering has dried many a father's tear. It is not the law of God that our people must live eternally in despair and that their children should die of disease and want. Our people demand a better life for themselves and for their children; they want food and clothing, employment and protection." Here was the Sufi prisoner of conscience turning to God for help in answering people's prayers because the ears of government were deaf and the hearts of officialdom cold. "Deny them their rights," Zulfi warned, "and they will find a redeemer and if none is available they will redeem themselves. No plan for change is needed when the people seek it. The mood of the people is the plan. But arrogant functionaries, oblivious of this, want only to find solutions for the regime's perpetuation."[53] He saw himself now as the people's "Redeemer," no longer constrained to listen to his secular manifesto draughtsmen, like J. A. Rahim, or socialist pundits, like Mubashir Hasan, for he—Zulfikar Ali Bhutto—was attuned to the "mood of the people," sensed it in his heart, his spirit, through the mystic bond of love he felt for all the people of Pakistan, and which they, especially the poorest, neediest, wisest, most devout among them, felt for him.

> The classical excuse of colonial masters, whenever subject people have risen against them, has been that all the troubles are due to a handful of political agitators. . . . This Government would like to make the world believe much the same sort of thing. . . . Stripped of the maze of prejudice and fabrication, the truth, radiant in its clarity, stands as my witness when I say that neither I preached violence nor hatched a plan to instigate the students. . . . This regime, which has slandered the word "revolution" in describing its *coup d'etat,* celebrates a Revolution Day each year, but has the temerity to punish people for uttering that word. . . . [D]emocracy . . . exists . . . like the fragrance of a spring flower. It is a melody of liberty, richer in sensation than a tangible touch. But more than a feeling, democracy is fundamental right, it is adult franchise, the secrecy of ballot, free press, free association, independence of the judiciary, supremacy of the

> legislature, controls on the executive. . . . Under the canons of this regime, the printed word is in disgrace, the franchise limited to individuals subject either to intimidation or allurement, the body of the law contaminated by arbitrary acts . . . and the right of assembly in ashes in the furnace of Section 144. . . . This is the depressing reality but this does not necessarily mean that a change is not possible without violence.

Popular opposition to Ayub and his regime had spread during the last months of 1968 and in early 1969, both in the East and West of Pakistan. A general strike closed down Pindi for a day in late November, and police continued to clash with students and workers, while the army watched warily from its barracks. Little more than a week later another general strike paralyzed Dacca on 7 December. This time police fire drew young Bengali blood, and radical Maulana Bhashani, whose National Awami Party was ideologically modeled on Maoist doctrine, issued a call for the complete shutdown of East Pakistan in mid-December. Mujib's Awami League endorsed Bhashani's call, and soon all the small shops and businesses of East Bengal closed their doors.

Ayub saw that his days in power were numbered. He could speak nowhere in public without getting shot at or causing a riot. Nor had his once-robust health returned. Still, he had the army, and though it might not be powerful enough to beat India, its strength was more than sufficient to quell street disturbances and political opposition. But none of the heads of the three top services—Army General Yahya Khan; Navy Admiral S. M. Ahsan (1910–89); Air Marshal Nur Khan—was willing to order troops to fire on civilians, advising instead negotiated settlements. Former Air Marshal Mohammad Asghar Khan (b. 1921) entered the political fray in Lahore on 17 November 1968, announcing his support for "people's rights" to the press on that day, and thereafter addressing bar association meetings throughout the West, calling for the restoration of democratic rights and freedoms. His sober and prestigious voice had, as he later put it, "an electrifying effect on the public mind."[54]

Zulfi was released from prison during the second week of February 1969 and returned to Larkana, where he was to remain "under house arrest" inside his Al-Murtaza compound. And Ayub accepted the unanimous advice of his three chiefs, and agreed to call a roundtable conference of political leaders from the East as well as the West to attempt to resolve what had by now become a virtual state of undeclared civil war against his regime, reducing Pakistan's economy to a shambles.

At about this time the Democratic Action Committee (DAC) was formed. It included the conservative Jamaat-i-Islami and Nizam-i-Islam religious parties from the West, and Mujib's Awami League as well as Bhashani's more radical National Awami Party of the East. The committee agreed to talk with Ayub about the future of Pakistan's Constitution and the upcoming elections, but only after he lifted his state of emergency with its invidious Defense of Pakistan Rules. For much of January rioting bloodied the streets of Dacca, Karachi, Lahore, and Pindi as police continued to battle students and striking workers stationed behind bus and rickshaw barricades. Ayub broadcast a personal appeal to parents and teachers, asking them to "use their influence" to help restore calm and peaceful order to Pakistan's streets. But the students kept shouting, "Ayub must go!" and hurling brickbats as well as fireballs at police, whose ammunition was live.

On 14 February, Zulfi announced in Al-Murtaza that he, Mustafa Khar, Mubashir Hasan, and several other People's Party stalwarts would "fast until death"[55] to effect the lifting of Ayub's draconian martial laws. It was the only time Zulfi ever followed in the footsteps of Mahatma Gandhi, yet his announced fast appealed to many of his humblest, devout Muslim followers as proof positive of his saintly Sufi powers, his willingness to sacrifice his life for the people and for Pakistan. The timing was excellent: three days later Ayub ended the state of emergency, releasing Bhutto and his leading lieutenants from detention. All of Larkana turned out to celebrate, as did most of Sind, indeed, most of West Pakistan, for Wali Khan was released on the same day from his Frontier prison. "Jiye Bhutto!" (Long Live Bhutto!) filled the air above Sind. Zulfi was hailed both as Quaid-i-Awam and Shaheed, having risked his all, hurling shots from his sling at the giant, bringing down the longest-lasting, once seemingly strongest dictator of them all.

Ayub hoped to continue at least nominally in control. By releasing Mujib as well as Bhutto, and inviting them both to a roundtable conference with all other political leaders of note, Ayub seems to have imagined that he might retain his much-diminished presidency a bit longer. Zulfi would have none of it. He did not accept the invitation to meet with Ayub or any of the others, launching his own final campaign for the presidency from Karachi a few days after his release. "We will take a bullet on our chest, but will bring a revolution!" promised their Quaid-i-Awam to his adoring Sindhis. "Give us bread and clothing; otherwise vacate the chair!"[56] he shouted at Ayub. Hundreds of thousands roared back, "Jiye Bhutto!"

On 21 February Ayub announced his decision to remove himself from the campaign for reelection. The very next day the government dropped its charges against Mujib, who was released from jail and found an ocean of Bengalis awaiting him in Dacca. There were many more Bengalis than Sindhis, of course, and Mujib was as popular among his people as Zulfi was in Sind and most of Punjab. Shortly after his release Mujib agreed to attend Ayub's conference, held in Pindi from 10 to 12 March. Bhutto and Bhashani both boycotted that meeting, however, and though Wali Khan of the Frontier National Awami Party (NAP) was there, as were other less popular Western leaders, nothing was achieved. Strikes and student violence proliferated everywhere. The cities were unsafe and political murder was not unusual. By late March conservative leaders of the West accused Bhutto and Bhashani of leading Pakistan "into anarchy."[57] On 25 March 1969 Ayub stepped down: "I cannot preside over the destruction of my country."[58] He handed his powers over to General Agha Mohammad Yayha Khan (b. 1917), who immediately proclaimed martial law throughout Pakistan, and assumed the title Chief Martial Law Administrator in addition to those of President and Army Commander in Chief.

Yahya Khan announced his resolve to preside over the "smooth transfer of power" to "representatives of the people elected freely and impartially on the basis of adult franchise." He asserted that he had "no ambition" to retain power. But Asghar Khan, who knew Yahya well, immediately condemned this "betrayal of democracy," insisting that the little general was, in fact, "a highly ambitious person."[59]

Zulfi also knew Yahya, who had hunted with him in Larkana and drunk with him before and after each shooting. Yahya's principal staff officer, moreover, Major General (later Lieutenant General) S.G.M. Peerzada, was a boon drinking companion of both men, but a heart attack in 1964 had forced him to become more temperate than either. Until his heart attack Peerzada had served Ayub as military secretary, the most important and powerful secretarial post in Pakistan; the military secretary organized the president's daily calendar—deciding with whom he should meet, when, and usually for how long, attending most meetings himself—thereby acquiring detailed information and an understanding of the president's mind better than any ministers or other high officials. A clever military secretary could become almost as powerful as the president he served, were he so inclined. Peerzada was a man of sharp intellect and political ambition, self-effacing, patient, able to achieve under a less-than-sagacious Yahya Khan the position of singular influence and power he had aspired to but was obliged to resign from by Ayub at the first sign of his physical frailty. Peerzada could "never forget" his ouster from the Eden of Ayub's Presidential House, and after Zulfi Bhutto was also sacked in 1966, he and Peerzada "developed a close friendship based on their common animosity towards Ayub," as G. W. Choudhury accurately observed.[60] Peerzada and Zulfi saw eye to eye in their negative assessments of Ayub, and Zulfi hoped to use Peerzada as his swiftest access to the top of Pakistan's martial pole of power. There were, after all, several cases still pending against him in various courts of law, for though Ayub had stepped down, his lower minions lingered on, in Sind's provincial offices as well as at the head of the district police of Larkana and Naodaro.

In mid-May 1969 Zulfi wrote Larkana's superintendent of police, Raja Sarwar Khan, to complain about his deputy, one Mr. Kadri, whom he viewed as the "chief instrument" of Ayub's "harassment" against himself as "the former President's principal target."[61] "Mr. Kadri committed every wrong in order to fulfill the task assigned to him . . . among other devices, he picked up two or three notorious goondas from my home-town Naodaro and coaxed them to cause me and my people as much irritation and annoyance as possible." *Siasat!* By the same post, Zulfi also wrote Colonel Dost Mohammad, then martial law administrator for Sind's Khairpur Division, informing him of "the kind of trouble and harassment that the police continue to give us in Larkana," though he pointedly added, "I know that this is not the policy of the present regime but these officials are bent upon creating friction."[62] Zulfi sent Mumtaz Bhutto personally to "explain in greater detail" what he found "intolerable" about Kadri's "misbehaviour."

Two days later, after meeting with Mumtaz, the colonel replied, "I will carry out investigation myself and make sure that NO victimisation of any sort takes place."[63] Zulfi was "thankful" for the colonel's "interest" in keeping close watch over Kadri, adding "not that I am bothered by the malicious activities of such individuals," but "I merely wanted to draw your attention to the continuing mischief of Mr. Kadri only because I believe that times have changed and that the present administration does not intend to continue the polices of the former regime in this particular connection."[64] He was quite right. Peerzada had promised him as much, and Peerzada's word now was law.

Bureaucratic underlings, however, knew how ephemeral orders issued from the top of Pakistan's central government generally proved to be. Even military leaders were quick to displace one another. Zulfi found many thorns and sharp stones strewn in his path on every backroad he or his new party traversed, even in Larkana. Deputy Commissioner Masud Mufti had the temerity to write him in early August of 1969 "to draw your attention" to Martial Law Rule No. 21, "which forbids the holding of public meetings and processions of a political nature."[65] Members of the People's Party had "committed a breach" of that regulation "during your journey from Moenjodaro to Larkana," Zulfi was informed, and ordered to "please direct the members of the People's Party to desist from such displays in future." It was intolerable to Zulfi to be nattered at this way by men who, whatever their titles might be, were little more than fleas to his ego. He replied to "Mr. Mufti" in Larkana, and sent a copy to Peerzada in Pindi:

> I am not a drawingroom politician. Whenever I travel in this country and wherever I halt people are bound to assemble to meet me. If the car in which I sit is preceded or followed by other motor cars carrying friends and party members happening to go the same way, you call that a procession. . . . I am not the former President of Pakistan to need weapons to protect me from the affection of the people. You surely cannot expect me to rebuke my friends, my party members and the thousands of sympathisers, and rudely tell them that they are not wanted, that they should not have waited for hours in the sweltering heat to meet me. . . . I care for the people of Pakistan, for the humble folk whom the administration grown callous by years of unchecked power has been trampling underfoot.[66]

In "marked contrast" to his own modest and restrained political activity, Mujibur Rahman "strides across the land of East Pakistan as a mighty colossus preaching his hymn of hate without let or hindrance." The growing popularity of Mujib's Awami League, with its Six Points demanding virtual autonomy for Pakistan's "Bangladesh" wing, was now viewed by Zulfi as a treacherous threat and challenge to Pakistan's integrity. Generals Yahya and Peerzada saw Mujib and his party in much the same Western light, leading Bengali "separatist forces that will strike at the roots of national unity."

"The aims of my party, the nature of my struggle are to be found in the attainment of fundamental principles involving freedom, national sovereignty and not based on irrelevant matters having no bearing on the nation's future."[67] This would not be the last time Zulfi Bhutto was careful to point out to Yahya (through Peerzada) his own position and that of his party in contrast to Mujib and his. "Do not think Mujibur Rahman alone commands influence and is able to attract attention and get warm ovations and that for this reason he should be treated exceptionally. Let me tell you Mr. Deputy Commissioner that this state of affairs cannot continue," Zulfi warned Mufti.

> Mr. Mujibur Rahman's Six Points which spell the destruction of Pakistan have gained support in Sind and this is ominous. It can cast the shadow of secession over other regions of Pakistan. . . . The enemies of the country both within and outside are active in their mischief to undo Pakistan. . . . The Jamaat-e-Islami which wants to exploit our dear religion only to spread its baneful influence is not capable of overcoming the crisis. . . . Pampering such parties led in the past to the Ahmadi riots and the imposition of Martial Law in 1953 . . . the Jamaat is sharpening its knives and preparing demands for the declaration of Ahmadis as a minority and for the segregation of Shias. In this way they will turn Muslim against Muslim and bring chaos to Pakistan. It is a fallacy to think that a simple and expedient tilt towards such forces serves the country's interest or even that of a regime.

Here, indeed, Zulfi was quite right, but unfortunately he would soon forget the wisdom of this cogent analysis in later desperate attempts to shore up his own collapsing regime's support. He contemptuously dismissed the leaders of all other parties in the West as either too "weak" or "vacillating," labeling the PDP "a collection of uninspiring fossils." Turning to the major parties of the East, Zulfi castigated Bhashani's National Awami Party as "oriented to either Moscow or Peking," and charged that Mujib's Awami League "wants the dismemberment of Pakistan."

Zulfi insisted that his People's Party was "the only national party in the country that inspires the confidence of the people and is both dynamic and vigorous enough to meet the test of our times. . . . A blow to this party is a blow to national solidarity," he cautioned.

> We are patriots, we have served Pakistan courageously and have won the affection and esteem of the people. We do not want to create problems. Trouble this time would mean the end of Pakistan. We want to co-operate in the task of overcoming the constitutional dilemma but co-operation cannot be one-sided. We do not seek any privileges. All we expect is to be left alone to organize our party and to prepare for the future. Whatever the circumstances I cannot fail my people. Please do not suffer from the complex of power. Do not repeat the mistakes that were committed not so long ago by a despot the like of whom Pakistan will not see again. The situation is brittle, it is fragile, indeed—it is delicate! . . . If another crisis erupts in the near future, take it from me that this time you will find Pakistan in two or more pieces. . . . I believe that I am among the few who have the capability and the conviction to hold this country together and to make it march to progress and glory. . . . I played with my life and risked death to free the people for a better Pakistan.[68]

The passionate closing section of the letter to the deputy commissioner of Larkana, reveals much about Zulfi Bhutto's complex mind and the motivation underlying most of his major decisions during the next decade. He considered himself not only a Pakistani patriot but the one leader truly capable of "saving" his country and its people from impending disaster. He believed it was his "destiny" to do both. Zulfi thought he best understood his people's passions, their insatiable "thirst" for democracy, too powerful for any dike to restrain. He alone could hold back or unleash those torrents, turning them on or off whenever he chose with his mighty rhetorical powers. He had done it in Lahore, in Pindi, in Karachi, in Larkana, even in remote villages all along the Frontier. "Whatever the circumstances I cannot fail my people," he wrote, literally meaning it, believing himself *infallible,* endowed with destiny's mantle for unique greatness and power, and for suffering and sacrifice as well, potential martyr that he was. "Please do not suffer from the complex of power," Zulfi wisely cautioned Masud Mufti, who might well have responded, "Physician, heal thyself!"

"He really believed in whatever he said," Husna recalled, "Because he was very *intense.* If he said something he believed in it, even if he didn't start off by believing. . . . He was like that . . . very single-minded, very intense, and very total about everything. . . . He was human, of course, he made mistakes, and he admitted them to me. In fact, he would always laugh, and say, '*You* never let me get away with it, do you?' . . . I became sort of his conscience . . . bare, you know, naked, the naked truth."[69]

8

Free Elections and the Birth of Bangladesh

(1970–1971)

"We are the people of Pakistan," Zulfi shouted, launching his party's first national election campaign in Karachi on the eve of his forty-second birthday, 4 January, 1970. "The people are my roundtable conference. The Roundtable of Ayub Khan comprised only a few individuals who were conspiring against the people, shutting themselves up in rooms. This is why we did not join the Roundtable Conference. Our politics is the politics of the masses. It is politics of the open. The Roundtable Conference was a deep conspiracy against the people. Ayub Khan wanted to sabotage the people's struggle. . . . He did not want to surrender power. . . . The capitalists, the feudalists, the army and the police were supporting him, but the people were not supporting him."[1]

Might it still have been possible in those final hours of Ayub Khan's administration to fashion some political formula for saving Pakistan from the violent disintegration that would sever the two wings in 1971? Had Zulfi agreed to attend the Roundtable Conference, was there any reason to believe that he and Mujib could sit patiently listening to one another? In which language? For how long? The catalogue of Bengali grievances condensed into the Awami League's Six Points, after all, is too well known to require repetition. The majority of Pakistan's population, who lived in the overcrowded East wing, felt not only exploited economically by the West but oppressed politically and militarily, much the way most South Asians had felt under British imperial rule. To patriotic West Pakistanis, such feelings were incomprehensible, or betrayed deep-rooted treachery that could never be accepted. Zulfi Bhutto, patriotic West Pakistani that he was, viewed Mujibur Rahman and most of his Bengali followers as little better than traitors to the Land of the Pure, for which Quaid-i-Azam Jinnah and his own father, Sir Shah Nawaz, had lived and died. Like Yahya Khan, he was willing to give them more posts in the higher services, even in the army, and a more equitable share of the foreign exchange that their jute exports earned the world over. He had no objection, moreover, to their speaking, reading, and taking civil service examinations in Bengali, but what more did they want? Were they Pakistani Muslims? Or were they traitors? How much autonomy could any single region expect and demand, before that region became an independent nation? As a good Sindhi, Zulfi well understood the powerful tug of regional roots, yet Sind was a part of Pakistan. As was East Bengal. And Punjab, and the Frontier. All of them combined made Pakistan what it was, a great Muslim Nation, much more than the sum of its provincial parts. As for Ayub, how could Zulfi sit at any table over which that old fool, or fox, might preside? No, the roundtable called by Ayub was a "deep conspiracy against the people," Bhutto believed.

"Time has shown that our decision was in the interest of the people," Zulfi continued in his election campaign kickoff at Nishtar Park. "If we had joined the Roundtable Conference, the movement would have failed. Ayub Khan would not have decided to hold elections. Nothing could be gained . . . neither elections nor the decision on adult franchise. He had only conceded the parliamentary system and the other leaders wanted that I should call off the movement. He was a hunter but I am a better hunter. He could trap others but failed to trap Zulfikar Ali. I escaped the trap because I did not want the people to be trapped. I shall always remain with the people."

The roundtable would, of course, have demanded national elections, a few months earlier perhaps, and the franchise monopoly long enjoyed by Ayub's Basic Democrats would also have been scrapped because the universal adult franchise demand was one agreed upon by all major parties, East and West. Zulfi may, however, have been right in so proudly asserting that he was a "better hunter" than Ayub; he had, after all, done much with his verbal shots to bring down the field marshal he had supported and shored up those eight long ministerial years before. Now he continued to attack the fallen dictator, insisting that "there was a time when Ayub Khan told me that if we could defend East Pakistan, our armies could break through to Delhi. He repeated this many times. He told me that this was the only hitch in the way otherwise we could launch a struggle for Kashmir. Later, when we went to Washington, Mr. Robert McNamara talked to me at a dinner in our Embassy and asked me how we would defend East Pakistan in case of war with India. He admitted that we could advance in the West. . . . I told him that if East Pakistan could be defended then perhaps we could face India. He refused to believe that."

Zulfi was now ready to reveal all the "secrets" he knew about Ayub's fears and weaknesses, appreciating how much Peerzada, if not Yahya, would relish such anecdotes. He had, of course, often told these stories at private parties, with his brilliant mimickry of Ayub and the others, but now he was speaking to the people, as their self-appointed tribune. "I will not contest the elections on trifles. . . . I had told Ayub Khan that the defence of East Pakistan was not difficult. . . . Everybody knows that our valiant forces did teach India a lesson in the Rann of Kutch. . . . Ayub Khan thought I was sentimental. I told him that I was not sentimental and that he should allow me to carry out my policy. But he accepted a cease-fire and took the matter to the International Court of Justice. He declared that he was a friend of India. The result was that we lost territory which rightly belonged to us." Zulfi believed that, as did most of the Pakistanis who heard his hoarse but powerful voice. Like him, they *wanted* it to be true, wished it were true, that Pakistan's smaller army *could* beat any Indian army, no matter how large and well-equipped. One Muslim Pakistani soldier with a rifle and bare bayonet was worth at least ten Hindu Indian soldiers, most of them were convinced. Was it not true?

"Our military commanders had requested two divisions," Zulfi revealed, spilling secrets. Since the 1965 War many of his listeners were less positive that religion and national origin alone created supersoldiers. Maybe numbers played a small part. Now Zulfi was explaining to them why the numbers had been denied, and the victory march into Delhi aborted before it began. "Our Finance Minister had enough money for industries but refused to allocate funds for the raising of two divisions demanded by the army. Later, even Ayub opposed the move. If the two divisions had been raised, the situation would have been different today."

Zulfi returned to Larkana for his annual hunt, and birthday celebration, but was angered and dismayed to be facing some of the same old petty problems in his home district, where Ayub Khuhro was set to run against him in one of three Sindhi constituencies Zulfi would successfully contest that year. Even though a new impartial superintendent of police had been assigned to Larkana, Zulfi felt obliged to write him on 10 January concerning a call by an inspector in connection with an alleged case of cattle theft. He suspected "some hidden malice and meanness"; it was part of a conspiracy instigated by "my petty and inconsequential political opponents."[2] *Siasat!*

To an immense audience that filled Pindi's Liaquat Gardens a week later, Zulfi spoke of religion. "Islam is our religion. We are Muslims and we are proud of that. You know I have not only served Pakistan but I have also served Islam in the Middle East. *Allah* will decide on Judgement Day how best I have served the cause of Islam."

"Our second principle is democracy. . . . [I]n a democracy the people rule. We insist on democracy, for it is provided in Islam. . . . [and] socio-economic equality or *Musawat* has been given the highest priority. . . . We shall, therefore, bring about *Musawat.* No power on earth can stop us." Zulfi's fervor turned his campaign appearances into old-time religious revival meetings, where those who watched him and heard his invocations of the Prophet's name wept as they cheered and screamed, some fainting from the ecstasy of being in his presence. "We have no hidden personal motives in our struggle for *Musawat,*" Zulfi answered his free-enterprise critics. "We are working for the good of the people at large. . . . If countries like Iran and Egypt can introduce free education and medical care, why should a great Islamic country like Pakistan not be able to do that? Why can't the people educate their children? Why can't the sick get admission to hospitals? . . . Is it because the 22 families are in possession of the country's wealth?" Ayub's family was one of that cohort named by Dr. Mahbub ul Huq, the former chief economist of the Planning Commission, in his economic analysis of Pakistan's ills. Zulfi told the crowd that "the people had been fed on false promises for the last 22 years, insisting" "there is an end to patience." One part of Zulfi Bhutto believed his humane words. The poor people who listened also believed them, and many thought that by voting for this good and devout man their poverty would soon end and the gross inequities of the economy would disappear. Zulfi promised that the introduction of "Islamic Socialism" to Pakistan would do away with the myriad prevalent disparities. "I am prepared to sacrifice everything . . . even the lives of my children. . . . I am not afraid of being killed," Zulfi shouted, breaking open his shirt. "Come on, fire bullets at me. I am prepared to die for the sake of the people." Such histrionics dazzled the West Pakistanis crowded round him. Zulfi Bhutto never feared death. A martyr's death, after all, would assure him an eternity in heaven. The slings and arrows of police officers and rival waderos bothered him more than the prospect of a hero's death in the service of his people, more and more of whom virtually worshipped him for saying aloud what many of them felt, yet lacked the courage or ability to articulate.

Then on the subject of Kashmir, Zulfi ridiculed Ayub's "cowardly" leadership. "India pulled out all her troops from occupied Kashmir" to counter the Chinese attack in November 1962, Zulfi revealed. "Ayub Khan was busy sightseeing in Hunza . . . riding a mule. . . . Kashmir had no troops at all. . . . Any action by Pakistan would have ended the Kashmir issue forever . . . in conformity with justice. . . . We lost a golden opportunity."

From Pindi Zulfi drove to Peshawar, where he addressed a mammoth meeting in Jinnah Park. He mocked the fallen Ayub, who had been shot at in Peshawar two years earlier, when he rose to speak there just a few days after Zulfi had launched his new party's crusade against the dictator. "When the shots rang out, our great Field Marshal fell flat on the ground and took cover. My dear friends, I believe that life and death are in the hands of God but I can assure you that if Bhutto had been involved in such an incident, he would never have fallen flat on the ground."[4] The Frontier audience cheered wildly, some firing guns into the air, for here was a man worthy of leading this nation of tribal braves. Despite his Sindhi birth, Zulfi Bhutto spoke like a true Pathan. He contemptuously ignored his Western political opponents, landlords like Punjab's Mian Mumtaz Daultana, who led the Council Muslim League during this campaign, and warriors like former Air Marshal Mohammad Asghar Khan of the Frontier, who had just recently joined the political fray, called for an early restoration of democracy to Pakistan. "Everyone is posing as a lion," Zulfi roared, dismissing such political opponents as "jackals. Now they claim that they fought for democracy. Some of them were sitting in England and others had locked themselves in their houses. . . . The people fought against Ayub Khan, and I was with them all through the struggle." "Bhutto Zindabad!" echoed from thousands of lusty lungs across the Khyber toward Afghanistan's border.

Multitudes came from every remote village of West Pakistan to hear him, for Zulfi Bhutto was fast becoming a legend in a land where courage and self-sacrifice are weighed against the purest gold in the hearts of Muslim men and their begums. Zulfi's passionate, hypnotic powers affected women as well as men. Few West Pakistanis were immune to his charm and messianic appeal.

At Mardan a month later Zulfi promised his followers, "I shall always side with you and die with you. Zulfikar Ali Bhutto will be with you at every step, at every turn and on every front."[5]

At Gujrat in March, Zulfi shouted "This is not my party. . . . This is your party."[6] Wherever he stopped to speak on the campaign trail, despite the heat and dust, which fatigued many of those who accompanied him, Zulfi seemed to thrive, growing stronger on sleepless days and weeks, from the adulation, the masses. His people, bright-eyed though barefoot, smiling and waving though often unwashed, were devoted to him, for was he not their new savior?

Bread-and-butter issues were not neglected by the energetic candidate, who reminded his Listeners that "Articles of daily use like milk, meat, salt and cloth are becoming dearer, but there is no corresponding increase in the income of the poor people. The capitalists and the millowners are making a lot of money . . . Why are the workers in jails? Why are not the feudal lords and capitalists in jails in spite of indulging in smuggling and black-marketing?" Zulfi's cry was "*Roti, Kapra, aur Makan*" ("Bread, Cloth, and Shelter"). It soon became the perennial slogan of his People's Party, for those basic needs would remain unsatisfied after each interlude of his party's rule.

One part of Zulfi believed his campaign slogans and promises, but another Zulfikar Ali Bhutto understood and appreciated all the temptations and pitfalls of administrative power. He had worked closely with Ayub for eight years and with Mirza before him, hence who knew better than this Bhutto, after all, the pleasures of cosmopolitan high life? His silk suits were all handmade, his Havana cigars handrolled, his Scotch whiskey the oldest and best. Part of him hated the others for enjoying it all as much as he did. So out on the hustings, facing the barren rocks and dust-filled fields of the Frontier, and the ill-fed, underclothed, homeless masses who turned out in their tens of thousands to hear his words of hope and cheer his promises, Zulfi meant what he said, even though his brain knew how impossible it would be to keep any of his grandiose promises.

Again the largest crowd gathered at Lahore's Mochi Gate to hear Zulfi attack Ayub's weakness, explaining how the Kashmir War of 1965 would have been won, had he been in charge. "When Indian forces were busy committing aggression against Pakistan, Ayub Khan was playing golf in Swat. . . . "My friends and brothers, our brave forces were ready to advance. . . . General Yahya Khan, Air Marshal Nur Khan, and the present PAF Chief, Air Marshal Rahim, all of them wanted to teach India a lesson. But Musa Khan and Ayub Khan were frightened. And right at that time I was saying at the Security Council that we would fight for a thousand years."[7] The crowd went wild. Zulfi Bhutto was their hero, a military strategist, a diplomatic genius, an unmatched orator. Ayub, who was twice his size, clearly had less than half his courage or brains. "After the war, Ayub Khan went to Washington and met President Johnson who frightened him, threatened him and forced him to his knees. I was pained to see the President of a brave nation being so easily browbeaten . . . extremely pained."

On 30 March, Zulfi and his party left Shahdadpur after breakfast for Sanghar, Pir Pagaro's Hur territory, where bright triumphal arches decked in the colors of the People's Party had been erected a few days earlier but "hacked down with hatchets" the night before. District officers and the police joined the procession of cars just a few miles out of Sanghar. The local police had had word of an ambush and urged him to stop at the government rest house en route. He accepted the cautionary invitation, but before the cars reached the rest house, "they opened up," and "a regular battle" ensued "with guns and rifles. Our opponents were occupying bunkers and were sitting in trees with their rifles," Zulfi reported after reaching Karachi. "I was in a motor car but I came out because I wanted to face death bravely. I started walking forward and told my opponents . . . I was Zulfikar Ali Bhutto and they could shoot me. Then they started firing again. My friends fell upon me and I had a narrow escape." In describing his experience twelve days later to a crowd in Karachi's Lyari district, the heartland of PPP support, Zulfi boldly proclaimed, "Whatever the conspiracies against us, whatever the plans against us, we are not going to be scared off. . . . The whole of Pakistan has been turned into a Sanghar. . . . [W]hatever the number of murder threats or murder attempts, we shall continue to fight them with the help of the people. This is my solemn promise to you."[8] The Sindhi assemblage roared out in one voice, "Jiye Bhutto!"

"The Pakistan People's Party is the party of the masses. It is your party and it represents your feelings. . . . The people of Pakistan from Karachi to Khyber support the Pakistan People's Party because this party belongs to the common man. If one Bhutto is assassinated, a thousand Bhuttos will arise. If I am killed, the people of Pakistan will carry on this democratic struggle." Even Zulfi did not dare to lay any claim, however, to the support of East Pakistan's people, who constituted the majority of enfranchised voters in the nation as a whole. Mujib daily addressed much larger crowds, and they, in their millions, cried, "*Joy Bangla!*" ("Victory to Bengal").

Mujib's popularity throughout the East was far greater than Zulfi's in the West, for there were 10 million more people in the East and all of them spoke and understood Bengali. And Mujib was as stirring an orator in Bengali as Zulfi was in Sindhi, English, or Urdu. "Sheikh Mujibur Rahman unleashed hatred against West Pakistan," Zulfi later declared in his pamphlet *The Great Tragedy*. "He used every means to mobilise the people of East Pakistan. Riding on the high crest of Bengali nationalism, and with a number of jail terms to his credit, the Bengali leader raised the emotions of his people to a frenzied pitch. . . . He went about like a Messiah telling the poverty stricken people of East Pakistan that their salvation lay in Six Points. . . . He cleverly concealed his true intentions in an atmosphere of hatred. The language and the methods were of fascism. . . . Sheikh Mujibur Rahman was a spell-binding orator. He used his political skill with a mastery that no Bengali leader to this day has rivalled or surpassed."[9]

Given the provincial and linguistic fragmentation of the West and the large number of opposing parties, Zulfi knew that he had no chance of beating Mujib in a nationwide universal-franchise election. He had, however, powerful allies in Yahya and Peerzada, who, like him, feared Mujib's "secessionist" plan. But as Suhrawardy's leading disciple, Mujib was clever enough to understand the forces allied in West Pakistan against him. Since his release from prison he had appeared singularly flexible, even accepting Ayub's invitation to the Roundtable Conference, which Zulfi had rejected. And Yahya had promised to be an "impartial referee" in policing the elections, so Mujib had good reason to hope that he would peacefully emerge as Pakistan's next premier especially if he were willing to permit the little general to remain Pakistan's nominal president. Such an arrangement was anathema to Zulfikar Ali Bhutto.

Bhutto saw Yahya socially as well as officially during this preelection period, for Yahya Khan was Shi'i by faith; his ancestors were Iranian-born. Zulfi attended Yahya's daughter's gala wedding in April, and drank every toast to the beautiful bride's health, as did her father, who convivially told Bhutto, "You fellows must now take over."[10] But Zulfi knew that by "fellows" he meant all contending politicians, and was wary of Yahya's seeming generosity and apparent willingness to relinquish power. In announcing elections the past November, Yahya had insisted that they be held within a Legal Framework Order (L.F.O.), whose final definition he had issued at the end of March. It required adherence to certain basic principles by all parties; faith in "Islamic ideology"; adherence to the "integrity of Pakistan," to the holding of "free and fair elections," and to the enjoyment of "provincial autonomy" within a centrally integrated federal system; and the promising of equable "economic progress" for all provinces. Zulfi suspected and feared collusion between Yahya and Mujib, and between Yahya and the fundamentalist Islamic parties of the West. "Let us hope that the electoral race will be impartially supervised and that the same set of rules will apply to all."

On 4 April 1970 Yahya flew to Dacca to meet with Mujib and his advisers, as well as with other Bengali political leaders. He was certain that the overwhelming majority of Bengali leaders had accepted his L.F.O. without major reservations. Mujib, like Bhutto, had a more complex mind than the general, who was soon "shocked" to learn from tapes played for him by his intelligence officers that Mujib had confided to his closest colleagues, "My aim is to establish Bangladesh; I will tear the L.F.O. into pieces as soon as the elections are over. Who could challenge me once the elections are over?"[11] Zulfi doubtless understood Mujib much better than Yahya did; which is to say that he never trusted his populist counterpart of the East, rarely believing anything Mujib said. Zulfi knew that, like Zulfi himself, Mujib felt constrained to say one thing one day, to one audience, and often its opposite the next time he spoke. Politics, after all, was the "art" of compromise, was it not?

Zulfi understood, of course, as he admitted in a speech given that April that "a constitution will be framed but the constitution alone will not solve your problems; elections will also be conducted but they alone cannot serve all purposes. . . . [E]lections have been held in our neighbouring country [India], but the plight of the people in that country remains as deplorable as here because they also have a capitalist system."[12] Dr. Mubashir was Zulfi's economic adviser at this time, and he needed a scapegoat that the poor could blame for their misery, something that could be summed up preferably in a single phrase. The "capitalist system" filled the bill. To reinforce the choice of target, knowing how simplistically inaccurate it was, he stated that "when Quaid-i-Azam struggled for Pakistan he aimed at undoing the capitalist system." Nothing could be further from the truth. Barrister Jinnah remained a singularly wealthy, property-acquiring capitalist till his last days in India, continuing to purchase valuable property in Bombay and Lahore almost to the very eve of Pakistan's birth, the subcontinent's Partition.

Zulfi's argument that "Quaid-i-Azam did not want the capitalist system of India to prevail in Pakistan," because it was "a system being practised by the Hindus," was, moreover, a total distortion of Jinnah's liberal secularism. Not only did Jinnah fall in love with and marry a Parsi, and work closely with Hindu lawyers, bankers, and other business leaders in Bombay during his mature life, but for years before Partition and the birth of Pakistan he served as a managing director of Tata Enterprises. He was neither anticapitalist nor professionally anti-Hindu, though some of his foremost political adversaries were, of course, Hindus. Zulfi, however, wanted to believe that Great Leader Jinnah would have agreed with him because, as he rightly noted, "The Muslims of the subcontinent made innumerable sacrifices" during the long and painful struggle for Pakistan, yet what had they achieved? What good had it done for the common man, the average Pakistani? "You can see that the system which is a legacy of the British still prevails. . . . The troubles of the people have increased. . . . We have given you in writing our three principles. You can see that there is no scope for cheating." What made him think about cheating? "Those who have cheated you have been exposed. . . . You have to compare their achievements with our performance."[13] That opportunity was still to come.

Zulfi dressed like Mao on the campaign trail, wearing a plain green jacket and a Mao cap, which gave him instant identity both with China and a "people's" revolution. It was the closest he came in trying to compete with the Awami and National Awami Party leadership of Bengal. At times he was more specific in identifying his party and people's movement with Mao, though he was always careful to deny any imputations by his opponents that he was a communist. "We need a long march," Zulfi told his cheering audience at Abbottabad. "When I say that we would undertake a long march I am dubbed as a communist. I say that the first long march in history was undertaken by Imam Hussain."[14]

Zulfi was so ceaselessly attacked by conservative Islamic leaders for his notorious personal behavior and exotic tastes that he often underscored by such allusions to Islamic tradition that he was first and foremost a Muslim. "We shall wage a 'jihad' for the cause of Islam, not only in Pakistan but anywhere in the world, if required. . . . If Muslim blood is being mercilessly shed in India, you cannot just wring your hands. If atrocities are being committed on the Muslims in the Middle East, you will have to do something about it. If the People's Party had been in power and if there had been a people's government in Pakistan, it would not have allowed India to get away with the killing of Muslims in Ahmedabad." Hindu-Muslim riots had recently taken many lives in the capital of Gujarat, yet here was another grand promise Zulfi would find impossible to keep when he gained power over Islamabad's administration.

Zulfi's bold tactics on the stump, and his tireless speechifying seemed to "charge his batteries" rather than deplete his energy. His ability to attract crowds and keep them in thrall started to worry his old friends in Pindi and Lahore, the landed barons, the general officers, who once thought they knew and could trust this "wadero-actor." Many began to wonder if Ayub wasn't right about the "madman," after all. "Our opponents are joining hands against us," Zulfi warned an audience at Kohat. "They are conspiring to oppose a new party . . . because we have come to you with a message that rings true. They know that our voice is your voice. They know that we are not going to betray you. . . . We shall have to launch a long march because we want to bring about a fundamental change."[15]

Yahya started to feel more anxious about dealing with Zulfi than about Mujib, who often sounded more reasonable. Yahya's minister of information, General Sher Ali, who directed Pakistan's powerful National Press Trust, did his best to keep news of the People's Party and its fiery leader's increasingly popular speeches, out of the daily headlines. Zulfi charged that the press had imposed a "blackout" on his activities and speeches, though he also insisted that the government-controlled press "distorted" what he said on the hustings. The Pakistan People's Party had only a few small papers of its own. The daily *Musawaat* (Socialism) and the monthly *Nusrat* (Victory) were two—both of which were named by Hanif Ramay of Lahore, one of Zulfi's most creative and talented early followers, a poet and artist as well as a journalist-editor. Neither paper, however, had a circulation even nearly approaching Pakistan's dailies, *Dawn,* the *Pakistan Times, Morning News,* and *Evening Star*. Zulfi's ability to win mass support was less hindered by negative or negligible press reports in West Pakistan, however, than might have been true elsewhere, even in neighboring India. The chances that most Pakistani readers of major English dailies would have voted for him would only diminish the more his speeches were accurately reported.

The Information Ministry did control Pakistan's radio as well as its press, and there Zulfi had better reason to feel aggrieved because radio news was heard by people in all major towns and village bazaars. "I have been touring this area and there have been processions and meetings of historic proportions," Zulfi protested angrily in Peshawar, "but not a word about them over the radio. This is not fair play. Our government claims impartiality. Is this impartiality?"[16] Then Zulfi called on Minister of Information General Ali "as a gentleman to submit his resignation," contemptuously adding, "but you cannot expect this from him. He has always been in service. He knows only how to work for a salary. . . . He should face bullets like we did in Sanghar."

In Baluchistan that June of 1970 Zulfi promised his Quetta audience, "This is your party and it will never betray you. This is my word of honour to you."[17] Zulfi now charged that Liaquat Ali Khan "was assassinated because he wanted to introduce Islamic Socialism. His murder was the result of a great conspiracy. That is the reason also why Zulfikar Ali Bhutto was fired upon in Sanghar."[18]

At the end of July, Yahya appointed an election commission, chaired by Bengali Justice A. Satter, who ordered new electoral rolls drafted throughout Pakistan. It was an enormous job, but Yahya promised elections before the year's close and everyone hoped they would be held before December. By the end of

August more than 31 million East Pakistanis were named eligible voters, with little over 25 million in the West. Zulfi knew that challenge of numbers in the East was more dangerous to his chances of winning political power than the combined forces of all parties arrayed against him the West. Still, he insisted that August, "I am confident that we shall . . . by the grace of god, emerge as a majority. No one can stop us."[19]

In September the first natural disaster of that year of catastrophes struck East Pakistan. Devastating floods inundated that low-lying land, leaving hundreds of thousands homeless. Rural damage was estimated in hundreds of millions of rupees. Polling was to have started in early October, but because of the floods, Yahya postponed elections till 7 December.

A number of People's Party activists and leaders were now thrown in prison as the campaign heated up. Zulfi sensed another "conspiracy." "Goondaism of the opposition is rampant and is increasing daily," Zulfi wrote Larkana's deputy commissioner, O. M. Qarni, on 16 August, 1970, complaining that "nothing is done to punish the miscreants," though "action is taken against my party workers."[20] His People's Party was being singled out and "victimised" because "we are against the economic system of the capitalists, bankers, feudal lords and exploiters. Secondly, we do not want a compromise with India."[21]

A thousand miles to the east worse disasters struck: a cyclone and a tidal wave took an estimated half a million peasant lives within a few days. Mujib now insisted that unless his Awami League's Six Point program was fully implemented under Pakistan's new Constitution, he would "launch a mass movement." He and his followers viewed the approaching elections as their "last chance" to live in dignity as citizens of Pakistan, instead of remaining "slaves in their own soil."[22] Though he spoke in a different tongue, Mujib used many of the same images as Zulfi, charging that "the few privileged families had sucked the blood of Bengal in such a way that the very backbone of the economy . . . had been shattered." Further, "a conspiracy was being hatched to purchase Bengalee members of the assembly to weaken Bengal's unity."

One of Zulfi's five contests would be won in Lahore, where he drew his largest Punjab crowds. Just a month before polls opened, he declaimed that his People's Party "has come into being only for the poor people, workers and the peasants . . . other parties are the stooges of the capitalists. . . . They have sucked the blood of the people and have mercilessly looted them."[23] On 18 November, Zulfi addressed his largest national audience, asserting over radio and television that "it is our moral duty to lift the people of Pakistan from the quagmire of poverty."[24] The "awakening of the people and their increasing determination to resist exploitation" endangered "vested interests," who were "exploiting religion" to divert popular attention from the poverty and oppression found everywhere. He went on to say, "The unresolved problems are legion. The cancerous growth of corruption has to be arrested. We promise that we shall tackle the problem of corruption vigorously . . . We promise to the people of Pakistan that we will give them a clean administration. We pledge to you that our Government will consist of men of integrity . . . We will come down with a heavy hand on crime and violence. We will . . . control rising prices. We will provide fair wages . . . We will remove the ghettos and clear the spectre of slums. We will restore the freedom of the press; . . . We will electrify villages and we will recognise student power, the fountainhead of our future . . . We will fortify our defences and increase the wages of Government servants at the lower echelons . . . We will protect the rights of the minorities. We will abolish black laws . . . We will seek to give free medical aid to the poor. We will take flood control measures. Heavy industries will be owned by the people . . . I make this pledge to you solemnly as a Muslim and as a Pakistani."[25]

Zulfi's Pakistan People's Party won 62 of the 82 seats in Punjab, and 18 out of the 27 in Sind, and 1 seat in the North-West Frontier Province, capturing 81 seats, just over half the total won by Mujib's Awami League, which won 160 seats, all from East Pakistan. A few days after the results were announced, Zulfi thanked his supporters at a rally outside the Punjab Assembly in Lahore and congratulated Mujib. "We respect the majority," but "both Punjab and Sind are centres of power. We may or may not form a government at the Centre but the keys of the Punjab Assembly Chambers are in my pocket." In the other "lie the keys of the Sind Assembly and . . . no central government can run without our co-operation." He then openly warned Mujib as well as Yahya: "If the People's Party does not support it, no government will be able to work, nor will the constitution be framed. The Centre needs our co-operation. . . . When I say that we must get our due, I mean that. . . . I am ready to die for my people."[26]

Mujib saw no reason to strike a deal with Zulfi Bhutto before taking up the reins he had so clearly won in a free and fair election. "No power on earth would be able to frustrate the legitimate demands of Bangla Desh," Mujib insisted on 17 December 1970. Yahya agreed, flying to Dacca early in January 1971 to meet with Sheikh Mujibur Rahman, to whom he referred openly as "the next Prime Minister of Pakistan." Shortly after returning from Dacca, Yahya along with Peerzada flew to Larkana, where they stayed with Bhutto on 17 January.

"The President informed us of his discussions at Dacca in which he told Mujibur Rahman that three alternatives were open to the Awami League, namely, to try and go it alone, to cooperate with the People's Party, or to cooperate with the small and defeated parties of the West Wing; and that, in his opinion, the best course would be for the two majority parties to arrive at an arrangement. For our part, we discussed with the President the implications of Six Points and expressed our serious misgivings about them."[27] Zulfi then deputed G. M. Khar, his party's Punjab general secretary, to Dacca to meet with Mujib and "lay the groundwork" for his own flight for a summit meeting there on 27 January 1971. Zulfi was disappointed to find Mujib "intractable," reporting later that Mujib insisted that "he was not in a position to deviate one inch from Six Points."[28] Zulfi countered that "public opinion in the West Wing was against Six Points," the general impression there being that "Six Points spelt the end of Pakistan." He pressed Mujib to "agree to a reasonable delay" before convening the National Assembly, but Mujib was firm on 15 February. Zulfi feared that once the assembly met, it would immediately elect a speaker from the East, and then "impose a Six Point Constitution." His strategy, therefore, was to boycott an early meeting of the National Assembly in Dacca. He met his party lieutenants on 2 February in Lahore, two days later in Karachi, and the following week in Multan, ordering them to stay away from Dacca at all costs. Zulfi met with Yahya again in Pindi on 11 February and urged him to postpone convening the assembly for at least another six weeks. Two days later Yahya announced that the assembly would meet in Dacca on 3 March "for the purpose of framing a Constitution for Pakistan."

Zulfi's response to Yahya's announcement was that his party would "not attend" the session. "We cannot go there only to endorse the constitution already prepared by a party, and return humiliated." Zulfi warned his newly elected delegates to the National Assembly that he would "break the legs"[29] of any party member who dared attend. All agreed to hand him their written "resignations" before the end of February in order to strengthen his bargaining position with Mujib and Yahya.

"The country is passing through a grave crisis," Zulfi warned on the Punjab University campus in Lahore late in February 1971. "When I went to East Pakistan, I told Mr. Mujib that it was his Awami League that had been elected on Six Points, not the People's Party. I told him that we had been elected on the basis of our stand on revolutionary changes in the economic system and on evolving an independent foreign policy, that we would try our utmost to cooperate with him. . . . However, there is a limit. If we went too far it would lead the country to disaster."[30] Zulfi thought he could pressure Mujib into backing down. He hoped to convince Mujib that he could control the East wing of Pakistan, but only if Mujib agreed to leave Bhutto in charge of the West.

"Mr. Mujib quotes me as saying that if need be I would go to Dacca for talks," Zulfi told his student audience in Lahore. "Well, I am prepared for two, three or even ten rounds of talks. . . . We have great respect for the people of East Pakistan just as we have for the people of the Punjab, NWFP, Baluchistan and Sind. Their interests are our interests. But it is painful that slogans based on provincial prejudices are raised. Why do they not raise slogans for the whole of Pakistan? . . . If we are to serve our country, our nation and our religion, we will have to strike off the shackles of exploitation. The success of the People's Party here and the Awami League there has been made possible by the people."[31]

Here was Zulfi's "solution," put as diplomatically yet as bluntly as he could phrase it, as he believed it, wishing to serve his country, his nation, his religion, and eventually "to strike off the shackles of exploitation." "*Idhar hum, udhar tum!*" was the Urdu formulation he used, "Us here, you there!" The People's Party would control West Pakistan, the Awami League would run the East. In personal terms that meant Zulfi over the West, Mujib over the East. Or as Zulfi now put it, more delicately:

> In both parts of the country only the people have emerged victorious. That is why we avoided a confrontation. In fact, we kept retreating. . . . We did not play up our differences. We went to East Pakistan to explain our stand. . . . I held conferences with my colleagues in Lahore, Multan

> and Karachi. We retreated so much that people began to ask what had happened to Bhutto. But it is regrettable that Mr. Mujib remained rigid. . . . The President of Pakistan went to Dacca and announced that Mujib would be the Prime Minister of the country. Both said they had had satisfactory talks, so it was presumed that the constitution had virtually been framed. But we have a duty to those millions who elected us.[32]

Thus Zulfi reminded Yahya as well as Mujib of his singular sensitivity and unique strength. As if superior electoral numbers alone would matter to a true Pakistani Muslim. India had many more voters than all of Bangla Desh, after all.

"Mr. Z. A. Bhutto and the People's Party have suddenly started striking postures and issuing pronouncements, which reveal a tendency to subvert the constitutional processes by obstructing the normal functioning of the National Assembly," Mujib told the press in a special release on 24 February. "Haunted by the spectre of famine and denied the bare means of subsistence, the people of 'Bangla Desh' have been reduced to a state of total destitution. We can on no account allow this state of affairs to continue . . . It is unfortunate that Bengalis have in effect been dubbed as 'enemies,' in whose midst the representatives from West Pakistan would feel themselves 'hostages'. Unwarranted aspersions were cast on Bengali Members of the National Assembly when the Assembly was termed a 'slaughter house'. All these extravagant charges were made only because the Session of the National Assembly had been called at Dacca. . . . If the atmosphere is poisoned in this way, could not Bengalis legitimately question whether they should be called upon to go to West Pakistan?. . . . the conspirators and the vested interests and their lackeys are embarking upon the last desperate bid to frustrate the "adoption of a constitution by the elected representatives of the people and the transfer of power to them."[33]

Three days later Zulfi addressed a huge meeting in Lahore. "I propose two alternatives to resolve the present crisis—postponement of the National Assembly session or removal of the 120 day time limit for the Assembly to frame a constitution. If either of these alternatives is accepted, I shall go to Dacca tomorrow to meet Mr. Mujibur Rahman to resolve the pre-session deadlock. If the session of the Assembly is held on March 3 . . . without PPP's participation, I shall launch a popular agitation from one end of West Pakistan to the other. . . ."[34] The gauntlet was thrown more toward Yahya Khan than Mujib. "*I* shall go to Dacca tomorrow," he said—not offering even to accompany the president or to discuss his "solution" first with the powers who nominally ruled as yet from Pindi. Zulfi felt neglected by Yahya, offended by that little general to whom he had extended much hospitality in Larkana and had been especially considerate, thanks to Peerzada's adroit diplomacy, never attacking him as mercilessly as he had attacked Ayub Khan. Soon Yahya's turn would come, however, for he had had the temerity to speak of Mujib as "prime minister," simply because Mujib's party won the most seats in a National Assembly destined never to meet.

"I refute the allegation levelled by Mr. Mujibur Rahman that I have been conspiring to create impediments in the transfer of power," Zulfi went on. "It is unimaginable that I could be in league with bureaucrats or the capitalists or the regime, or any foreign power, because all of them have shown consistent hostility to the People's Party. . . . It is for the preservation of . . . Pakistan that I have been struggling. I would offer any sacrifice for it." Zulfi meant it. At least part of him did. "If the country disintegrates there will be nothing left to save."

Still more: "Exception has been taken to my statement that the PPP members would be 'double hostages' in East Pakistan and that the National Assembly would be a 'slaughter house' for them. I cannot afford to be away from West Pakistan for 120 days when Indian troops are massed on the West Pakistan borders. My duty is to be with my people when their security is being threatened. . . . If the National Assembly meets on 3rd March my party will launch a campaign of protest."[35]

On 1 March, Yahya capitulated. The National Assembly would be summoned at "a later date."[36] Mujib met with his party leaders immediately, angrily announcing that a six-day strike would start in Dacca and be extended throughout Bangladesh on the third of March. On 7 March he would reveal his "final programme" at a mass meeting planned for the Race Course Maidan in Dacca. "You will see history made if the conspirators fail to come to their senses," he promised. Mujib singled out Bhutto as the one leader of a "minority party," who had "always been acting in the most irresponsible manner."[37]

The next day Zulfi held a press conference in Karachi. "God is our witness that all our efforts since the election have been directed to finding a solution to the constitutional problem which would not only right the injustices done to the people of East Pakistan in the past but also enable all the people of the nation to live in harmony."[38] He had obviously not expected Mujib to launch a general strike. He had underestimated the depth of Bengali frustration, desperation, and disenchantment in his search to "negotiate" a better position for himself. "We demanded postponement of the National Assembly session only to provide the two major parties with an opportunity to have another dialogue," Zulfi explained. "We are willing to have this dialogue anywhere at any time that the Awami League would like."

The streets of Dacca were flooded with people by now, and Bangla Desh's labor pains had begun. Several unarmed young men were fired upon at Farm Gate, two of them killed on the first day of the strike. "They have been shot at because they along with the rest of the People of Bangla Desh had stood up to protest against the gross insult inflicted upon Bangla Desh by the powers that be," protested Mujib. "I . . . urge the elements who are seeking to co-confront the people with force to desist forthwith. . . . [F]iring upon unarmed masses amounts to genocide and is a crime against humanity."[39] Mujib called upon "every Bengalee . . . including government officials" not to cooperate in any way with "anti-people forces." All government offices and courts were closed, as were semiofficial transport operations, including Pakistan International Airlines and railways. The only planes that flew into Dacca were military planes in which every seat was filled with crack Punjabi and Pathan soldiers, dressed in civilian clothes. Throughout the month of March thousands of Pakistani troops would thus be airlifted into Bangla Desh in preparation for the final desperate attempt to hold by force the allegiance of 70 million alienated Bengalis. A regiment of Baluchi troops reached the port of Chittagong aboard the *Swat* on 3 March, but Bengali stevedores refused to lift its ammunition out of deep cargo holds. Barricades went up at major crossroads between that port city and the capital to its north, as "The Land of Bengal" braced itself for the most violent blows of all.

Admiral S. M. Ahsan, who before Partition had served on Lord Louis Mountbatten's staff, and then became military secretary to Governor-General Jinnah, was governor of East Pakistan at this time. He courageously refused to order Pakistani troops to open fire on striking Bengalis. He was relieved of his command on 1 March. Yahya initially replaced the good Admiral with Major General Sahabzada Yaqub Khan, but he too rejected genocidal orders, hence was replaced by Lt. General Tikka Khan, who had never shown the slightest reluctance to order troops to fire on anyone. On 6 March, Yahya announced to the nation his "latest step": he would convene the new National Assembly in Dacca on 25 March.[40] Mujib addressed an estimated million Bengalis, who filled the Race Course Maidan in the heart of Dacca, the next afternoon. "There is still time for us to live as brothers if things are settled peacefully. No one can stop us from realising our rights when we are prepared to lay down our lives. We have learnt how to shed blood."[41] His Awami League lieutenants were ready to carry on if Mujib were arrested or shot. Bangla Desh would, indeed, be born soon after the imminent bloodbath. The exodus of 10 million refugees to India would follow.

"Pakistan has never faced a more dangerous crisis in her 23 years than she faces today." On 14 March in Karachi's Nishtar Park Zulfi spoke to a much smaller audience than Mujib's had been. "Sincere efforts" would suffice to "preserve the integrity and solidarity of Pakistan."[42] Zulfi insisted that "only an intellectually bankrupt person" could accuse him of "colluding" with officialdom to "create" the current crisis. "Who is Bhutto colluding with?" It was not the capitalists; "they have been asking for our heads." It was not government officials; "they tried to persecute me after I left the Ayub Government." It was not the reactionaries; "the whole struggle of Bhutto has been against the reactionaries." And about the so-called secret meetings with the president, "We represent West Pakistan. . . . [He] is forced to meet with us." Moreover Yahya had been meeting with the majority party leader in the East.

Zulfi argued that the current political crisis was compounded by economic inequities in trade and balance of payments between the East and West. "The Awami League alleges that West Pakistan owes Rs. 31 billion to East Pakistan but I cannot accept the position that West Pakistan should agree to repay this amount. And there is a total of Rs. 40 billion loan in foreign exchange. The Awami League contends that West Pakistan will have to pay Rs. 38 billion. Now if I had compromised with them, . . . [y]ou would have questioned my integrity . . . and since the East Pakistanis are in a majority, I would have been unable to do anything."[43] That posed a problem in logical behavior: professing faith in democracy, Zulfi had no intention of accepting a decision taken by a "simple majority" of Bengalis.

"Please remember this crisis . . . was inevitable," Zulfi declared. "On the one hand, East Pakistan wanted virtual independence, and on the other West Pakistan did not want to be exploited . . . the crisis was destined to come . . . Collusions were made with the forces of darkness. A big upheaval took place . . . [T]he introduction of socialism . . . alone is the way to help the poor and to eliminate exploitation from both East and West Pakistan . . . On this there should be no question of majority or minority. If they are in a majority there, we are in a majority here. Pakistan consists of two parts. Both parts have to prosper equally. We want power to be transferred. There is no other way out."[44] His mind kept racing back and forth, frantically, seeking a "way out." Yet part of him knew it was too late. The die of martial decision had been cast.

"My dear friends . . . I have been weeping for my Bengali brothers. I am dying to be able to help them, because they are being killed. I cannot sleep these days. I have always served Bengal. Bengalis are our brothers. . . . Mr. Mujibur Rahman, I say to you; you have made a mistake. Let us sit together. Let us go to the Assembly and frame a constitution there, so that Martial Law is buried and power transferred to the people. . . . A constitution can be framed . . . But if you are to go on talking about a 'Bangla Desh', we too, in view of our majority, can talk about 'Sindh Desh' or 'Punjab Desh.' . . . in that case it would be asked where has the Pakistan of the Quaid-i-Azam gone?"[45]

The crowd cheered wildly, tearfully. Zulfi was skilled at pressing every button of West Pakistani emotion. The repeated promise to sacrifice his own life gave him the aura of a Muslim martyr. The reiteration of his faith in Pakistan and in the Quaid-i-Azam's dream added luster to his image. Even as his reminder that he alone had the strength to stand against both superpowers and India added muscle to his stature. He was braver than the brave, bigger and tougher than any soldier or general. How could it be said that he would join with anyone or anything to destroy Pakistan?

The day after Zulfi spoke, Yahya flew to Dacca to meet Mujib. Baluchi political leader Nawab Akbar Khan Bugti had been in Dacca and had urged the flight. Bugti believed that Mujib was willing to lead a unified Pakistan. Bugti blamed Bhutto for the current deadlock, insisting that he was "worse and more ruthless than former President Ayub Khan."[46] Many other minority party leaders in the West, who agreed on virtually nothing else, agreed in denouncing Zulfi Bhutto for "creating a crisis."

At a press conference in Karachi on 15 March, Zulfi announced his final solution to the deadlock.

> It is not we who have demanded a constitution based on two economies and virtually two countries. We have said that in framing a constitution for Pakistan the West Wing must be heard and protection for its legitimate interests incorporated. . . . We say that power should be handed over to the representatives of the people in both wings. We say that at the Centre power should be transferred to the majority parties of both the

> wings, and in the provinces to the majority parties in the provinces. Only such an arrangement will ensure the unity of Pakistan.[47]

It was Zulfi's boldest attempt to reach over Yahya's head to Mujib, inviting his Bengali counterpart to join forces in a "people's revolution." Mujib, who trusted Zulfi Bhutto no more than he trusted Yahya or any other West Pakistani, ignored the daring invitation.

Pathan Khan Abdul Wali Khan, president of the National Awami Party in the North-West Frontier Province, insisted that Bhutto had "no standing" in the NWFP and hence could not speak for that part, at least, of West Pakistan. The former air marshal Asghar Khan was equally negative about Bhutto's proposal, as were leaders of every other party in the West. Mian Mumtaz Mohammad Daultana, president of the Council Muslim League of Punjab, said that "since there is only one Pakistan . . . there can be no question of two majority parties in one country . . . running the affairs of the country."[48]

Mujib called for greater "non-co-operation," vowing that "the struggle" would "continue with renewed vigour until our goal of emancipation is realised."[49] Dacca and all other major cities of Bangladesh remained paralyzed by the strike, which was remarkably effective in the face of martial law and the army. More civilians were shot, however, as soldiers opened fire in Joydevpur. Yahya met Mujib for an hour and a half on 19 March in the President's House in Dacca, and they agreed to meet again the following morning. Asked by reporters how the talks had gone, Mujib responded glumly, "I always hope for the best and prepare for the worst."

On 21 March Zulfi flew to Dacca with his party's team of advisers. "On seeing the green fields of Dacca when the plane was descending," Zulfi recalled, "I was overcome by an indescribable sensation. . . . I could not believe that in the last few years so much resentment had grown that our brothers and sisters should revolt against the country for which so many had shed blood. . . ."[50] A military escort took him to Dacca's Intercontinental Hotel, where the top-floor presidential suite had been reserved for him. An angry crowd of Awami Leaguers shouted "abuses and indulged in hooliganism" as he entered the hotel lobby and waited for the elevator that had been slowed down by Mujib's strike. When the elevator's gold doors finally opened, Iqbal Ismay stepped out with one of Mujib's closest lieutenants. Zulfi, by now tired of waiting and hearing abusive Bengali, pointed an angry finger at Mujib's colleague. "I know where you're off to, and I want you to tell *him* [Mujib] that *I* am the Destiny!"[51] Mujib's friend only smiled.

That evening Zulfi met with Yahya at the President's House, and over drinks and dinner Yahya briefed him on his meetings with Mujib. Mujib had "proposed" the immediate withdrawal of martial law and the convening of the National Assembly "divided *ab initio* into two Committees," one for West Pakistan, the other for Bangla Desh. The committee for West Pakistan would meet in Islamabad; that for Bangla Desh in Dacca. Both committees would "prepare their separate reports" and "discuss and debate . . . ways and means of living together." An "interim arrangement" would give East Pakistan "autonomy on the basis of Six Points," and the West wing would "be free to work out their quantum of autonomy according to a mutually acceptable procedure, subject to the President's approval." Yahya would "continue running the Central Government" until a new constitution could be approved by the assembly. Yahya had tentatively agreed to the proposal "subject primarily to my agreement," Zulfi recalled in his *Great Tragedy* recapitulation of those dark and fateful late-March days.[52]

"The scheme was fraught with danger," Zulfi felt. "I wanted a little time for reflection." He told Yahya that he could say "nothing" without the "benefit of advice" from his "colleagues," who, upon his return to the hotel "expressed their misgivings and suggested that I should not accept the proposal as it contained the seeds of two Pakistans. I was relieved to get their reactions which were similar to my own."[53]

The next morning at eleven o'clock he returned to the President's House for his first meeting with Mujib and Yahya. After the usual small talk, Mujib "turned to the President and asked him if he had given his final approval to the proposals of the Awami League. The President reminded him that it was necessary for me also to agree and for that reason I was present. . . . On that Mujibur Rahman remarked that the proposals had been communicated to the President and it was for the President to convince me; and went on to say that once Mr. Bhutto agreed in principle to the proposals, they could hold formal discussions, but until then the discussions were of an informal nature. . . . The President replied that this was not good enough, but Mujibur Rahman remained adamant."[54]

Was Yahya simply using Bhutto's recalcitrance as his excuse for rejecting Mujib's offer? Why else, after all, would he say that it was "necessary" for Bhutto to "agree" before he could express his own "final approval" and that "informal" discussions among the three were "not good enough"? Had the president-general and his military colleagues already decided upon their course of action, and was all this political and "constitutional" haggling merely a delaying tactic? Or had Yahya himself, in fact, fallen hostage to Zulfi's power play? Had their dinner tête-à-tête the previous evening given simple-minded Yahya the courage, thanks to Zulfi's shrewd analysis of Mujib's proposal and his outspoken advice against capitulating to such a "Bengali ultimatum," to reject Mujib's scheme? Whatever their motives and feelings may have been, those three men who held the fate of Pakistan in their hands that morning of 22 March 1971 might still have averted the imminent tragedy of genocidal war had they but reached some agreement; instead they parted swiftly, without resolving anything.

Zulfi followed Mujib out of the president's office. "I was a little surprised," he later wrote, by what he called Mujib's "sudden change of attitude" as soon as they were alone in the next room—Mujib had asked the military secretary to "leave" so that they could "talk." According to Zulfi, "He grasped me by the hand and made me sit next to him. He told me that the situation was very grave and that he needed my help to overcome it. At this point, thinking the room might be bugged, we walked out to the verandah towards the back of the house and sat in the portico."[55] Zulfi then reports that Mujib "told me that he now realised that the People's Party was the only force in West Pakistan and that the other politicians of West Pakistan were wasting his time. . . . [I]t was essential for the two of us to agree. He told me I could do whatever I wanted in West Pakistan and he would support me. In return I should leave East Pakistan alone and assist him. . . . He suggested that I should become the Prime Minister of West Pakistan and he would look after East Pakistan."[56] Even retrospectively Zulfi could not quite bring himself to write of Mujib as a potential prime minister of East Pakistan, only that Mujib would "look after" that wing! But how easily "I should become the Prime Minister of West Pakistan" flowed from his pen. That, of course, was Zulfi's own scheme by now. *Siasat!*

Zulfi's remarkably creative little work, appropriately entitled *The Great Tragedy,* continues:

> According to him this was the only way out of the impasse. He cautioned me against the military and told me not to trust them: if they destroyed him first they would also destroy me. I replied that I would much rather be destroyed by the military than by history. He pressed me to give my consent to his proposal and to agree to the setting up *ab initio* of the two Committees. . . . He told me that he would like to meet me again and that he would arrange a secret meeting between the two of us. In the meantime I should ask Mr. Ghulam Mustafa Khar to keep in touch with him. . . . I explained to Mujibur Rahman that my request for the postponement of the National Assembly had been made in good faith, and that his reaction was unnecessarily violent. . . . I told him that I would naturally give my most careful thought to his proposal and do everything possible to arrive at a fair settlement.[57]

Yahya had watched the two men as they walked round the garden path through the parted curtains of his window, and sent for Zulfi after Mujib departed. "When the President expressed his surprise at what he called 'the honeymoon between the two of you', I told him that such dialogues are a part of politics," Zulfi recorded. He then also conveyed his "considered opinion" of Mujib's proposal. "I could not be a party to the proposed scheme as it inevitably meant two Pakistans. This was my main objection . . . but it also contained other serious defects."[58] Provincial autonomy for the West wing would be "difficult to work out," Zulfi reasoned, and because "Martial Law was the source of law then obtaining in Pakistan" lifting it would remove all "legal authority" from the president and central government. "There would thus be a vacuum," Zulfi noted, and bringing his never-forgotten Latin into play, added that "the government of each province could acquire *de facto* and *de jure* sovereign status." Ever-suspicious of Mujib's motives, "In such a situation nothing could stop the secession of East Pakistan." Thus he armed Yahya with "legal" and subtle political reasons for justifying the military "solution" that the latter naturally favored. Even more than Zulfi, no doubt, Yahya was sufficiently intoxicated to believe that a "show of force" should suffice to bring Mujib and his Bengalis to their "senses."

"I suggested to the President that he should exert his authority and influence to make Mujibur Rahman agree to a compromise," Zulfi wrote. "Surely, it was reasonable to expect Mujibur Rahman to make some sensible adjustment that would preserve national unity and at the same time give him the substance of his demands."[59] Yahya then "requested me to discuss all these questions in detail with his experts," for the little president himself had no need to waste any more of his precious time on these petty political squabbles and endless, useless arguments. His military mind was now resolved on its next moves. He had never studied Latin, but he knew his duty when faced with a party preaching treasonous "secession."

The twenty-third of March 1971 was celebrated throughout West Pakistan as the thirty-first anniversary of the Muslim League's adoption of its Lahore Resolution, annually remembered as Pakistan Day. But that morning most Dacca newspapers brought out a special supplement headlined "Emancipation of Bangla Desh." It included a message from Sheikh Mujibur Rahman that "ours is the right cause and hence victory is ours," ending with "Joy Bangla!"[60] The same day Yahya announced another postponement of the convening of the National Assembly, giving no new date. At a press conference in Dacca on that day Zulfi said, "We want a democratic arrangement for the future of the country and to the satisfaction of the people of both the wings."[61] Yahya sent a Pakistan Day message to all loyal Pakistanis and troops, East and West: "Today we must rededicate ourselves to the principles enunciated by the father of the nation; namely, unity in our ranks, faith in our destiny and discipline in the conduct of our affairs."[62]

Mujib addressed a huge crowd in front of his house at 32, Dhanmandi Road in Dacca on the evening of 23 March, which he called "Resistance Day." Under the flag of Bangladesh, he declared, "Our people have learnt to shed blood for the achievement of their just and legitimate cause and any intimidation, coercion and application of force to resist them will simply be an exercise in futility."[63]

By 24 March reports that the military was firing on civilians were pouring into Awami League headquarters in Dacca. Tajuddin Ahmed, general secretary of the party, stated that "any attempt to frustrate the efforts of arriving at a political solution would be reckless." He urged "those concerned to desist from such actions," and called upon "the people to be vigilant and . . . ready to make any sacrifice to defeat the conspiracies of anti-people forces."[64] Mujib reaffirmed that night that "we will free the people of Bangla Desh . . . Either we shall live like men or we shall go out of existence fighting for our cause."[65]

Zulfi met with Yahya and Peerzada again on the morning of 24 March and told them that "the time for a settlement seemed to be running out. . . . I informed the President that I had sent back some of my party leaders as I felt that their presence was more necessary in West Pakistan."[66] But Zulfi and Khar, who had received a secret "message" from Mujib, stayed on a bit longer. Khar met with Mujib that night and reported he found him "perturbed" over the firings in Chittagong. The next morning Zulfi returned for another talk with Yayha and Peerzada at the President's House, bringing along J. A. Rahim as well as Khar. They were all booked to fly out of Karachi the next day, on the fateful morning of 26 March.

On the night of 25 March, Zulfi and his friends "went to our rooms," after they finished dinner at about 10:30. "An hour later we were awakened by the noise of gun-fire," Zulfi wrote. "A number of my friends came to my room. . . . We witnessed from out [the] hotel room the military operations for about three hours. A number of places were ablaze and we saw the demolition of the office of the newspaper 'The People'. This local English daily had indulged in crude and unrestrained provocation against the Army and West Pakistan. With the horizon ablaze, my thought turned to the past and to the future. . . . Had we reached the point of no return—or would time heal the wounds and open a new chapter in the history of Pakistan?"[67] Zulfi tried to phone Yahya soon after Operation Search Light started but learned, to his surprise, that the president had already flown home. Even as Zulfi watched the fatal fire light up Dacca's midnight sky, Yahya was "sipping his scotch and soda, at 40,000 ft. over Ceylon."[68]

U.S. M-24 tanks led the Punjabi-Baluchi assault upon student dormitories on the campus of the University of Dacca. Iqbal and Jagnath halls were filled with sleeping students and faculty when the tanks opened fire and continued shooting for at least five minutes. Soldiers crouched behind the tanks then charged into the shell-battered dorms with fixed bayonets and killed all persons still alive: students, professors, caretakers, and servants. Tikka Khan's troops were thorough.

Zulfi flew home on the morning of the twenty-sixth, and later recalled:

> On our way to the airport we saw the flags of Bangla Desh coming down from the house-tops and . . . barricades on the streets. . . . I was again haunted by thoughts of the future. I prayed that this turn in events should not degenerate into a protracted internecine conflict. I hoped that the patriotism of the common people would reappear in vigour and that the nightmare of fascism would disappear. . . . A tumultuous crowd welcomed us at Karachi Airport . . . and insisted that I make a speech; but I was in no mood for speeches. I nevertheless did manage to say: "By the Grace of God Pakistan has at last been saved". In my heart I hoped and prayed that I was right. The future will tell whether Pakistan has been saved or lost. But this much can safely be said that if the regime had not acted on the night of the 25th, on the following day the Awami League would have declared the independence of Bangladesh.[69]

For most of the world, however, the genocidal massacre unleashed by Pakistani forces on 26 March 1971 was a much louder and more memorable proclamation of the independence of Bangladesh than any proclamation to that effect that Mujib might have made on the radio. Clearly, Zulfi agreed with Yahya and Tikka's "final solution" to their Bangla Desh "problem," and like them, he obviously considered Mujib's Awami League demand and the hoisting of Bangladesh flags atop buildings all over Dacca a "nightmare of fascism." The rest of humankind would see the massacre of innocents as every foreign correspondent and most foreign diplomats in Dacca did: a dreadful example of the weakness of military terror and the limits of power when it comes to repressing a popular demand for "national" freedom. Consul-General Archer Blood, senior U.S. diplomat in Dacca, cabled Washington at the time to report the "mass killing of unarmed civilians, the systematic elimination of the intelligentsia, and the annihilation of the Hindu population."[70]

With Dacca still burning the next morning and Bangladesh emerging bloody but unbowed, Yahya went on Pakistan radio to declare that "Sheikh Mujibur Rahman's action of starting his non-cooperation movement is an act of treason. He and his party have defied the lawful authority for over three weeks. They have insulted Pakistan's flag and . . . have tried to run a parallel Government. They have created turmoil, terror and insecurity. . . . The Armed Forces, located in East Pakistan, have been subjected to taunts and insults of all kind. I wish to compliment them on the tremendous restraint that they have shown. . . . I am proud of them."[71]

Mujib had been taken from his home by Pakistani troops the night before, after the firing had started on the Dacca University campus. He was kept in prison on the Dacca Cantonement till 1 April, and then flown to West Pakistan and placed in solitary confinement in Mianwali jail—in the same "VIP" cell that had held Zulfi Bhutto years earlier. "I was in the house," Mujib told Zulfi at year's end while still a prisoner in Pindi. "I knew everything that was happening there but I did not escape. What for? Only to save the country. No violence! I could have put up a fight but I did not. So many people would have died." He had as yet been allowed to see no news report of the massacre of civilians in Bangladesh that started just before he was arrested. "But I had no doubt, Yahya wanted to arrest me. . . . I had asked all the political leaders to leave the house before my arrest. I will court arrest because it is my country. They started a machine gun in the house. Tell me what is the position of my country?" Mujib asked Zulfi, who had just taken over from Yahya as president of Pakistan, nine months later, and visited his Bengali prisoner on 27 December 1971.[72]

"Position of the country is . . . ," President Bhutto replied, nervously, evasively. "I told him [Yahya], 'You must immediately bring about political settlement and not let this bloodbath take place. You must restore rule to the elected leaders.' He said, 'Yes, yes, we'll do it very soon.'"

Three months earlier, however, in his *Great Tragedy* tract, Zulfi had written in support of the military action of March, denouncing all critics of Pakistan the world over as "disappointing, if not unfriendly," thanking only China for strong support. "The attitude of the British and the American press has been, to say the least, deplorable. In general the Western press has unashamedly supported the secessionist movement. . . . Prompted by India, the Soviet Union . . . chose to forget her own history, her own military interventions for self-preservation even beyond her borders in Hungary and Czechoslovakia."[73]

One of the Bengalis massacred in the first wave of late March was the husband of Husna Ahad, who had been a senior partner in the firm of Orr, Digham in Dacca, and a friend of Mujib's.[72] Husna herself had by then long since left her husband and was living in Karachi, in a house near 70, Clifton, which Zulfi rented for her. Zulfi received many other early and accurate reports of what was happening in war-torn Bangladesh. One of his closest Dacca friends, Sheikh Kaiser Rasheed, wrote from his Dhanmandi Road home (near Mujib's house) in April: that

> The enormity of the tragedy here beggars description. . . . I am afraid that every drop of blood outside the most rigid minimum requirement of military operations will leave a trail of abiding bitterness. . . . [P]lease contact [Yahya] and impress upon him the need not only to issue the strictest possible instructions in this regards but also to demonstrate their resolve that no infringements will be countenanced. . . . I am sure you are deeply concerned . . . I would urge you, however, to desist from making any sanctimoneous statements. In the context of the situation here they sound so utterly heartless, it is not true! I would recommend that you come across and see things for yourself.[75]

By the end of April nearly a million terrified Bengalis had fled to India, where they remained in refugee camps ringing Calcutta all the rest of that blood-drenched year. The flood tide of refugees stood at over 50,000 a day for nine months.

Yahya's former constitutional adviser, the Bengali Dr. G. W. Choudhury, returned to Dacca in late May of 1971:

> It was the worst experience of my life. Everywhere I went, I heard the same story: one person had lost a son; another a husband; many villages were burnt. . . . My next meeting with Yahya took place in Rawalpindi. . . . Yahya's first question was what I had seen in Dacca. My prompt reply was that no single foreign newspaper had exaggerated. . . . I also told him that it was not only the number of deaths but the manner in which innocent persons had been killed and women raped that had destroyed our cherished homeland for which the Muslims of the subcontinent had sacrificed so many thousands of lives. . . . He looked vacant and seemed unable to talk to me. He knew my devotion to the concept of a united Pakistan and he also knew that I had never supported Mujib's veiled secessionist plan. He could not, therefore, dismiss my account as that of a "typical secessionist under the influence of India." . . . Then I began a round of visits to a number of other members of the ruling elite. It was a shocking experience to see their attitude: "The rebels must be crushed; then we can talk of any political settlement." . . . A group of the junta headed by Hamid, Peerzada and Omar, thought that the problems of "East Pakistan" had been solved forever by force. . . . Justice Cornelius was asked to prepare a constitution giving "autonomy" to East Pakistan but "within limits," . . . Yahya told me frankly. The Cornelius constitution . . . was being prepared to deal with Bhutto who had been demanding that his party should be given effective power. . . . Yahya was not willing to yield to Bhutto's pressure. . . . Moreover, Bhutto was the most hated

person in East Bengal as the Bengalis believed, not without justification, that he, in collaboration with the hawkish generals, had precipitated the crisis.[76]

President Nixon said nothing against the Dacca massacre, but sent his White House security adviser, Henry Kissinger, to Pindi in early July. Kissinger met with Yahya, who briefed him on his own secret meetings with Mao and Chou, having recently served as Nixon's middleman in helping to "open" China to U.S. "recognition." On 8 July 1971 the press reported that Kissinger's departure from Pakistan had been "unexpectedly delayed" by "intestinal affliction"; actually, he had flown to Peking for his first meeting with Chou En-lai. With Yahya so cooperative and helpful in facilitating secret U.S. diplomacy, it could hardly to be expected that the White House would publicly rebuke him for attacking what Kissinger would later refer to as an "international basket case," meaning Bangladesh.

India's Prime Minister Indira Gandhi moved closer to Moscow at this time, and would sign the Twenty-Year Treaty of Peace, Friendship, and Cooperation with the Soviet Union on 9 August 1971, after being rebuffed by the White House and State Department in her efforts to win U.S. support against Pakistan. Zulfi, like Yahya, and Nixon, had never liked Indira Gandhi. All three men often joked about her "arrogance," referring to her by a number of deleted expletives, least offensive of which were "cow" and "bitch."

Yahya sent Zulfi to Teheran in July to seek more support from the shah for the war in the East, which had become more costly and militarily exhausting than any of his generals had anticipated. The Democrat-controlled U.S. Congress had imposed an embargo on shipments of arms to Pakistan as news of the genocidal murders reached the West. Iran was one of the closest major storehouses of U.S. weaponry that could be tapped without publicity. The shah, moreover, was especially friendly to his Shiite neighbors Yahya and Zulfi. From this time at least Zulfi became a close friend of the shah, inviting him regularly to Larkana, feeling he could as a rule rely on this new Persian "King of Kings" for all sorts of help, financial as well as military.

"Of course, the Shah could be charming," Zulfi later reflected on this "friend," who was in some respects much like himself. "He possessed a certain modesty in his vanity. He had a very good memory. He was superstitious. . . . I am not very sure about his courage. . . . He was impeccable in manner. . . . He was soft spoken and courteous . . . punctual to the dot, hard working but with mixed up priorities. He was both easy to please and difficult to please. The slightest thing could cause him to misunderstand or misconstrue. He was replete with imaginary grievances."[77]

Yahya hoped to reach early agreement with Zulfi on a constitutional scheme that could allow them to share power. On 29 July, Yahya invited Zulfi to bring his PPP "team" along to confer with him and Peerzada and Justice A. R. Cornelius, their constitutional adviser, in Karachi's Government House. Zulfi assured the president that his party "had no intention to bring about confrontation with the regime as this may lead the country to disaster." He also told

Yahya, "We believe in the declarations made by you as President," adding a more personal note, "we know you and in a personal regime it is better to deal with the man you know." But, "If our opponents are put in power, we will oppose them and destroy them. They have no mandate from the people." Sweetening again, he graciously assured Yahya, "You can be the Head of State in the next Government." Zulfi himself, of course, would be prime minister. He told Yahya, "You may consult other parties, we do not object," but refused absolutely to sit down with the others at any roundtable conference. He had promised the people a great deal, which posed, as Zulfi now delicately put it, "some basic problems" because "we cannot give these up or resile from them." Nonetheless, he hoped a satisfactory constitution could be adopted within 120 days of the convening of a new constituent assembly. J. A. Rahim, the party's secretary-general was there at Zulfi's side, and remarked, "Even if Mujibur Rahman's treason had not taken place the crisis would not have ended with the framing of the Constitution. The crisis arises out of property relationship. . . . If we come to power we would introduce changes. . . ." Rahim himself would be one of Zulfi's thorniest "problems" in years ahead. His mind was as sharp as his tongue, and his memory was as good as Zulfi Bhutto's.[78]

Yahya spoke more slowly than either Zulfi or Rahim, as he replied.

> Good to deal with people you know—I know you. I am a soldier but we have not come by a coup. Power was thrust on us, we have not come because we wanted to. . . . I am anxious to get the Country back upon the rails of democracy. I am myself frightened by my own blind power—Power is with God Almighty. The Constitution is not an ordinary law and how can you with your majority in Punjab and Sind deny to the people in NWFP or Baluchistan their freedom to decide on what conditions they agree. . . . There can be no majority in Constitution-making. You objected to Mujib's majority bull-dozing—Others would object to your bulldozing them.
>
> U.K has no written Constitution. . . . The more important thing is good government. We must have a Constitution that is workable and looks after the needs of the people. In meeting you I am discussing with representatives of the people. I will discuss with other parties but I cannot consult all 120 millions. . . .

"Don't talk of confrontation with me," he said softly, turning to focus his heavy-lidded eyes on Zulfi's tight-lipped face. "I am not your Opponent, but while I am in power I shall continue to do the job that I am required to do by my position. . . . I want to transfer power to you but I cannot say who will be in power in the Centre."[79] Yahya then suggested that the People's Party go East and "win the seats" in new elections for the many "vacancies" that had been left there since the fighting started. He planned to call elections in East Pakistan within "30 days."

A week after that confrontation in Karachi, Zulfi's appendix flared up. He was rushed with a high fever on 5 August to a Karachi hospital, where Dr. Nasir Sheikh's skillful surgery saved his life. The inflamed appendix had just started to burst, spewing its poison throughout his abdominal cavity. One week after the operation he was able to walk and dictate letters, though it would be another fortnight before he could return to his normal fast pace.

On 25 August he met again with Yahya and Peerzada to discuss the situation in East Pakistan. Yahya informed Zulfi of his reasons for wanting to appoint the unpopular Dr. A. M. Malik as civilian governor of East Pakistan a week later. "I told the President that his decision vindicated the position I had taken right from March 27th when I met you in Karachi," Zulfi wrote Peerzada, "that military action was inadequate unless accompanied by a political initiative. However, I wish to reiterate that the appointment of a civilian Governor is not the same thing as the introduction of the democratic process which I have called for."[80]

Zulfi feared that Yahya was planning another "conspiracy" against him, much as Ayub had tried to do by inviting Mujib to the Roundtable Conference in 1969. He warned Peerzada on 2 September, "I have in the national interest refrained from enlarging upon my views in public and to the press, in order to avoid inflaming public opinion at such a critical juncture." But time was running out, he insisted, sending the letter with G. M. Khar, whom he asked not only to deliver it but to set the time for Zulfi's next meeting with Yahya on 9 September. Bhutto had wanted the meeting in Karachi, but Yahya invited him to Islamabad. Zulfi politely declined, sending another, longer letter through Khar directly to Yahya in Rawalpindi on 8 September. In it he called for "clear-cut decisions and initiatives" to be taken "immediately." "An anguished and disturbed Nation wants to know its fate," Zulfi wrote. "History is an exacting judge. Without prejudice or rancour all of us must address ourselves to the central task, the survival of Pakistan. We simply cannot act in pique. . . . Destiny has placed in your hands and in ours an onerous responsibility. Being answerable to God and our people . . . we are determined to arrive at a lasting solution acceptable to the people of a united Pakistan."[81]

Zulfi addressed a meeting of People's Party workers in Hyderabad that same day, making public his growing frustration with Yahya. "The President should take an immediate decision. I am 43 and I have the strength and courage to organise a movement." Zulfi labeled those who opposed immediate transfer of power "toadies and conspirators," and cautioned them to "cease their conspiracies. We will defeat all enemies of the people."[82] Three days later he addressed a larger crowd at the mausoleum of Quaid-i-Azam Jinnah in Karachi, declaiming:

> Brothers, as Zulfikar Ali Bhutto, an individual, I am nothing. You are my power and my strength. No government in the world can solve its problems without the co-operation and support of its people. No problem can be solved through oppression, through bullets, through force. A few months ago when I said this first, a number of sycophants and defeated politicians said that this amounted to treachery. Had democracy been restored then, we would not be in this sorry state today. . . . What happened in East Pakistan, can happen in West Pakistan as well. . . . But, I

> warn the powers that be that we will not permit bloodshed here. We will not permit another Jallianwala Bagh. The Quaid-i-Azam made Pakistan with the sacrifices of the people. . . . Oh! my Quaid, did you dream of the Pakistan that we are living in today? Was it your concept, your dream? . . . He was a democrat and he respected the democratic process. . . . He said that the constitution will be made by the elected representatives of the people. . . . Speak, speak my Quaid, when will this night of oppression end? . . . For God's sake, restore democracy.[83]

On 18 September after a long meeting with Zulfi in Karachi, Yahya "pledged" that before year's end a "Constituent National Assembly" would be convened, once by-elections were completed in East Pakistan, if that were possible. Then "constitutional governments" should be formed "both at the centre and in the provinces," Zulfi announced at a subsequent press conference, pretending to believe that both wings of Pakistan could remain under a single government. "We believe the main task before us today is to save the country," Zulfi said on 29 September. "Nobody wants war, but if there is war it will be total war, which must mean the total participation of the people." Zulfi was still ready for a "thousand year war" with India, and believed himself better qualified than Yahya to lead Pakistan to "victory." He was not totally unaware, however, of Pakistan's perilous state, and the weakness of its shattered economy as well as polity.

> East Pakistan is in flames; the whole country is in near ruin. How will Pakistan be rescued if new fires are to erupt in other parts of the country? And erupt they will if the people's rights are not recognised. . . . [O]ur demands on behalf of the people are for the solidarity of the country and the well-being of the people as a whole, and totally different from those of the Awami League whose leadership sought to wreck Pakistan. We simply cannot be tarred with the same brush. It would be a monstrous injustice to the people of Pakistan to equate the two. Patriots and traitors cannot be equated . . . the PPP is not in league with India, Pakistan's mortal enemy, to destroy our nationhood.[84]

By September Yahya's new Inter-Services Intelligence chief, Major General Jillani, reported sizable Indian troop movements toward the Bengal border, and heavy Soviet shipments of arms into India. Yahya also knew, as Indira Gandhi did, that India was paying more every month simply to feed and shelter the millions of Bangladesh refugees squatting around Calcutta than she had for the entire war in 1965. To seize the moment in late September, Yahya conveyed a secret peace plan promising, in effect, immediate implementation of all Six Points to Indira Gandhi through her new ambassador in Islamabad, Jai Kumar Atal. But now that Indira had her Soviet backing, she did not bother to respond to such clandestine overtures. Once the monsoon stopped and river levels dropped, India's assembled forces would begin an irresistible move across the East's border.

The moderate Bengali elder statesman Nurul Amin was appointed "Prime

Minister" of Pakistan by Yahya on the eve of the Indian invasion. Yahya remained president, and offered Bhutto two cabinet posts, deputy prime minister and minister for foreign affairs. Zulfi refused at first, insulted at the thought of being number two to a Bengali less popular than Mujib. "I told the President that his advisers were misleading him. I told him they were like cobras," Zulfi announced at a public meeting in Multan that October. "And you must have read in the papers that they found cobras in the President's House, Karachi!"[85]

"Every war is fought on a political level," Zulfi argued. "If the people of Pakistan are not with the Government how will it face the enemy successfully? . . . They are talking of a total war. If it is going to be a total war, then the people cannot be ignored. . . . We are their representatives. We need time to consolidate the country, at least six months."[86] Pakistan had less than two months left, however. By mid-November, Indian military sorties began crossing the Bangladesh border, probing Pakistan's soft defenses. With less than 100,000 fighting men in all of the East Wing, the Pakistan Army was barely able to control the disaffected millions of unarmed Bengalis. It proved no match for India's forces, led by Lt.-General Jagjit Singh Aurora under the overall command of General (later Field Marshal) Sam Manekshaw, and their Soviet weaponry.

Indira Gandhi flew to Washington that November, but heard nothing from Nixon or Kissinger to convince her that Pakistan's leaders deserved more time to work out an acceptable constitution. "We are told that the confrontation of troops is a threat to peace," she said on the eve of her departure from Washington. "Is there any peace when a whole people are massacred? Will the world be concerned only if people die because of war between two countries and not if hundreds of thousands are butchered and expelled by a military regime waging war against the people?"[87]

Zulfi accepted Yahya's urgent invitation to fly to Peking as Pakistan's special envoy to request Chinese military support, should India invade in the East. He was promised "everything" by Mao and Chou, but when Islamabad's desperate calls went out two weeks later, China delivered virtually nothing. On the morning of 22 November 1971 India's tanks and troop-filled trucks crossed the border north of Calcutta, rolling toward Jessore. Pakistan's ambassador to the United States, General Raza, called a press conference in Washington to announce the "attack," and when asked by one reporter if war was "imminent," the ambassador irately replied, "It is not only imminent, but it is on."

Two days later Zulfi received a top secret message from one of his confidants in the president's secretariat, informing him that "any day now you should become the Prime Minister to our great delight," and going on to advise him that "once the back of Indian forces is broken in the East, Pakistan should occupy the whole of Eastern India and make it a permanent part of East Pakistan. . . . This will also provide a physical link with China. Kashmir should be taken at any price, even the Sikh Punjab and turned into Khalistan."[88] Zulfi thought enough of this letter to keep it in his Karachi Library, though the anxious author, postscripted the request "N.B. Kindly destroy this letter immediately after you have seen it."

The next day, 25 November, Zulfi met with Yahya, who now took him fully into his "confidence," hoping to remain the president in a new government, with Zulfi as prime minister. As soon as the war ended old Nurul Amin would be displaced, of course, because Yahya and Zulfi knew he was mere Bengali window dressing for the outside world. What Yahya himself did not as yet understand was that he too would soon be jettisoned by his ambitious new "ally."

"I did not consider it necessary to go into certain matters at the meeting," Zulfi wrote the tottering president a few days later in prefacing comments on a proposed constitution, some features of which he believed were unwarranted. A constitution "should be rooted in reality" and that meant that a parliamentary federal system was unavoidable and imperative. A principal cause of Pakistan's calamitous situation was its "quasi-unitary system" and a decade of presidential dictatorship. "In any Parliamentary system the main executive authority is vested in the Prime Minister. You now propose to give Constitutional sanctity to the office of the President, combining it with that of the Commander-in-Chief of the Army, endowing the President, moreover, with Emergency powers and to some extent Martial Law powers too."[89] Foolish Yahya thought he might be able to work in harness with Zulfi, for were they not hunting buddies and strong drinking Shi'i friends?

"With so much power to be vested in the President, who could have so wrongly advised you to top it all by conferring upon the President the disproportionate and provocative special responsibilities for the preservation of the integrity and ideology of Pakistan and for the protection of fundamental rights. This provision . . . casts an aspersion on the patriotism of every elected representative and puts the whole nation under a cloud of suspicion. It arrogates to an individual the sole authority and wisdom to determine the loyalty and affection of the whole populace to the motherland. . . . We do not want a dud President; we want an effective President but not, as under such a scheme, a virtual dictator. Democracy has its own rules, and once it comes into play such a travesty can only drive a thwarted nation on to a collision course . . . the scheme will not work . . . I therefore urge you to delete this provision before the Constitution is promulgated . . . to avert a major crisis." Zulfi had other criticisms. "Governors should be appointed on the recommendation of the Chief Minister concerned. You, Mr. President, have said that the power to appoint Governors must rest with you as you want to appoint non-party men to this position. I can only say that no person is above party. . . . Man is a political animal."[90]

Yahya had hoped to remain president for "at least two terms," under the new Constitution, but when Zulfi demurred—"this would simply not be acceptable to the people of Pakistan"—he agreed to "continue to hold office for the first fixed term," and then stand for election to a second term. Now Zulfi cautioned him that "even this one-term proposal" was subject to the approval of "my Central Committee." A week later, on 2 December, Zulfi wrote Peerzada to report that the Central Committee had discussed the "situation," and "urged me to draw the regime's attention" to many new problems. "The people have lost faith in a government riddled with contradictions, sensitive to the dictates of foreign powers, and which in no way represents them." "Today's position is that a war is on and . . . [m]assive mobilisation is required. Instead, pitiful mistakes are being made . . . the impression is that India is giving us a kicking in East Pakistan and we are on our knees begging foreign powers to come to our help."[91] The next day Yahya invited Zulfi to join, as deputy prime minister, a coalition government with Nurul Amin. Bhutto responded, "It goes without saying that if the Pakistan People's Party accepts this invitation it would be making a very great sacrifice in the supreme national interest." Zulfi then went on to spell out "our understanding" that the coalition would consist of seven East Wing ministers and six from the West, none of whom should be members of the religious Jamaat-i-Islami party, which had consistently attacked Bhutto and his party as "anti-Islam." Five of the six ministers from the West would be PPP ministers from Sind and Punjab, and would control the Foreign Affairs, Home, Establishment, and Information and Broadcasting ministries, in addition to the deputy prime ministership.[92]

"The crisis facing Pakistan today is total," Zulfi summed up. "Although the National Assembly is, according to the time-table put forward by you, not to meet till the 27th of December, we can join the Coalition Government with immediate effect, provided our minimum terms are fulfilled. The critical situation does not brook delay." Zulfi was back in harness, returning to a military government post similar to the one he had left five years earlier. Only this time the military president and Pakistan itself were both about to fall.

9

President Bhutto "Picks Up the Pieces"

(December 1971–July 1972)

As the sun set on 3 December 1971, Pakistani fighter-bombers attacked airports in Amritsar, Agra, Srinagar, and half a dozen smaller Indian cities, and the army advanced across the international border from Sind into Rajasthan. The "total war" Zulfi had been pressing Yahya to fight was at last begun now that Zulfi was back in power. He advised Yahya much the way he had advised Ayub during the 1965 war. The F-86 Sabre jets and Mirages hit few of their targets, however, making some runways and hangars inoperative but hardly destroying India's air force on the ground. Nor did Pakistan's demoralized army get very far, for Indian air power was ready to immobilize Pakistani armor, quickly bringing Pakistan's western offensive to a halt. In the East, Jessore fell on 7 December, and India's long columns of Soviet-built trucks towing 105 mm artillery, Jeeps mounted with recoilless rifles, and troop carriers kept rolling on toward Dacca.

Zulfi flew to New York on 8 December 1971, hoping to recoup on the diplomatic front some of the ground Pakistan had lost in Bangladesh. "We will not rest," he assured his cheering countrymen who had come to Karachi's airport to see him off, not if it took "a thousand years" to clear "Indian aggression from the sacred soil of Pakistan." He rested a bit at the Pierre, however, before going to the Waldorf Towers for breakfast with Henry Kissinger in UN Ambassador George Bush's suite on 11 December. "Chinese wallpaper and discreet waiters made one nearly forget that eight thousand miles away the future of my guest's country hung by a thread," Kissinger recalled. "Elegant, eloquent, subtle, Bhutto was at last a representative who would be able to compete with the Indian leaders for public attention. The legacy of distrust engendered by his flamboyant demeanor and occasionally cynical conduct haunted Bhutto within our government," wrote Zulfi's American counterpart. "I found him brilliant, charming, of global stature in his perceptions. . . . He did not suffer fools gladly. Since he had many to contend with, this provided him with more than the ordinary share of enemies."[1]

Kissinger advised Bhutto that "Pakistan would not be saved by mock-tough rhetoric. 'It is not that we do not want to help you; it is that we want to preserve you. It is all very well to proclaim principles but finally we have to assure your survival.' . . . The next forty-eight hours would be decisive. We should not waste them in posturing for the history books. . . . Bhutto was composed and understanding. He knew the facts as well as I; he was a man without illusions, prepared to do what was necessary, however painful, to save what was left of his country."[2]

The following evening Foreign Minister Bhutto told the UN Security Council, "Time is running out." Swaran Singh, Indira Gandhi's foreign minister, had just finished speaking when Zulfi rose to note that he knew him "very well." Recalling what Kissinger told him, he said, "I am not going to indulge in glib rhetoric or semantic contrivances because the situation is far too serious. The fat is in the fire, and the time has come for us to act. . . . Either we act individually or we act collectively." He also promised not to "assume a sanctimonious attitude," admitting "we have made mistakes. Man is not infallible. Mistakes have been made everywhere . . . by the Roman Empire, by the British Empire, by every state in the world. But states are not penalised for their mistakes. . . . We are prepared to rectify those mistakes in a civilised spirit, in a spirit of understanding and co-operation."[3] Thus Zulfi launched upon one of his longest sanctimonious speeches. He retraced the history of Pakistan and its many conflicts with India, declaring, "Pakistan is an ideal. It will last even if it is physically destroyed. We are prepared to face that physical destruction. We are prepared for the decimation of 120 million people. We will then begin anew and build a new Pakistan." The speed with which he could discard the tragic errors of his past and plan optimistically for the future was part of his potent appeal to Pakistan's populace, to Punjabis as well as Sindhis, and to many tribespeople along Pakistan's North-West Frontier who faced death and adversity every day with courage and faith in God.

"We are too poor. There is too much misery," Zulfi said. Compassion was also part of his popular appeal. "It is unfortunate that today we should be pitted against each other and one of us should dream semi-barbarically of the liquidation and annihilation of another. . . . It is simply not possible, because then India will be pitted against 120 million people, valiant people with a great past, fighting for their independence, fighting for their dignity. . . . So I offer a hand of friendship to India."[4]

Even as Zulfi was talking in New York, the new military commander of Pakistan's forces in the East at this time, General A.A.K. ("Tiger") Niazi, strode briskly toward the Intercontinental Hotel in Dacca. Then pausing to answer the questions of some foreign journalists, he perched himself on the chrome-plated seat of his shooting stick, pearl-handled pistol strapped to his ample waist. Asked "what the Pakistani Army would do if the Indian Army and the Mukti Bahini attacked Dacca from all sides," Niazi replied, "What you fellows don't know is our hidden strength. I tell you, things are going to happen very quickly, amazing things. By tomorrow or the next day the whole situation will have changed. . . . It doesn't matter whether we have enough men to defend the city. If you stay around, you'll see our men dying gloriously. . . . We know what we are dying for. What does the enemy know? Remember, every Muslim soldier is worth ten Hindus. We shall give a good account of ourselves. Gentlemen, the great battle for Dacca is about to begin!"[5]

"Bangladesh exists on the lips of the Government of India. Bangladesh exists in their mind; Bangladesh does not exist in reality," Zulfi shouted in New York. "We are prepared to die. We are not afraid to die. Our people are brave. We and India have shared 5,000 years of history. . . . Believe me, Mexico might occupy the United States, Denmark might occupy Germany, Finland might occupy the Soviet Union—but Pakistan will not be occupied by India in any circumstances. Remember that. . . . We shall fight, and we shall fight for 1,000 years as we have fought for 1,000 years in the past. . . . We can continue."[6]

Nixon ordered the nuclear-armed aircraft carrier *Enterprise,* of the Seventh Fleet, to steam with its escort convoy into the Bay of Bengal, reportedly just to "help evacuate" U.S. personnel if Dacca fell. New Delhi recognized such nuclear weapons rattling for what it was, however, and Mrs. Gandhi resolved at this time to go full ahead with India's own plutonium-production plans that would in two and a half years trigger India's first nuclear explosion under the sands of Rajasthan, close enough to be felt in Sind.

"Today we are pitted against India and a great power," Zulfi charged in the Security Council. "India is a big country. . . . But today it is standing on the shoulders of a big power to look bigger. If it did not . . . it would not have been arrogant enough to defy the will of the General Assembly and the whole world expressed in a resolution calling for a cease-fire, the end of hostilities and the withdrawal of forces. . . . Today, the Soviet Union has openly and brazenly come out in support of India. . . . Otherwise the blockade of the Bay of Bengal would not have taken place."[7]

Kissinger had assured Zulfi of the White House's resolve to defend West Pakistan against any possible joint Soviet-Indian attack, but it was now clear to all diplomats at the UN that nothing could keep Bangladesh under the sovereign control of Islamabad for many more days. General Niazi had, in fact, already signaled India of his intention to surrender. There was painfully little Pakistani resistance to the steady multipronged advance of India's gigantic army and air force that had moved big guns and bombs within range of Dacca's suburbs.

"Take East Pakistan for five or ten years, we will have it back," Zulfi challenged Swaran Singh, turning to aim his barbed words at the turbaned head of his bearded adversary. "Believe me, Mr. Foreign Minister . . . this is the lesson of history from the beginning of time—what belongs to a people will go to that people. . . . East Pakistan is part of Pakistan—you know this. Remember this well."[8] That would remain an ambition and frustration: to lure Bangladesh back to its pre-1971 status. "Today, the Indian Foreign Minister said that the problem could only be resolved if a representative of the so-called Bangladesh government, which was created by India, was represented in the Security Council," Bhutto continued. "Are you going to permit this kind of precedent, when provincial parties and those who are clients of larger countries should have representation before the Security Council? . . . sometimes the Sikhs will come to the Security Council; sometimes the Punjabis . . . We can also bring some people from India. We have not indulged in that kind of mischief. But give us some time. . . . Listen, Sardar Singh, Golden Bengal belongs to us not to India. . . . We will fight to the bitter end. We will fight to the last man."[9]

In every village and hamlet, as the Indian troops entered, joyful Bengalis rushed out to greet them with garlands of marigolds and tearful cries of "*Joi Bangla!*" The flag of Bangladesh was everywhere—over huts and trees, draped on Soviet guns, attached to the antennas of Indian tanks. The words of *Gurudev* (Divine Teacher) Rabindranath Tagore's anthem "Amar Sonar Bangla" (Our golden Bengal") rang out from the lungs of millions who had waited in silent terror for nine months for this birth-dawn of freedom. "O Mother! During spring the fragrance of your mango groves maddens my heart with delight. O Mother! During autumn in your blooming fields I have heard your sweet murmuring laughter. What beauty, what shade, and what love, what affection I perceive beneath the banyan trees. . . . My life passes in the shady village homes filled with rice from your fields. They madden my heart with delight. O Mother! . . . I am poor, but what little I have, I lay at thy feet. And it maddens my heart with delight."

Bhutto spoke again the next day in New York about how bravely Dacca was being defended, and how fiercely Pakistan would continue to fight. But the two-week Bangladesh war was almost over. "Tiger" Niazi was ready to turn his pearl-handled revolver over to General Aurora, abjectly touching his forehead to the head of the stalwart Sikh hero of Bangladesh within days of Zulfi's sanctimonious threats and promises.

On 15 December, Bhutto addressed the Security Council for the third and final time, full of bravado:

> So what if Dacca falls? So what if the whole of East Pakistan falls? So what if the whole of West Pakistan falls? . . . We will build a new Pakistan. We will build a better Pakistan. . . . Mr. President, you referred to the "distinguished" Foreign Minister of India. . . . How is he distinguished when his hands are full of blood, when his heart is full of venom? . . . I extended a hand of friendship to him the other day. . . . I am talking as the authentic leader of the people of West Pakistan who elected me at the polls in a more impressive victory than the victory that Mujibur Rahman received in East Pakistan. . . . But he did not take cognizance of it. . . . I say what Cato said to the Romans, "Carthage must be destroyed." If India thinks that it is going to subjugate Pakistan, Eastern Pakistan as well as Western Pakistan—because we are one people, we are one state—then we shall say, "Carthage must be destroyed." We shall tell our children and they will tell their children. . . .

Eighteen-year-old Benazir, down from Cambridge to be with her father, was sitting just behind him.

> We will fight for a thousand years. . . . India is intoxicated today with its military successes. . . . So you will see . . . this is the beginning of the

> road. . . . Today, it is Pakistan. We are your guinea pigs today. But there will be other guinea pigs. . . . You want us to lick the dust. We are not going to lick the dust. . . . I am not a rat. I have never ratted in my life. I have faced assassination attempts. I have faced imprisonments, I have always confronted crises. Today I am not ratting, but I am leaving your Security Council. I find it disgraceful to my person and my country to remain here a moment longer . . . legalise aggression . . . I will not be a party to it. We will fight; we will go back and fight. My country beckons me. . . .

"You can take your Security Council," he thundered as he tore up his notes on the resolution, tossing the paper aside. "Here you are. I am going."[10] Stiff-legged, Zulfi strode out of the chamber. Benazir and the rest of the delegation followed. "My father was very upset as we walked and walked, seeing the devastating repercussions ahead for Pakistan," Benazir recalled. " 'Now Pakistan will have to face the shame of surrender to India. There will be a terrible price to pay,' " he agonized.[11]

Niazi surrendered on 16 December, signing a formal instrument in Dacca's race course the next afternoon and giving up for his entire army, 93,000 of Pakistan's best soldiers, and supporters. It was the most ignominious defeat in the nation's history, one that even Zulfi Bhutto's rhetoric was unable to conceal from the millions in West Pakistan, who had hoped and prayed and desperately tried to believe every brave lie he told them.

"Piles of dead soldiers lie about the roads and ditches at the entrance to Dacca," Gavin Young, the only British correspondent left inside Dacca at this time, wrote for London's *Observer* on 19 December. "With the advance guard of the Indian Army, the first Indian general into the city, Major-General Gandharu Nagra, said: 'Casualties are severe. Very messy.' I asked him if he had already met General Niazi, . . . 'Oh yes. He said he was very happy to see me. We knew each other at college.' "

> That first morning of the cease-fire . . . [o]fficers who had been comrades at the same staff college . . . stood looking at one another wondering who to blame. The Indian troops—dusty in buses, jeeps, or trucks—took the garlands and embraces, the cries of "Joi Bangla" from leaping near-hysterical Bengalis. . . . Niazi told me: "We surrendered because otherwise we'd have had to destroy the city between our two armies. We would have had to surrender in the end, so what was the point in continuing?" He gave a pale smile. . . . He is not at all a brilliant man, but he is dedicated—not a political thinker. The war would have ended a week earlier, but President Yahya Khan convinced Niazi that China and the United States would intervene on Pakistan's side.[12]

From New York Zulfi flew to Key Biscane, Florida on 17 December to meet with Nixon and Kissinger on Bebe Rebozo's yacht, where he was assured of ample U.S. military and monetary support. He boarded Pan Am flight 106 on the evening of 18 December for the long voyage home to the much-diminished nation he was now to lead, stepping down in Rome shortly after noon the next day. Air Marshal Rahim Khan had ordered a Pakistan International Airlines Boeing to be sent specially for Bhutto from Karachi. The flight to Islamabad allowed Zulfi to stopover in Teheran to confer with the shah before finishing the last lap of his sadly triumphal trip home.

The People's Party organized mass meetings from Pindi to Karachi as word of the Dacca debacle cast its shadow across West Pakistan. The cry of "Death to Yahya Khan" alternated with "Long life to Bhutto!" Yahya had wished to hang on, but when his chief of staff, Lieutenant General Hamid Khan, briefed most of the junior officers at the National Defence College in Pindi on the morning of 20 December on the recent events in the East, he was met with angry questions and epithets. "The younger officers were shouting 'Bastards', 'Drunkards!', 'Disgraceful!' and 'Shame!' . . . Lieutenant-General Hamid Khan's composure and that of the generals of the front row had completely collapsed. . . . Yahya Khan had played his last card. The game was up."[13]

Zulfi landed at Pindi soon after the well-staged drama in the NDC had convinced Yahya that he had lost more than a war. Mustafa Khar was waiting in his blue Mercedes inside the airport gate, and drove with Zulfi directly to the well-guarded President's House. The building surrounded by high barbed-wire walls with spiked steel gates would soon become Zulfi's residence. Yahya was nervously waiting inside with a broken Peerzada and shaken Hamid, sipping whiskey without any soda. They conferred for an hour alone, Pakistan's two presidents, who "knew each other well," and had shared so many of the same pleasures. Yet despite his Persian ancestry, Yahya was never really a political animal, only a military one. Now all his dreams of glory lay shattered behind him. Zulfi's lay ahead.

Information Minister Roedad Khan had been called to the President's House by Yahya earlier in the day "to make preparations" to broadcast news of the transfer of power that was about to occur. "I went there with a very heavy heart," Roedad Khan recalled, for by then his elder brother, Khan Abdul Khalik Khan, the only PPP candidate from the North-West Frontier Province to win a seat in the National Assembly (from Mardan) against Wali Khan's ruling National Awami Party, had quarreled with Bhutto. "He never forgave my brother," Roedad explained, and actually "got up from his chair and threateningly approached him . . . going red in the face. He [Bhutto] had a very bad temper."[14] Roedad Khan and the then Cabinet Secretary Ghulam Ishaq Khan—now president of Pakistan—walked "up and down on the lawn" outside the President's House as the "new order" and the "new people" arrived: Mumtaz Ali Bhutto, Dr. Mubashir, J. A. Rahim, and G. M. Jatoi. Finally, Yahya and Zulfi appeared. "Then the transfer of power took place," Roedad remembered, chagrined that Bhutto had not so much as nodded to him. As he started to leave, however, "Mr. Bhutto called my name and said, 'You'll be hearing from me!'" Roedad returned to his Information Ministry office and waited for the phone to ring. Within the hour it did. He picked up the receiver and heard as harsh and loud a voice as he'd ever recalled bark, "Zulfikar Ali Bhutto!" "*Sir!*" he snapped in reply. "I want you to do the following," Zulfi ordered, "I want to address the nation tonight, and I want you to outline some points, and come to see me at 5 this evening."

That evening Roedad learned he had been sacked from Information and Broadcasting, replaced by Zulfi's barrister, Hafeez Pirzada, "who understands our aims and objectives," Bhutto explained. Pirzada, whose birthplace in Sind was fewer than fifty miles from Larkana, had since 1958 shared Mr. Dingomal's Karachi chambers with Zulfi and Mumtaz. After Zulfi left Ayub's cabinet, Hafeez defended him in many cases, including the famous tractor-misappropriation case, and the confiscation of some 115 of his personal "hunting and shooting weapons," which "I got . . . quashed," Pirzada recalled. "Then we became very, very close friends."[15] After the quashing of all other charges Ayub's legal advisers had filed against Bhutto, Hafeez was appointed to Zulfi's party's Central Committee, and would remain his barrister and a member of his cabinet for the next five years. Hafeez and his lovely first wife, Sadia, to whom Zulfi was strongly attracted, often dined with President Bhutto.

Bhutto had been driven to Pindi's Punjab House after having been "sworn in" as the new president of Pakistan and chief martial law administrator, the two jobs he took over from Yahya on 20 December 1971. Yahya was left to pack up and vacate the President's House as soon as he could. Lieutenant General Gul Hasan had been called to Punjab House, and thought, in fact, that he was going to meet Yahya there as he ascended the narrow stairs, flanked by two six-foot-six Presidential House guards. Hasan asked the military secretary as he got to the upper floor, "What's the Chief doing here?" "You'll see," was all that the General replied. He found Zulfi Bhutto waiting inside the drawing room with Air Marshal Rahim Khan. "Bhutto embraced me and he started to cry, and all that, all put on," Gul Hasan recalled, later shocked to learn how "creative a liar" Bhutto could be, and what "a vicious man" he was. "Totally unbalanced."[16] Bhutto asked Gul Hasan to take over Yahya's job of army commander in chief, for that was one position Zulfi knew he could not fill. As current chief of the general staff, Gul Hasan knew all the general officers personally, having trained many of them, including one obsequious "dark horse," Major General Zia ul-Haq (1924–88). Gul Hasan asked for time to "think about it," still surprised to find himself with a new president.

After the defeat just sustained by Pakistan's army, the job of commander in chief must have sounded more like a crown of thorns than a high honor. But Zulfi could not wait, insisting he had to address the nation that night, and "the cornerstone of his speech" was "going to be my appointment!" General Hasan felt flattered, but to give himself some moments to consider the offer, excused himself and went into a bedroom to mull it over. Half an hour later he accepted on three conditions, the first being that he wanted no promotion, preferring to remain a lieutenant general. Zulfi agreed, and that night Gul Hasan listened to the broadcast, which didn't start until several hours after it had been scheduled, about 10:30 P.M. "'We've appointed Gul Hasan as Commander-in-Chief,' [Bhutto] said," the now-retired general recalled. "'He's a good soldier, popular with the men, and he hasn't dabbled in politics, but I'm afraid he'll have to serve in the same rank, because we can't afford to fatten people unnecessarily.'

It was an out and out lie! So next morning I rang him up and said, 'What the hell are you doing?' He said, 'This is politics!' "[17] A few months later the outspoken Gul Hasan would be packed off to Austria, where he served as Pakistan's ambassador for the next three years, whereupon he was posted to Athens till his retirement in 1977.

"My dear countrymen, my dear friends, my dear students, labourers, peasants . . . those who fought for Pakistan . . . I have come in at a very late hour, at a decisive moment in the history of Pakistan," Zulfi began, speaking over radio and television in English.

> We are facing the worst crisis in our country's life, a deadly crisis. We have to pick up the pieces, very small pieces, but we will make a new Pakistan, a prosperous and progressive Pakistan, a Pakistan free of exploitation, a Pakistan envisaged by the Quaid-i-Azam. . . . [T]hat Pakistan will come, it is bound to come. This is my faith . . . but . . . I need your co-operation. I am no magician . . . without your co-operation I simply cannot succeed. But with your co-operation . . . I am taller than the Himalayas. . . . You must give me time, my dear countrymen, and I will do my best . . . I have been working round the clock.[18]

He apologized for not speaking in Urdu, but had opted for English because "the world is listening." "I wish I were not alive today," because of the way the Indians were "gloating" over their victory, but this was "not the end"; rather, it was a new "beginning" for Pakistan.

> I have been summoned by the nation as the authentic voice of the people of Pakistan . . . by virtue of the verdict that you gave in the national elections. . . . I would not like to see Martial Law remain one day longer than necessary . . . one second more than necessary. I want the flowering of our society. . . . I want suffocation to end. . . . This is not the way civilized countries are run. Civilization means Civil Rule . . . democracy. . . . We have to rebuild democratic institutions. . . . We have to rebuild hope in the future. We have to rebuild a situation in which the common man, the poor man in the street, can tell me to go to hell. . . . We have to make our Government accountable. . . . But because we are in such a terrible situation, you will have to give some time to me.[19]

It was the sort of speech Pakistanis needed that cold December night, after a year of bitter disappointments and nightmares. Precious little of what Zulfi Bhutto promised would ever be brought to fruition, yet just hearing him say what he did in so strong a voice helped lift the fallen spirits of millions who had felt, less than a week earlier, that their nation had died an ignominious death by surrender. "I want to tell our gallant armed forces who have fought in East Pakistan that our hearts are with you," he said. "We will not rest . . . till we have redeemed your honour. . . . [T]hese are not empty words." What made him think they might be construed as empty words? "We will stand by you, we are with you. If you go down, we all will go down together. . . . [R]emember my words . . . we are doing everything in our power. . . . [w]e will bend backwards to see to it that not a moment is wasted for the correct results. I do not want to spell them out because that may give indication to other people."[20]

He talked on for more than an hour, insisting that he was prepared to work for "a *modus vivendi*" in the East. He was also ready to "give the country the rule of law," but not quite yet. "Soon, very soon," he promised, "but please give me time. . . . The lack of accountability has left us in a very bad plight. We have not lost a war. We have not failed. . . . Our soldiers fought valiantly. We are the victims of our system. . . . The Western press has been against us."[21] Not the United States this time, not the Soviet Union, not even India, but the Western press. It was a new level of creativity. Shortly after, "I am afraid the nation has been fed on lies. Deception has been the order of the day."

"For economic and social justice, I will move as fast as is necessary, to see the burden of the common man lifted. . . . I am a man who works 24 hours a day. . . . I expect the bureaucracy to do the same." Zulfi also vowed to finish "corruption and maladministration." There would be no "vindictiveness" toward any "bureaucrat who has misbehaved with me in the past." Yet before firing Roedad Khan, he had ordered the arrest and incarceration of Altaf Gauhar. The police were ordered to "put an end to [the bureaucracy's] tyranny," its "*Zulm*." He would indulge in no nepotism, show no "favoritism" to friends or relatives, no *sifarish*. "I am not going to permit my relations to say that we are related to Mr. Bhutto. I have no relations, I have no family. My family is the people of Pakistan. My children are the people of Pakistan."[22]

To help him "rebuild democracy," Zulfi appointed Mumtaz Ali Bhutto as governor of Sind, and his old wadero friend and People's Party patron, Mir Rasool Bakhsh Khan Talpur, descendant of the pre-British ruling Talpur mirs of Sind, Mumtaz's "senior advisor," the martial law equivalent of chief minister. Mumtaz's appointment made Jatoi "hysterical," Zulfi recalled; he had promised Sind's chief ministership to Jatoi, yet now resolved instead to keep him "with me in the Federal Government, holding a very important portfolio and later, he could go to Sind."[23] As chief martial law administrator, President Bhutto could appoint anyone he wished to govern any province or manage any ministry. He made Jatoi minister for political affairs, communications and natural resources. Not only Jatoi, of course, but his playful wife, Maria, thus remained at Zulfi's side, always within reach.

Zulfi brought J. A. Rahim into his cabinet as minister for presidential affairs, culture, town planning and agrovilles, and made Dr. Mubashir minister for finance, economic affairs and development. He kept the foreign affairs, defence, interior and provincial co-ordination ministries in his own busy hands, bringing his former foreign secretary Aziz Ahmad back into service from retirement. He also used the considerable experience of Oxford-trained Rafi Raza to advise him on primarily military matters.

One week after Zulfi took charge in Pindi, he had himself driven after sundown to a well-guarded army bungalow near the military airport. Mujib had been flown there the day before by helicopter from his prison cell, where he had seen no newspaper, heard no radio since his arrest in Dacca in late March. "When they brought me here, I was thinking that another disaster was coming," the tall, haggard prisoner said, faintly smiling at the sight of his old "friend," though as yet uncertain why all the armed guards were in that room as well. Then Zulfi told him of the transfer of power in Pindi. "I am happy," Mujib responded. "Believe me I am happy. Tell me what is the condition of Bengal? I am very perturbed."[24]

Bhutto explained that India's army was "occupying" Dacca, had moved in with Soviet guns and trucks, under cover of Soviet warships and planes. "That means they have killed us," Mujib cried. "Bhutto they have killed us. . . . You'd better get me to Dacca. . . . [Y]ou have to promise one thing to me that I will be having your help, if the Indians put me in jail. You will have to fight for me."

"We will fight together, Mujib," Zulfi promised.

"No, no," Mujib replied, knowing full well with whom he was dealing, and knowing, of course, that every word he said was being taped in this well-wired sitting room. "I have to settle matters there. The occupation army is there, that I am sure. Awami League cannot move them, that I am sure. The world will understand . . . I want your help. . . . Believe me, I never wanted the Indians. You don't believe me?"

"I believe you," Bhutto said. "Otherwise, do you think that I would have immediately taken this action? The first thing I did was to order your release from there and coming here."

"Well, you have taken the charge, I am happy," Mujib repeated, though he did not sound happy, nor did he feel happy. His furtive eyes darted around that strange "safe-house" room. He wondered if he would ever be permitted to see his beloved Dacca again, or his family and friends. "You have done away with the military?"

"Now it is a secondary matter," Zulfi replied evasively. "Military will not come back again. I just want to see the unity of my country on whatever terms you want." He still dreamed of undoing the past, using this prisoner, *his* prisoner now (even as "retired" Yahya was), to replay history's tape, to put Humpty-Dumpty Pakistan back together again. Why not? Was he now not both president and chief martial law administrator of Pakistan? And wasn't this poor Bengali fool his mouse to play with, to tease and tickle as he liked? "I just want to see the unity of *my* country," he said, not *ours,* for he still distrusted Mujib almost as much as Mujib mistrusted him.

"No, no terms," Mujib told him, fearing the worst as he looked into those bloodshot eyes. "Tell me what is the position of my country."

"Mujib *bhai* [brother], if there is any possibility that we can live together?"

"I will hold a public meeting in Dacca," Mujib said. "I'll meet my people, discuss with them and inform you. That is my promise . . . I will speak drop by drop . . . they will believe me. Inshallah we'll do something. I have not to say about this outright. You understand my position." He was terrified at the thought of his Awami League comrades learning that he was actually negotiating here with President Bhutto. Yet if he did not speak softly to this man who held the keys to his life, what would become of him? How could he survive? When would he be sent home?

Zulfi played with his Bengali prisoner for another eleven days, leaving Mujib to simmer and sweat, to tremble and wonder what on earth would happen to him, and what had befallen his once-"Golden" Bengal? Till 7 January 1972 Zulfi left him there alone, returning that night, after his birthday celebrations in Larkana were over. He was more anxious now to woo Mujib, to bring him and all his Bengali millions back under his protective West wing. What a coup that would be! What Ayub and Yahya had lost he, Zulfikar Ali Bhutto, would recapture. Quaid-i-Awam Shaheed—Savior Bhutto!

"On the 27th you said that we can have two or three things together," Zulfi reminded Mujib as soon as they were alone again, "defence, foreign affairs and currency. . . ."[25]

"Before that you have to allow me to go," Mujib replied nervously.

"But I don't mind Confederation [however] loose it is," Zulfi insisted, pressing him for some minimal commitment.

"No, no . . . I have already told you that we can wait. . . . Some sort of arrangements we can make. I told you it will be Confederation. This is absolutely between you and me. I know that you have not told anybody."

"No, why should I tell," muttered Bhutto. "I'll keep it absolutely confidential."

"You leave it to me," Mujib insisted, "absolutely leave it to me. Trust me."

"I trust you," Zulfi told him. They "trusted" each other.

"I will tell these things to the millions of people of Bangla Desh when I first go there . . . that Yahya before handing over the charge to Mr. Bhutto wanted first to kill me. I will demand from there Yahya's hanging. You understand: trial, trial. . . . I'll say, 'First give trial to this man, then I'll talk to you.' . . . You understand. I have to take in confidence my people. The wounds are serious."

"Who do you think was the bigger villain—Yahya or Peerzada?" Bhutto asked him.

"Both."

"I know," Zulfi agreed, content, it almost seemed, to sit all night, chatting. They were all three his prisoners now. He alone had emerged from that last mad year of blood and terror, not only free but on top, comfortably secure at Pakistan's pinnacle of power. Yet he was so eager to win back Bangladesh that he told Mujib, "Listen, Presidentship, Prime Ministership, whatever you take I am prepared to retire to the country. I swear by Holy Quran that I am prepared to retire."

"I told you there is no objection," Mujib replied. "My idea was that we will live together and we will rule this country. This Haramzada [Yahya] started the fuss telling me this was what you had said . . . and . . . communicating to you. Then the disaster started in Bengal."

"But now we have to put the things right," Zulfi reminded him.

"We have to try," Mujib conceded. "But you know the route. I have to go. I have to cope. . . . [Y]ou know the occupation army is there. . . . [O]ur army . . . massacred . . . as in Indonesia. . . . Another point is there, I tell you frankly . . . the West Bengalees, . . . I have to take them out from Bengal and crush. . . . My difficulties you understand."

"Yes, it is very difficult," Zulfi agreed. He was faced now with similar outspoken opposition, and it would keep growing. Especially those who blamed *him* for losing most of Pakistan. "I am prepared to fly to Dacca."

"No, no," Mujib insisted, holding his head painfully. "I'll meet with my people . . . I want time. . . . I'll tell you from there. . . . You have become a hero. You will continue for the whole life."

"Nay, I don't want," Zulfi told him. "You should be our President." Zulfi also offered him "fifty thousand Dollars," but Mujib was too nervous to take any money from him, suggesting instead that he "contribute it for my chartered plane."

Shortly after midnight on 8 January 1972 Mujib flew to London. There were no direct flights from Pakistan to Bangladesh, and Mujib preferred London to Teheran or Ankara, the other options Bhutto had suggested. The PIA jet landed at Heathrow in gray fog at 6:30 that morning, and a much relieved, bedraggled Mujibur Rahman, the nation-savior hero of Bangladesh, was whisked off in a Rolls limousine to the elegant Claridge's Hotel in the heart of Mayfair. Before noon on that first day of his freedom, Mujib had spoken from his suite with Indira Gandhi in Delhi and Prime Minister Edward Heath in Downing Street. Heath authorized release of an RAF plane to fly his fellow Commonwealth prime minister home the next day. Mujib stopped first in Delhi, where he received a hero's welcome from Indira and her entire cabinet. In Dacca, an adoring ocean of Bengali faces awaited him, looking up to Bangabandhu (Nation-Unifier) Sheikh Mujibur Rahman for answers to all the tragic needs of his young nation, born after so much trauma, left in such desperate poverty. The "difficulties" confronting him were far greater than Mujib had imagined.

President Bhutto's own problems proliferated as well, though he pretended to be on top of everything, jauntily telling an American journalist from the *Baltimore Sun,* "If you Americans think Franklin Roosevelt had an amazing first hundred days, watch us."[26] The reporter was impressed, informing his readers of how hard Pakistan's new president worked, "sleeping only three or four hours each night." He had to travel much, of course, dividing his nights between Pindi and Karachi when he was not in Larkana or Lahore.

With the new year, Zulfi announced his immediate nationalization of ten "categories" of major industries, including iron and steel, basic metals, heavy engineering, heavy electrical, motor vehicles, tractors, basic chemicals, petrochemicals, cement, and public utilities. Capital equipment worth an estimated Rs. 1 billion (c. $100 million) was brought under state control, and the new Board of Industrial Management, chaired by Dr. Mubashir Hasan, was appointed by Bhutto to supervise some thirty major industries. A debilitating flight of capital began soon after this martial-law socialist policy was initiated, and would continue throughout the period of Zulfi's People's Party government as well. The sudden and unexpected nationalization, of course, plummeted Pakistan's international credit rating in London as well as in New York and Washington. Zulfi turned more and more for financial support to North African friends, especially to Muammar Qaddafi of Libya, to his good hunting friend the sheikh of Abu Dhabi, and to Communist China.

"We must dedicate all our energies to ensure maximum production and highest quality," Zulfi told the nation in announcing the nationalization move. "The workers will now have a real stake in the success of these undertakings. . . . There is no substitute for hard work. . . . There is now a people's Government, and the people are the ultimate masters."[27] On the next day Zulfi addressed a public meeting in Karachi, speaking in Sindhi:

> You must never forget that I am one of you, one from amongst you. . . . When I was summoned from New York . . . the country was very close to civil war. . . . I was needed by my countrymen and was wanted in my homeland. But I did not know that things were so bad, because as truth was never told to you and as you were fed on nothing but lies, I was also being supplied with incorrect news. . . . Yahya Khan had asked us to accept power. . . . He said to me that I must accept the Government which belonged to the people. . . . I told him that the ship of the nation is afire. . . . I and my Party had declared in April and appealed in the name of God that power should be transferred immediately to the people. . . . To that Yahya Khan and his colleagues replied . . . that whoever talked of the transfer of power was a traitor. . . . Now it is for you, the people of Karachi, to give your verdict as to who proved a traitor. Were they the traitors or were we the traitors? What did these people do? They had beaten the workers . . . They lashed the peasants . . . They ruined the country so much that we have now reached our present tragic plight.[28]

Now it was Yahya's turn to be Zulfi's target, as Ayub previously had been. Zulfi told his people that Yahya only intended to "lead the politicians to their doom," that he was "a liar, a drunkard, and a fraud" who conspired against the people. As proof positive of Yahya's treachery, Zulfi noted that during the recent elections "my opponent in Larkana [Khuhro] was promised support and was given money. . . . 'Over my dead body I will transfer power to Bhutto,' Yahya Khan said! Listen, listen my comrades, listen carefully."[29]

People had become restive, and many were angry, weary, fed up though underfed, heartsore and depressed. Prices kept rising and everything was scarce—food, shelter, clothing, all the things Bhutto's party had promised them. "You cannot have all the reforms in one day," he cried out.

> Even the most sacred reforms . . . took 23 years . . . and they were brought by our Holy Prophet. I am not even worth the dust of his feet. . . . On the one hand you are asking for democracy and on the other hand you want reforms. . . . Confusion is being created in the minds of the people. No compensation has been given (they complain). This is sheer high-handedness (they say). Now on the one hand people are saying that it is hardly anything which I have done and that it is not enough. . . . Yes, agricultural reforms are also on their way. They are accusing me of being a landlord, a feudal lord. . . . Lands are of no consequence to me. . . . As for the educational reforms, . . . this is only the first phase of our reforms. . . . Every day they are taunting me about my manifesto. . . . But who can digest in one day all that the manifesto contains? . . . [w]ho is going to be

> benefited by the industrial units which we have taken over? . . . [T]hese are your factories, your industries. . . . [N]o Government of this kind had ever come to power. . . . Very soon I would bring labour reforms. Give me some time. . . . The promise of democracy is a true promise, a genuine promise, but those who are shouting every day for democracy to be ushered immediately . . . are in reality the henchmen of the capitalists. Beware of them.[30]

Workers who were not paid went on strike. Students left their classes and joined in protest rallies, demanding more reforms and cheaper food and rent. Nothing had as yet changed for the better, only for the worse, it seemed, because the economic picture was so bleak. For most of the 1960s Pakistan's economy had been relatively robust, especially in the West wing, but now all the indicators were moving down. Some kept falling for the next half decade, till Zulfi Bhutto was toppled from his seat at the top.

Zulfi tried intimidation: "If you don't work I will have to bring another Government, perhaps the same . . . type of people who used to rule over you. . . ." He tried remonstrance: "Are you such ignorant children who cannot understand that in two days or even 10 days we cannot do everything? Don't you know that your country is a backward country? Don't you know that in one day I cannot nationalise everything?" He cast himself as victim: "Why are you doing all this to me?" He pleaded: "I appeal to you, my dear workers, to stop, for God's sake, this labour unrest and all such things. Start working. Get down to your jobs . . . I swear by God and his Prophet that we will be on your side." Then: "But don't press us. . . . We also have our breaking point."[31]

Undeniably, the problems facing Pakistan were monumental. They would have challenged the most experienced, level-headed administrator, or the capacity of any team of experts in many fields. But Zulfi Bhutto really had no solid grounding in administration, and few capable men he trusted to assist or guide him. He tried to "run" Pakistan as he had run his estates in Larkana, or the way he managed his personal affairs, like the feudal lord he still was, with alternate threats and promises, with carrots and sticks, with bribes and hunting rifles, curses and tears, and solemn oaths to God. And part of him meant each contradictory move, each conflicting word or phrase, for he was many Bhuttos, all wrestling inside him. His greatest weakness as president, and later as prime minister may have been his inability to attract to his service men of greater integrity or probity. Perhaps no leader ever feels quite comfortable or "safe" with assistants superior to himself, yet some middling minds have the talent and capacity to recruit high-level ability, even genius.

Zulfi was always suspicious of everyone around him. He usually feared some conspiracy or other, and was himself so astute at realpolitik that he suspected his closest colleagues and allies of plotting against him. He scolded his closest associates a great deal; perhaps he thought that just pointing out their weaknesses would suffice to make them discard abhorrent habits of a lifetime. He also seems to have hoped that by working inordinately long hours, he—Zulfikar Ali Bhutto—would set so high a standard for others that intense dedication would in itself solve Pakistan's most intractable problems. He counseled new governors and ministers to keep "uppermost in mind" that they must "remain in touch with the people, with their sentiments and aspirations. This can best be achieved by travelling among the people as much as possible. . . . [O]fficers must work and not be distracted. . . . Pomp, show and ceremony do not represent respect. Respect must be earned . . . by hard, honest work. This standard of service I am not only setting down for you but fully intend to maintain myself as an example."[32]

Zulfi promised to convene provincial assemblies on 23 March 1972, a promise he made two months earlier. "Pakistan has been made weak because the people have been denied their rights. . . . The wishes of the people can be properly manifested only through institutions."[33] It was his best advisers who had convinced him of that truism of all good government. He asserted now his realization that "Governors cannot single-handedly run the entire political machinery of the Provinces," though his first two appointees in Sind and Punjab, Mumtaz Bhutto and Mustafa Khar, came as close to trying to do so as Zulfi himself did at the center. "It is being asked why I continue to act under the cover of Martial Law," Zulfi noted, first arguing in response, "I have a mandate from the people, a mandate without parallel in our history," then adding contrarily, "The title of Chief Martial Law Administrator I have inherited . . . [is] not our only inheritance. . . . Will the new self-professed mentors of democracy, who readily call for the immediate lifting of Martial Law, show us how simultaneously to lift the masses from their morass of miseries and injustice? . . . The powers of Martial Law have been used . . . for the sole purpose of bringing . . . basic reforms. . . . Once this first phase of reforms is over . . . the ground would be laid for the full flowering of democracy."

On the eve of flying off on a support-seeking tour of Arab countries, Bhutto invited leaders of Karachi's business and industrial community to meet with him at the airport's VIP lounge. "I am not your enemy," he told them. "In the previous regimes industrialisation went on without any rational framework . . . without any social context . . . any egalitarian context. . . . We have to go in accordance with the requirements of modern times. . . . As you sow so shall you reap and as we sow so shall we reap. . . . You say respectable people have been thrown into jail. You call yourself respectable. I call myself respectable. But what does the common man call us? He calls us parasites and blood-suckers."[34]

Zulfi withdrew Pakistan from Britain's Commonwealth of Nations and U.S.-dominated SEATO. Then he flew to China in early February, where he drank many toasts to the health of Chairman Mao Tse-tung and Chou En-lai, to "the further development of friendship and co-operation between Pakistan and China," and "to "Afro-Asian solidarity."[35]

Back home a week later, Bhutto announced his new labor policy on 10 February. Workers would share 4 percent of the annual profits of most of the larger companies, including all of those just nationalized. Elected shop stewards in all factories would "represent" workers' "interest and point of view" in all negotiations with management. Special labor courts would be obliged to adjudicate grievances within twenty days, and workers were promised "full protection" against "arbitrary retrenchment" or "termination of services." Cheap workers' housing and free education to matric (high-school) level for at least one child of every worker were to be provided as soon as possible by the state. Old age pensions, and group insurance for injured workers and the extension of social security benefits to domestic servants were also promised. It was the most liberal and comprehensive labor policy ever introduced in Pakistan, and though never fully implemented went a long way toward living up to the PPP manifesto promises in that vital area of social reform. Zulfi's minister of labor, Mohammad Hanif, who drafted the policy, had himself been a leading activist and revolutionary.

"We are not so naive as to think that a mere new set of laws will transform overnight the national economic life of our society," President Bhutto cautioned, in announcing the new policy. "But I am confident that if all of us try hard, and work hard, we will all benefit. . . . This is the only way to serve Pakistan. . . . All our yesterdays have seen failure. Let us strive for a better tomorrow, a morrow at the service of the common man."[36] Here was Bhutto at his best, idealistic, inspiring, yet practical in his sober warning, reaching out to help lift Pakistan's "common man" from hopelessness to a new life—or at least to its prospect for his children.

The feudal side of Bhutto, however, managed to subvert virtually all the idealistic good intentions he spoke so much about. "My dear Brother," he wrote to the governor of Sind's senior adviser, that same February:

> Habits do not change easily but men of responsibility have to change habits if they are to serve their nation. I have changed. . . . Before undertaking your present responsibilities . . . it was a different matter if you went with a bus-load of females to Shahbaz Qalander or wasted precious hours every evening at Sanjhi's. . . . I request you in the name of the country . . . to put a *complete* end to such things. . . . many tongues have already begun to wag. You know the mentality of our people. They do not spare anyone, including the saints.[37]

Was this last word of caution elicited by Zulfi's acute awareness of what people were always saying about him?

On 1 March, President Bhutto addressed Pakistan's "Citizens" and "Haris" (peasant serfs) on nationwide radio and television: "Tonight is your night, as I am speaking to you on land reforms. . . . From its inception, and throughout its struggle against injustice, the Pakistan People's Party has been committed to the eradication of the curse of feudalism. . . . The reforms I am introducing . . . will bring dignity and salvation to our rural masses who from today will be able to lift their heads from the dust and regain their pride and manhood."[38] Rightly noting that the "majority of our peasantry have since time immemorial been suspended in the vicious web of abject poverty and servitude," he promised, "We shall not allow this abominable *status quo* to continue." Few of his listening *haris,* however, understood either English or Latin, which was perhaps just as well because most of the changes he promised hardly alleviated their misery. Landlord ownership ceilings were lowered, from 500 irrigated acres to 150, and from 1,000 unirrigated acres to 300. The limits, however, applied only to individual landlords, not to entire families, and most large holders did precisely what Zulfi and his wadero cohorts had always done: parceled out excess acreage among relatives. Ayub's land reforms, Zulfi explained, "were a subterfuge . . . reforms in the name only." The "upper limit" of landholdings had not included such areas as "orchards" or lands "gifted" to one's heirs or a "family's female dependents." "The runaway scheme did not stop here"; there were also special exemptions for "Shikargahs" (hunting fields) and stud and livestock farms. "One of the exempted 'Shikargahs' stretched over more than 100,000 acres," he said, thinking of Jatoi's; his own was slightly smaller.

"We are not permitting such exemptions or concessions," President Bhutto declared righteously. "All 'Shikargahs' will be resumed and the land distributed to the peasants, except for those historical 'Shikargahs' which will be run by the state."[39] Naturally, both the Bhutto and Jatoi hunting fields were "historical." Zulfi did not, of course, mind having his holdings run by his own state. Provincial governments were to determine how "resumed lands are to be utilized," and former owners would have the "first right" to lease lands that were made available for lease. "In fairness," Zulfi added, "we cannot ignore investments made by existing owners in installing tubewells, purchasing tractors, breaking new lands and adopting costly modern techniques of agricultural production," as he himself had recently done on thousands of his Larkana acres. "We are compelled in equity to allow an existing owner to have an additional . . . 3,000 units," if he did any of the above. Large tracts of "Government land" in Baluchistan, in the Pat Feeder area, had been "encroached upon by several influential persons and there has been a protracted controversy over entitlement to these lands," Zulfi noted. Those "influential" Baluchi sardars had long been opposed to Bhutto's new party, and now "to end confusion and provide lands to the tillers, the Government has decided to resume all lands in the Pat Feeder area . . . without any compensation. The resumed lands will be granted to poor farmers of the region." For each enemy thus made, Zulfi won at least a thousand friends, though Baluchistan would soon be engulfed in a protracted and costly war against his central government.

During Ayub's "decade of development," many officers and civil servants most of whom were Basic Democrats, took advantage of options to buy fairly large tracts of irrigated lands in Punjab and Sind at remarkably low prices, given the potential value of such lands once fully brought under cultivation. President Bhutto now called that "the most disgraceful abuse of power," for during that period many Punjabi officers acquired holdings in Sind. "This has created a new class of absentee-landlords," Zulfi pointed out in a tone of appropriately righteous indignation, ordering the immediate confiscation, without compensation, of all lands in excess of one hundred acres that had been so acquired by "any Government servant"—except for members of the armed forces. All such lands would be parceled out to landless tenants or peasants with "below-subsistence" holdings. Here again, for every powerful enemy Bhutto made, he won at least a hundred votes, in some cases many more.

"He thought, let's do something quickly for the poor people," barrister

Yahya Bakhtiar, who was to become his attorney general and would represent him during his final appeal for his life, recalled. "And he thought the land reforms, nationalization and all that will give him money, and enough security to do something."[40] Chou En-lai had asked Zulfi during their recent meeting in China why he was in such a hurry about his reforms, Bakhtiar remembered from having accompanied Bhutto to China. Because they were "*very* good friends," Chou's words of caution elicited this very frank reply from Bhutto: " 'You are right, but do you realize that perhaps the Army will not give me a chance to consolidate my position? So I have to rush in a hurry to do something for my people.' " Chou advised him to build a "People's Army," explaining that was what China had done. "The idea took root and remained in Zulfi's mind as one of his top priorities."

"I have kept my pledge with God and man," Zulfi Bhutto said as he concluded his land reforms broadcast. "This day marks the beginning of a new saga in the annals of Pakistan. This is no prank with history. It adds a golden chapter to its volume on liberty. The hour has struck and we must rejoice on hearing the shackles break. . . . Friday the third of March shall be a public holiday to commemorate the infinite blessings of this day, the beauty and splendour of its promise."[41]

That holiday Friday he addressed his nation again, reminding his listeners that when he had taken charge the previous December, "we were in a cesspool and the nation had to be picked up. . . . We have no time to lose."

> I know there have been scarcities of essential commodities. The prices of *atta* [wheat flour] have risen. . . . The price of sugar is increasing. I am aware of all these developments. They hurt me very much and we are taking measures to rectify the position. What could I do if the granary was not to be found; if the food stocks were not available; if the wheat crop had been completely depleted by its faulty distribution? What could I do? . . . Even when a person takes an aspirin, his headache does not disappear with the taking of the pill. . . . [I]t takes a little time. But here in Pakistan we have had to undergo a major surgical operation.[42]

He then announced his decision to replace Lieutenant General Gul Hasan with Lieutenant General Tikka Khan (b. 1915) as the new head of the army, and Air Marshal Rahim Khan with Air Marshal Zafar Chaudhury. "From today we will no longer have the anachronistic and obsolete posts of Commander-in-Chief. Every wing of the Armed Forces . . . will be headed by a Chief of Staff. This is the practice in many countries. . . . So we have changed the colonial structure of the Armed Forces of Pakistan and injected a truly independent pattern into this vital service."[43] It was Zulfi's first step toward creating his People's Army. As commander in chief and president himself, he now felt more secure against any possible coup.

Zulfi trusted Tikka Khan more than he had Gul Hasan. As Lahore's corps commander and martial law administrator of Punjab during Yahya's period, Tikka had met Zulfi quite often in Lahore, "And I was totally neutral," the general (then governor of Punjab) recalled. "I remained out of politics and I didn't let anyone under me in the army take part in politics."[44] Were it not for Tikka Khan's "neutrality" in Lahore in 1969–70, the mass People's Party rallies at which Zulfi spoke and received such spectacular receptions might have turned into military massacres. Tikka Khan had been an important asset to Bhutto, at least by what he did *not* do, if not by what he did, yet he also proved helpful in other ways. After 6 March 1971, when he was sent as governor to Dacca, "I got letters [from Bhutto] 'Please look after so-and-so,' 'Look after so-and-so,' and then . . . Mr. Bhutto came round about 20th March. . . . I remember we had made all arrangements to receive them at the airport and take them to the Inter-continental Hotel. . . . I came back first of September 1971, and met Mr. Bhutto again here, in Islamabad-Pindi, and then we came closer. . . . [T]hen Yahya resigned . . . pressure from Gul Hasan's side, Army side. . . . And the Army was not happy. . . . Then he took over."[45] Tikka, though senior to Gul Hasan had not resigned but had gone to the Multan area as corps commander, for with "India in occupation of the East" it was still an "Emergency" for Pakistan, as he so strongly testified before the army's commission investigating the Dacca debacle in late February. Then on 2 March he was invited to see President Bhutto at his house, and they sat on the lawn together as the sun was setting, "And he said, 'I believe you have given a very good statement to the Commission,' and he would like to have a copy of that . . . then we discussed the '71 War and so forth, but I knew nothing about what was going to happen to me," Tikka insisted, for Bhutto had not as yet tipped his hand. The next day he received a letter from Bhutto, appointing him chief of staff, and ordering him to return to Pindi at once. "I was driven to his office, and we had tea together," Tikka Khan recounted. The president wished him "best of luck," and explained there had been "difficulty" with the "previous C-in-C . . . people were not cooperating."

General Akbar, who had trained the Kashmir "liberators" before launching Operation Gibraltar in 1965 and was now Bhutto's minister of state for internal security, had "secured" three generals in his office. General Gul Hasan was being "kept in 'protective custody' by Mustafa Khar."[46] Bhutto feared a military coup at this time. The three generals held by iron-fisted Akbar were brought to Tikka Khan's office in Pindi next morning, and the tough new chief told them, "You have to cooperate with the civil government. You are under the civil government . . . and so-on-and-so-forth, and after that I just disposed them off! I knew this Army fairly well," Tikka explained, smiling tightly.[47] The next day Tikka addressed some 300 of his officers at GHQ and told them "what they should do"—much as he had told the troika of generals earlier. A police strike had started over a week before on the Frontier and moved into Punjab, and when the president ordered the army and air force to "put it down," Gul Hasan and Rahim Khan had "not cooperated." Tikka Khan, however, always obeyed orders; as did General Akbar.

"We are determined to have a new vigorous institution of the Armed Forces," Zulfi told his nation on 3 March:

> We are absolutely determined to have . . . invincible Armed Forces. . . . It must again become the finest fighting machine in Asia. This we must do. This is a sacred task. . . . [T]he structure has been changed and the heads of the three services are dedicated individuals. . . . [T]he people of Pakistan and the Armed Forces themselves are equally determined to wipe out Bonapartic influences from the Armed Forces. . . . Bonapartism is an expression which means that professional soldiers turn into professional politicians. So I do not use the word Bonapartism I use the word Bonapartic [a subtle distinction only a Bonapartist like Bhutto would bother to make] because what has happened in Pakistan since 1954 and more openly since 1958 is that some professional Generals turned to politics not as a profession but as a plunder. . . . [T]hese Bonapartic influences must be rooted out, in the interest of the country, in the interest of Pakistan of tomorrow, in the interest of the Armed Forces and the people of Pakistan.[48]

Zulfi gave much thought to military reforms, and had read as much as he found time for on how Hitler and Mussolini had dealt with their rebellious generals, as well as how Napoleon had accomplished all that he did. He had been close enough to Mirza and Ayub and Yahya and Peerzada to feel he knew the "mind" of the Pakistan military and its temptations and blind spots, its singular strengths and weaknesses. He understood, of course, that the army had brought him in, and that the army assumed it could always throw him out if he misbehaved too much or started to take extreme liberties, as it had thrown out Liaquat Ali Khan and Khwaja Nazimuddin and Suhrawardy and Mujib, all those once-popular politicians who had stepped over the line of acceptable leadership behavior, whether in negotiations with India over Kashmir, in their Bengali separatism, or in treacherous secessionism. The army was Pakistan's protective wolfhound, always on duty, powerful enough to keep any enemy at bay or to destroy its "master" if he forgot the proper password or feeding hour. What Zulfi focused on now was how to retrain or rearm his army guard so as to make it his ever-loyal instrument rather than a clear and potent threat.

Rafi Raza and other bright military procurement specialists, including General Imtiaz Ahmad, who served as military secretary, briefed President Bhutto extensively on how best to update Pakistan's outmoded armed forces, bringing them from the depths of tank-warfare defeat into the modern missile age. Light French antitank SS-10 missiles were so accurate and relatively easy to handle that they seemed a sound investment, especially since five thousand of them would cost barely one-sixth the price of two hundred tanks, yet provide equally effective defense. Swiss Mosquito-Cobra 4 missiles were deemed even better for defense purposes, and the SS-11 and Swedish Bantams were longer-range offensive missiles. Instead of purchasing more costly howitzers, moreover, Zulfi was advised to consider mortar bazookas, like those used in the Soviet army, which could best be employed against tanks in rugged terrain like Kashmir. With lighter equipment, a Soviet division was armed at from between one-fifth to one-third the average cost of a U.S. division. Redeye missiles were much cheaper than Swedish Bofors guns, but the Redeyes would not be available to Pakistan for at least another five years. French Mirage III E fighters were considered the best outside the United States, and could carry nuclear or conventional bombs, both of which were to become increasingly important for Pakistan's new air force. The only "essential vessels" for the navy would be minesweepers and minelayers. Destroyers were too much of a "luxury," and submarines, also very expensive, were too "vulnerable" to attack by U.S. antisubmarine weapons, like RAT ASROC, the Mk46. Norwegian torpedo boats might suffice, or British Sea-Cats. Greater use of missiles would "give us superiority over our immediate neighbours."[49]

In his speech to the nation on 3 March, Bhutto called the recent police strike "a mutiny and mutiny at a time when the Armed Forces are facing an enemy. . . . But we faced it . . . with your power my friends. . . . I congratulate you and salute the gallant and brave people of Lahore for having responded to the call of the Governor of the Punjab. . . . So . . . we want to close this chapter for all time."[50] He promised police reforms as swiftly as possible, and had begun to think of creating his own elite force—the infamous Federal Security Force (FSF) or what came to be called Bhutto's "Palace Guard"—which was to play a more important and expensive role in his remaining years at the top. The FSF would be started in 1973. With Tikka Khan as chief; however, Zulfi could now rely on the army to handle any strikes. He warned his own party enthusiasts as well as his angry opponents: "While the Government is ever ready and receptive to the people's grievances . . . it cannot . . . allow itself to be coerced and permit individuals to take the law into their own hands and to indulge in violence and incite people to commit murders. This kind of planned anarchy cannot be tolerated and the common man is truly getting fed up with it. . . . I am not threatening anyone. I am the last person to threaten our people. I am one of you."[51]

Bhutto hoped to "embark upon negotiations with India" soon, and with Mujibur Rahman as well. His "sincerest desire," he insisted, was to "live in peace, to bring an end to hostility," and to bring all the Pakistani prisoners of war home. But Indians, he warned, might want to "extract the last drop of blood out of Pakistan," which was why this "curse of the Martial Law" had to continue. "[W]e want to lift it as soon as possible . . . I am sure you will not think that I am lying to you. Why should I lie to you? I have never lied to you. I don't want to lie to you. I am telling you the truth."[52]

Soon after this Zulfi wrote a top-secret memo to Tikka Khan, "Our Military Aim," to chart the course of Pakistan's military needs and diplomacy. "In the event of another Indo-Pakistan war," he felt it more than likely that Afghanistan would attack because there had been fears that it might during the 1971 War. Pakistan, therefore, had to be ready for a "twin attack." Though "not a soldier," Bhutto wrote,

> I have read quite a bit on military matters and warfare. My eight years in Government from 1958 to 1966 have given me sufficient experience, especially when I was Foreign Minister. . . . Recent events have raised our

> problems to an all-encompassing national challenge. I have for this reason arrived at certain conclusions based on the state of the country. . . . We have always wanted peace—but peace with honour and freedom. India, on the other hand, at the very birth of our country, thrust upon us violence, tyranny and oppression. She occupied Hyderabad by force, annexed Junagadh unlawfully after its accession to us, and for twenty-three years denied to the Kashmiris the plebiscite pledged before the United Nations. In 1971, she committed unconcealed aggression against us . . . bisected our country. . . . The threat from India keeps mounting. . . . So neither a few years of arranged peace, nor the present situation can possibly permit us to ignore the reality that there must inevitably, sooner or later, come another war.
>
> We must be clear in our minds now, what we intend to achieve in this new war when it comes. . . . There will no longer be an absence of clear thinking . . . on our side. . . . [I]n a new war the enemies around us will be depending largely upon material superiority and not the human factor, upon technique and not the force of an ideal. In this can lie their weakness and our strength. . . . Recent history has amply proved, in China, in Korea, in Israel and in Vietnam that a people's war can withstand innumerable ups and downs but that no force can succeed in altering its general march towards inevitable triumph.
>
> India is now estimated to have got some 975,000 troops, some 1050 combat aircraft and about 1650 tanks. . . . [But] let us not forget the whole background and history of India compared to our own. In the remotest of our villages the humblest of our people possess a self-confidence, a ready willingness to march forward into India—a spirit the equivalent of which cannot be found on the other side. . . . Great and terrible scourges have come to India from this side . . . every invasion from this side has defeated India . . . Thus the terror, fear and habit of defeat cannot be wiped out of their national memory overnight. . . . And we ourselves have ruled them for eight centuries. All this is not ancient history. It is current history—and the Indian masses are just now still struggling against the legacy of superstition, religious intolerance, caste system, racial animosities, . . . poverty, backwardness . . . ignorance, deceit and the unconquerable habit of submission and servility.[53]

Bhutto's memorandum went on to detail the possible military strategy and tactics that might come into play in his envisioned people's war against India: a northern "Panipat offensive" against Delhi, or a southern "Somnath offensive," thrust, cutting off Kathiawar and Bombay. The village-based people's army would be a "massive wave of raiders" that would always be ready to carry out V. P. Menon's warning to his Indian government in 1947: "The raiders are a grave threat to the integrity of India. Ever since the time of Mahmud of Ghazni . . . Srinagar today, Delhi tomorrow! A nation that forgets its history and its geography does so at its peril."[54] Whatever Tikka Khan's reply to Bhutto's memo may have been, Pakistan was never to be "prepared" for the people's war Zulfi Bhutto visualized against India—at least not in the remaining seven years of his lifetime, nor during the decade and a half that followed his death.

While drafting his secret plan for all-out war, Zulfi continued to practice the fine art of diplomacy he had studied at Berkeley. In late March he flew to Moscow for two days, meeting Premier Kosygin again and signing a five-year treaty that reopened a vital channel of international trade that had been closed to Pakistan since the Bangladesh War. Indira Gandhi flew to Dacca at this time, signing a twenty-five-year treaty of friendship, cooperation, and peace with Mujib. She promised to turn over to Bangladesh all the Pakistani prisoners, including General Niazi. Mujib announced that there would be no discussions with Pakistan concerning the return of the more than ninety thousand prisoners until Pakistan fully recognized Bangladesh.

"Brave citizens of Lahore," President Bhutto shouted in Urdu on 19 March 1973 inside the newly renamed Qaddafi Stadium to an overflow crowd of raucous Punjabis. Many among them were angry People's Guards—led by tough Mian Iftekhar Ahmad Tari, a close friend of Governor Khar—who had fought the PPP workers led by Hanif Ramay, who earlier had been put in charge of organizing the meeting.

> You must listen to your President quietly . . . please restrain yourselves and allow me to proceed. . . . Lahore is my city. . . . When I make an important announcement, . . . I come here to seek your consent. . . . Yahya's coterie had no vision. They were politically blind. . . . They were bound to fail. As for power . . . I wanted power from the people for the people, power for the worker and the peasant and the students and intellectuals. . . . Pakistan has a brave army. India must know that Pakistan has not been defeated. . . . It is only a . . . selfish, corrupt and dishonest coterie that has been defeated. . . . I had warned Yahya against war. . . . Then I declared, 'We will run Pakistan, we will save it. . . . In three months we have managed to salvage it. . . . We aim to do still more. . . . Five thousand years of exploitation cannot be remedied in a day. . . . In a year's time, *Insha Allah* [God willing], I shall be able to feed and clothe my people.[55]

By that time Bhutto had either imprisoned or retired many of Pakistan's bureaucratic "Brahmins," members of the vaunted Civil Service of Pakistan (CSP), the highest echelons of the administrative framework that had held Pakistan together from its inception, managing most of the ministries and departments of state. "People say that 2,000 Government servants have been retired," Zulfi told his by-now-cheering audience in Lahore. "But they were sucking the people's blood. They were parasites. There was a clamour for a purge. But when it took place, there have been complaints. We are in the midst of a revolution. . . . We must root out corruption. . . . Poor people will get justice. . . . Mao Tse-Tung said that sometimes the wind comes from the West, sometimes from the East. In Pakistan the wind blows from all sides."[56]

In early March 1972 Zulfi sought to convince the leaders of the North-West Frontier's dominant National Awami Party (NAP) and of Baluchistan's Jamiatul-Ulemi-i-Islam (JUI) to participate in the National Assembly for just a few day's session in April—to allow him to extend the state of martial law until 14 August. Bhutto was eager to negotiate a settlement over Bangladesh with India but did not want to embark on such discussions until he could claim to speak as head of a democratically elected and representative government, not just as the appointed civil martial-law successor to Yahya Khan. He managed to convince a skeptical Wali Khan, the secular-socialist head of NAP, of the cogency of his argument and the need for one "last" extension of martial law, even as he did Islamic fundamentalist Maulana Mufti Mahmood, the conservative head of JUI. Zulfi had appointed his strongest Frontier supporter, Hayat Mohammad Khan Sherpao, as governor of the NWFP, but to win over Wali Khan and Mufti Mahmud, had agreed to let them join forces in selecting new governors for the NWFP and Baluchistan after the assembly met. Then on the eve of convening the first brief session, he courted the Frontier's oldest and most conservative leader, Khan Abdul Qaiyum Khan—who was soon to become his home minister—forming a Frontier alliance between the PPP-and Qaiyum Khan's Muslim League (QML). When the National Assembly was convened in Islamabad in mid-April, Bhutto had pledges from 101 members out of a total of 142 in his pocket. His election as president of the assembly was thereby assured. It proved easy for him to push through his party's interim constitution, reinforcing many of the martial law powers and special privileges he had enjoyed—more than Ayub Khan had ever had.

Astute politician and statesman that he was, Zulfi soon saw that he really had no more need of his martial law crutches, which only weakened him morally, for his idealistic young PPP followers (as well as all opposition leaders) kept asking, "Why?" The legality of continuing martial law had been challenged in the courts, moreover, and was at this time pending before the Supreme Court of Pakistan. On the evening of 14 April as the Assembly met, a chorus of young voices kept shouting outside the handsome new Islamabad building, "Down with martial law! Bring back the rule of law!" Instead of ordering the arrest of the protesters, Bhutto ended his speech that night by dramatically announcing, much to the delight of the representatives, that martial law would be lifted on 21 April: "This demonstrates my commitment to democracy and my infinite faith in the people. We are the harbingers of a new order pulsating with pragmatic idealism and in tune with the symphony of the Third World."[57]

A week later President Bhutto took his oath of office under the interim Constitution at a lavish ceremony on the lawn of the President's House in Rawalpindi. "I wish to assure the leaders of the various parties who have assembled here that, if need be, I shall travel from Karachi to Khyber to strengthen national unity and to bring about greater co-operation among us."[58] To critics who had questioned "the need for this pomp and show and extravagant spending," Zulfi responded, "This is a people's gathering. There is no greater strength than the people. . . . It is the people's day, the poor man's day. . . . [T]he beginning of an era of democracy."

Under the interim constitution the provinces would enjoy "responsible" government led by chief ministers. The former governor of Sind, Mumtaz Bhutto, was made its chief minister and Mir Rasool Khan Talpur was appointed governor, a figurehead now, unlike the president, who retained supreme powers to appoint or remove such officeholders. "There should be no political vendettas," Zulfi wrote Talpur, in explaining the governor's obligations. "You must never forget that the Party belongs to the working classes . . . students and the down-trodden. Their interests come first. Try and improve their lot, try and make them feel happier, give them a sense of security . . . Corruption must be weeded out . . . Crime must be rooted out . . . Give particular attention to town planning, to health and cleanliness, to the inculcation of discipline, to the promotion of culture . . . The curse of Sind is to be found in the worst form of reactionaryism. . . . [D]o not give the slightest encouragement to reactionary forces . . . The President's House at Karachi has been transferred by me to the Government of Sind on certain specific conditions. I will not spell them out as these conditions are known to the present incumbent [Mumtaz]. However, please see to it that the wing reserved for . . . the Head of State [is] not used by anybody else. That should be kept clean and exclusively reserved."[59]

Zulfi and a lady "friend" often stayed overnight at the various Governor's mansions, and he was keen to ensure that the presidential wings, at least in Karachi's and Lahore's Government Houses, were kept spotless and always discreetly ready for him to use without prior notice.

A few days after writing Mir Talpur, he penned a similar letter to Mumtaz, in appointing him chief minister. "You have Almighty Allah to thank for the great honour bestowed on you. . . . The people expect much from you and . . . You should not fail them. Concentrate on the problems of the poor, remove the sub-human conditions. . . . Give water to places where its blessings have not yet reached. Electrify the Province. Make it a truly beautiful Province. Sind can be the California of Pakistan. . . . Karachi was once the pride of undivided India. Now it has become dirty and ugly. Put clothes on her and make her a beauty. . . . [T]o begin with, why don't you toy with the idea of making the stretch from the Airport to the intersection of Hotel Metropole into a model and modern European promenade, to impress all who land on the soil of Pakistan."[60] Zulfi hoped to transform Karachi into a prewar Beirut or Monte Carlo, but the now-abandoned concrete shell of the casino-hotel started shortly before his fall is a grim warning to future rulers of Pakistan of the danger of trying to erect a pleasure palace on the sands of the onetime capital of the "Land of the Pure."

Oriana Fallaci's interview with Indira Gandhi, who characterized Bhutto as "not a very balanced man," had just been published in Rome.[61] When Zulfi got wind of it, he immediately invited Fallaci to Pakistan. He spoke of Mujib In the subsequent widely read "Interview with History" as "mad, mad!" Reflecting on Yahya Khan, who was still under house arrest, Zulfi remarked, "What can you say of a leader who starts drinking as soon as he wakes up and doesn't stop until he goes to bed? . . . He was really Jack the Ripper. . . . The defeat we suffered is his . . . Yahya Khan and his gang of illiterate psychopaths."[62] He called Indira Gandhi "a mediocre woman with a mediocre intelligence . . . a diligent drudge . . . a woman devoid of initiative and imagination," but nonetheless he was ready to "meet her when and where she likes." That June they met at what had been the summer capital of the British Raj and is now the capital of India's Himachal Pradesh, Simla.

"We are going to India in circumstances which are but a part of the tragic legacy we inherited," President Bhutto informed his nation on the eve of his departure for Simla. "The war we have lost was not of my making. I had warned against it. But my warnings fell on the deaf ears of a power-drunk Junta. . . . Some disgruntled men of yesterday have the temerity to question why we are going to India. . . . Do they think that we should not seek the return of our prisoners? Do they suggest that we should allow the Indians to continue their occupation of two tehsils [sub-districts] in Sind and one in the Punjab?"[63]

President Bhutto flew east from Lahore on 21 June 1972, with his entourage of ninety-two officials, journalists, and friends, including his daughter Benazir. "Everyone will be looking for signs of how the meetings are progressing, so be extra careful," Zulfi cautioned Benazir on the plane that took them to Chandigarh. "You must not smile and give the impression that you are enjoying yourself while our soldiers are still in Indian prisoner-of-war camps. You must not look grim, either, which people can interpret as a sign of pessimism. . . . Don't look sad and don't look happy."[64]

Prime Minister Gandhi was waiting at Simla's eagle-perch airport as he stepped out of the Russian M-8 helicopter that had lifted them there from Punjab's capital. Despite what they thought and had said of each other, the two most powerful, popular South Asian leaders shook hands and managed to smile at this informal start of their summit.

Piloo Mody and his wife, Vina, had driven to Simla's hill station from Delhi the day. Soon after the Bhuttos reached Himachal Bhavan, the Governor's House at which they stayed during the summit, the Modys were brought over, and they chatted with Zulfi and Benazir for more than two hours, "recalling old days," Piloo remembered. "The next morning, as no Summit meeting was scheduled, I received a message to say that the President expected us at 11 o'clock . . . and would like us to stay over for lunch, which did not finish till 3.30 P.M."[65] Though Piloo recognized the pitfalls in Zulfi's "complex personality," as he put it, knowing how much of a "demagogue" his friend had become in the past few years, he still loved and admired him enough to hope that the summit might bring peace to South Asia.

"Bhutto, let us make no mistake, is above all a realist—although he may play his cards like a poker player—a great bluffer when he holds a poor hand, but always ready to arrive at mutually acceptable solutions if his opponent is big enough or shrewd enough not to push him into a corner," Piloo wrote of his oldest friend. "People who judge him only by his demagoguery and his occasional intemperate utterances fail to take into account how often he has tempered his demands or watered down his terms." Piloo's formula for breaking the summit deadlock, which seemed to have set in by the fourth day, was to lock Indira and Zulfi up in a small room at Simla and throw the key away "till the smoke came out of the chimney," as in papal elections.[66]

Piloo may have suggested that formula to Zulfi, for on the last day of the summit, the president hosted a dinner, and as soon as it ended left the table with Mrs. Gandhi. They went off alone into a small sitting room, and their attendant teams went into the larger billiard room on one side and the reception room on the other. "The two leaders would meet and then go back for further consultations and then meet again," Piloo recalled, noticing once when the billiard room door was left ajar, Mrs. Gandhi leaning over the green table "scratching away, obviously at the draft treaty."[67] Zulfi did not use the papal-election signal but devised a code of his own with Benazir, who kept coming down from her upstairs bedroom to ask if anything had happened. "If there is an agreement, we'll say a boy has been born. If there is no agreement, we'll say a girl has been born."[68] The Radcliffe-educated Benazir considered the equations "chauvinistic," but shortly before 1:00 A.M. the affirmative mantra rang out in Urdu from the floor below, "*Larka hai!*" (It's a boy!).

The agreement signed at Simla on 2 July 1972 stated that India and Pakistan had "resolved to settle their differences by peaceful means through bilateral negotiations or by any other peaceful means mutually agreed upon between them." Within thirty days of ratification by both nations, Indian troops were to be withdrawn from the roughly five thousand square miles of Pakistani desert soil they still occupied, and each nation agreed to "respect the line of control" in Kashmir resulting from the cease-fire of 17 December 1971. India's "final solution" of the Kashmir dispute appeared to be the permanent partition of that state along the new "border," to be "respected" bilaterally. No mention was made of a Kashmir "plebiscite" or of the ninety-three thousand Pakistani prisoners of war. Some Pakistanis labeled the agreement a "sellout" worse than Tashkent, insisting that it included a "secret agreement" by Bhutto to end Pakistan's claim to Kashmir, which reflected the new reality of Pakistan's much diminished size and status in a South Asia dominated by India.

"The Agreement that has been arrived at with India can be accepted or rejected by the National Assembly of Pakistan which represents you," Zulfi told the crowd of Pakistanis, who welcomed him at Lahore's airport on 3 July. "A people's Government can never agree to a Tashkent. Even our worst enemy cannot claim that this Agreement smells of another Tashkent," Zulfi insisted. "I had promised you that there would be no secret agreement. Now you can see that there has been no secret agreement. I tell you as a Muslim and I swear an oath—I swear in the presence of Almighty Allah that there has been no secret agreement. . . . On the vital question of Kashmir . . . we have made no compromise. We told them categorically that the people of Kashmir must exercise their right of self-determination. This was a question which could be decided only by the people of Kashmir. Neither Pakistan nor India had any say in this matter."[69]

What Bhutto had, in fact, agreed to at Simla was, to his mind, irrelevant as far as Kashmir was concerned: he had not accepted Indira Gandhi's position, no matter what was written on the piece of paper he may have signed. He was realist enough to understand how weak Pakistan was, how tiny, how lacking in power—for the moment. He never doubted that Pakistan would rise again some day to reclaim Kashmir, as he always believed it would do, for Kashmir was a Muslim state, after all, and thus "belonged," by definition, to Pakistan. No agreement could change that "reality," which was firmly rooted in the labyrinth of Zulfi Bhutto's brain. He had enjoyed the week's holiday in Simla, taking along the largest entourage any diplomat of South Asia ever took anywhere for an official meeting. And he had seen much of Benazir, eaten and drunk with Piloo, and breathed the bracing air British viceroys once had breathed. But he never for a moment meant to close off the Pakistani claim to Kashmir. He had needed the agreement primarily to prove to the rest of the world—doubting London, as well as skeptical Washington and Moscow—that Pakistan remained in the "great game," that its president was a shrewd diplomat cut from the same cloth as Talleyrand—for few were ready to believe that he was closer to Napoleon.

"Before my departure for Simla I had told you my task would not be an easy one. It was a very tough meeting," Zulfi told his audience at Lahore.

> They were negotiating from a position of strength because they had so many cards in their hands. But what did we have? . . . Apart from principles, we had our people with us. . . . In addition, I had the mercy and beneficence of Almighty Allah with me. . . . I talked to them [the Indians] firmly. . . . Finally, they agreed and said that on the question of Kashmir you may stick to your principles and we shall hold further discussions. . . . Now, as for the question of a "no-war-pact," you have seen that there is no such thing. What has been said in the Agreement is that we should avoid war. However, we shall continue our maximum efforts to uphold our righteous cause . . . based on the principles of justice.[70]

It was, in some ways, easier for Zulfi to say and mean one thing to one party and its opposite elsewhere, for he was many Bhuttos. He could be the suave diplomat, a sophisticated realist ready to concede whatever was necessary in his dealings with Mrs. Gandhi and Swaran Singh (and Piloo) in the rarified atmosphere of Simla's viceregal palaces, signing his name to a formal document as the cameras rolled, agreeing that "the basic issues and causes of conflict which have bedevilled the relations between the two countries for the last 25 years shall be resolved by peaceful means." Back at the Lahore airport, however, it was just as natural for Bhutto to swear and affirm by Almighty Allah that he had given away nothing, for he was among his people and they understood that their Quaid-i-Awam would never surrender Kashmir to a Hindu woman. They knew Zulfikar Ali Bhutto, almost as well as he knew himself—at least in this regard, when it came to matters of *Izzat,* of Pakistan's sacred soil, of divinely ordained justice. Only benighted Hindus could believe that any true Pakistani Muslim leader worth his salt would surrender Kashmir or agree to a permanent "no-war pact" with them! The war would go on for a thousand years, as it had for the past thousand and more, since the first troop of Arab cavalry, led by Muhammad ibn Kasim, galloped across Sind and brought *kafir* Hindus and Buddhists to their knees, offering them only Islam or death.

Zulfi flew the next day to Pindi airport, where another large crowd of admirers awaited him and listened to essentially the same message.[71] Once again the response was a rousing chorus of approval from people who viewed him not only as Pakistan's wise leader but as its heroic savior.

Though nothing was said in the Simla Agreement about Pakistan's prisoners of war because India had refused to discuss them without Bangladesh at the conference table, and Pakistan had yet to recognize Bangladesh, Zulfi promised to "negotiate further" on the matter when next he met with Indira Gandhi and with Mujib. He had at least won the promise of full withdrawal from the portions of Sind and Punjab still under Indian occupation, and vowed now that "Insha Allah we will succeed in finding a solution" to the POW problem soon. "We will remain friendly and peaceful towards those who do not commit aggression against us, do not occupy our territory and do not hold our POWs," Bhutto informed his cheering airport audience.

Ten days later he rose to address the National Assembly in Islamabad, ending its long debate over the Simla Agreement, during which he was accused, among other things, of having "betrayed" the people of Pakistan and every "pledge" he had made prior to his election. "If we have 'betrayed' every pledge that we have made, let the people of Pakistan determine it," Zulfi lashed back. "I believe, Sir, that we have unleashed a great revolution in Pakistan. I believe, Sir, that we have opened the floodgates of a golden bridge. I believe, Sir, that we have destroyed exploitation."[72] He trenchantly attacked all his critics, those within as well as outside his party, but then diplomatically appealed to them to "have some consideration. . . . It is not the future of my personal position that is at stake. At stake is Pakistan . . . the whole subcontinent. . . . The life and future and fortune . . . of seven hundred million people is at stake." He felt especially hurt by attacks from members of his party. He asked them to discuss their "grievances" with him "at some other place. . . . This is not a venue for the people to have a 'Kutchery'. . . . If in the National Assembly of Pakistan we are to take such an attitude I tell you quite frankly, I tell all friends on both sides of the House, that no individual can cope with the situation . . . then no person can save Pakistan."

He was quite right, of course. Yet how ironic for Zulfi Bhutto to voice that opinion in this highest of all elective assemblies in his country. "From the first day I took over I said that one person cannot make a country, but one person can destroy it. It is easy to destroy." Ayub Khan would have been surprised, had he been there, to find himself in total agreement with Zulfi Bhutto. Zulfi spoke for more than three hours that night because he felt compelled to answer virtually every disapproving word uttered in the four preceding days of debate. He was acutely sensitive to criticism but also loved political debate, relishing rhetoric and the tricks of logic or sarcasm. Few of his opponents were his match. "Do you think that Zulfikar Ali Bhutto of Larkana is responsible for the separation of East Pakistan?" he challenged them. "That this individual who was not even in the politics of Pakistan till 1954 . . . is the person who is finally, exclusively and completely responsible for the separation of Pakistan? If you want that, well, I say I can take this burden on myself. . . . I did it. Yahya Khan did not. Ayub Khan did not. The policy of exploitation did not. One thousand miles of separation did not. The fact that East Pakistan was in a majority and spoke a different language did not. . . . It is all my fault."[73] There was no responsive laughter, and few shook their heads or protested his claim to the blame, so he quickly added: "The crisis was in our stars. . . . It was boiling . . . nobody could stop it in 1971. It could have been stopped in 1950 . . . over the language issue. . . . I was nowhere on the scene. . . . Let us face the truth. If you want to hang me, hang me by all means, but the fact remains that I am not responsible for the separation."[74]

Did he intuit then that he was destined to hang? "Somehow he had an intuition that he didn't have enough time," Husna recalled. "He said, 'Either I'll die or I'll be killed . . . he was obssessed with it. . . . At times he used to say, 'I feel like giving up everything and going away.' . . . He was a man of intuition, but he always talked about death."[75]

In defending the Simla Agreement, Zulfi argued that there were only two other possible courses: the first was immediate war with India; the second, "to go under Indian tutelage" for all time. Neither was possible under his leadership. "My mandate is to build Pakistan. I will build Pakistan. . . . It is for the future generations to decide whether they want to make it a progressive, prosperous and happy Pakistan. It is for them to decide if they will go to war or to peace. . . . I cannot go to war. Not in the next 5, 10 or 15 years." He significantly added:

> We have to release great energies and we have also to unleash a great force. . . . I am not ashamed of what I have done to the people of Pakistan. I am proud of having galvanised them. I am proud that there is now a sense of dignity in the common man. I am proud that the *Hari* can tell his *Zamindar* to go to hell and that he wants his rights. If there is chaos in the wake it is productive chaos not negative chaos. . . . There will be problems but we must face them. . . . [w]e have let the giant sleep too long. . . . What we have done is to set a great movement of great ideals into motion. . . . I can mobilise the energies of the people of Pakistan and make them into a great, progressive and modern State. . . . Let the people of Pakistan then decide what is going to be their future with India.[76]

With Simla behind him, Bhutto could focus more of his time and thought on how to release the "great energies," unleash the "great force," part of which would be nuclear power. Zulfi's opponents were wrong in charging him with signing "secret agreements" with India; he continued openly to insist that "we believe the dispute of Jammu and Kashmir exists," and that it could be peacefully resolved only through acceptance of the "principle of self-determination. . . . There is nothing in the Simla document which says that we shall not go to the United Nations. It is true the emphasis is on bilateral efforts but it does not say that we cannot go to the United Nations."[77]

Bhutto's major secret agreements, however, were those he reached with Libya and China. And he continued covertly but assiduously to pursue dealings with France for raw materials and advanced technology. Pakistan required those French resources to build enough nuclear weapons to overcome the "temporary" military setbacks suffered in 1965 and 1971. It could then defeat India in the next and, it was hoped, "final" round in the South Asian war.

In August 1972, shortly after returning from Simla, President Bhutto ordered his three new service chiefs of staff to "put right" the "basic" weaknesses in Pakistan's armed forces: "We have spent many billions on defence with the result that this country is bankrupt and mortgaged to foreign debt . . . [but] we were unable to fight for more than 17 days in 1965 and for about 14 days in 1971." In a "top secret" memo, Bhutto noted that although Pakistan was always "threatened by aggression from India," its military leaders "do not seem to learn from mistakes. . . . In 1965 Ayub Khan did not think that India would cross the international border. . . . [L]ater he threw the blame on the Foreign Office. In 1971 Yahya Khan was convinced that India would not attack East Pakistan. . . . Now I want . . . everything to be organised systematically and scientifically. There must be no let-up. . . . We should start building air raid shelters in the right places. . . . There should be an underground Operations Room; there should be an underground arrangement for the Cabinet to meet. . . . Our aeroplanes should be better protected . . . Oil depots should be tucked away safely in the hills or underground. We should have an alternative oil pipeline from Iran to Pakistan. The camouflage of important buildings . . . should be of a permanent nature."[78]

Zulfi had told the National Assembly in defending the Simla Agreement "If you want the people of Kashmir to secure the right of self-determination they must fight for it. . . . [S]elf-determination cannot be achieved by proxy. . . . [I]f tomorrow the people of Kashmir start a freedom movement . . . we will be with them. We have not compromised anything. . . . We will fight if we want to fight. . . . This is an eternal position."[79]

By now all opposition had melted before the fire of his eloquence. The patriotic fervor of his message quickened each Pakistani pulse, awakening dreams all but abandoned.

> Sir, much mockery has been made of the policy of confrontation and a 1000-year war. . . . But it is a fact that the people of Pakistan accepted this concept. . . . What does a 1000-year war mean? . . . It means that the nation will never surrender its rights till eternity. . . . We have lived a thousand years in the subcontinent. A thousand years are nothing in the future. . . . [I]f the people of Pakistan . . . say we will fight for our rights till eternity, then [we] will fight. It may be a thousand years, it may be ten thousand years.

Cheers of acclamation arose. The agreement signed in Simla carried by an almost unanimous vote. All Bhutto stood for and said, and dreamed of and meant, echoed the deepest hopes of the Muslims of Sind and Punjab, and of the Frontier as well. Though the leaders who ruled in Baluchistan and the North-West Frontier were not of his People's Party, still their hearts beat in tune with his and their equally hot blood surged as his voice reached its febrile pitch of power.

> A defeated man cannot understand the meaning of a thousand year long struggle. Zulfikar Ali Bhutto has not been a traitor to the cause of Pakistan's struggle. . . . Victory and defeat are transitory events in the life of a nation. I can look down the vista of years and see the vision of victory after a 1000 years. . . . What can we do if we did not have men enough, if we had chicken usurpers, avaricious and corrupt individuals ruling Pakistan? Could not we in 1962 confront India and take Kashmir? But at that time the Field Marshal was hiding in Hunza. . . . Why do I admire Napoleon?—because Napoleon was the last professional soldier who went to war as a matter of policy. Professional soldiers do not go to war. It is only civilian leaders who go to war.[80]

At the end of his first six months in power, President Bhutto had managed to win back by diplomacy in Simla the tracts of Sind and Punjab lost to India. And he managed to buy precious time to rebuild his army and his people's spirits, with his skill at drafting ambiguous documents while professing faith in "eternal principles." He carried the often-rebellious National Assembly with him, moreover, by much the same force and mesmerizing blend of oratory and conviction that had won him such popular electoral support in 1970. He had proved more agile and resourceful, indeed, more brilliant, than his stoutest supporters had expected him to be. He surprised most of his party stalwarts first by retaining the cover of martial law while it was useful, then by discarding it with such alacrity once it became a political liability. He was never squeamish about shifting a position or stand, and his timing was almost the match of the ease with which he could debate both sides of any question. Yet even his genius at debate and his deft political maneuvering could not indefinitely hold at bay mounting provincial and socioeconomic pressures, frustrations, and discontent that kept Pakistan's most crowded urban centers and depressed rural backwaters boiling with ethnic and linguistic violence. The army proved only a temporary deterrent to the violence, which challenged Zulfi Bhutto's claims to political success for the rest of his last turbulent half decade in power.

10

Provincial Problems Proliferate

(mid-1972–early 1973)

By mid-1972 much of Sind, especially the poorest, overcrowded sections of Karachi, Hyderabad, and Larkana, were racked with deadly riots between Urdu-speaking *Muhajirs* (refugees) and their indigenous Sindhi-speaking neighbors. The riots posed a mortal challenge to President Bhutto's political powers and skills to resolve Pakistan's perennial ethno-linguistic-provincial problems. Sind was not the only province plagued with recurring ethno-linguistic violence, but it was the home base of Bhutto's power, and if Zulfi, with his "brother" Mumtaz as chief minister, could not control Sind, then what hope was there for his rule to remain effective over any part of Pakistan?

The roots of Sindhi separatism extended even deeper than those of the Bengali separatism that had so recently given birth to Bangladesh. Leaders of the "Sindhi Desh" movement, like Dr. G. M. Syed, traced the source of their claims to Sindhi "nationhood" as far back as Mohenjo-daro in the Third Millennium B.C. Long before the advent of Islam, at any rate, Sind was a kingdom ruled by Raja Dahir, who spoke Sindhi and issued royal decrees in that ancient language, and whose poetry and literature remain a most precious cultural legacy to his millions of Sindhi heirs. Urdu, however, the language of the Mughal "Army Camp" around Delhi, was chosen by Quaid-i-Azam Jinnah, rather than his native Sindhi, to serve as the national language of Pakistan, though in 1947 Urdu was as yet spoken by barely 10 percent of the entire nation's population. Most Muslim refugees from Delhi and Uttar Pradesh who fled to West Pakistan after its birth settled either in Karachi, the new capital, or elsewhere in Sind and were native Urdu-speakers. Many Urdu words were borrowed from Persian as well as Arabic, enriching Urdu's lexicon with teachings of Islam as well as the sublime poetry of Persia. Like most refugees, Pakistan's Muhajirs were generally hardworking, frugal people eager to recoup their tragic Partition losses, often finding more fecund fields for their talents and ambitions in their Islamic fatherland than they had in Mother India, whence they had fled. By 1972 some 30 percent of Sind's population was Muhajir, with almost 90 percent of bustling Karachi consisting of Urdu-speaking refugees. Throughout Ayub Khan's era of military rule, as well as during the briefer Yahya Khan interlude, Urdu remained the single language of required instruction in all West Pakistan schools, and the language of administration as well as justice. Now, however, Sind's government had introduced a new language bill that would bring instruction in Sindhi back to all public schools, even as Sindhi had already returned to regular daily use in Karachi's Government House and in its People's Party-dominated Assembly.

Leaders of Urdu-favoring religious parties, like Maulana Abul Ala Maudoodi's Jamaat-i-Islami, warned President Bhutto that Sind was "in flames," adding that "Your Governor [Talpur] and the Chief Minister [Mumtaz Bhutto]" were "held responsible by all quarters." Bhutto was urged to act swiftly to avert a disaster even worse than the loss of Bangladesh had been. "I am deeply grieved," Zulfi told his National Assembly that mid-July. "I have not struggled and toiled for this great country to make brother fight against brother, to engineer a situation in which the damage is irremediable."[1] Neither "Muhajirs" nor "Sindhis" were being killed daily in the streets of Karachi, he reminded his countrymen, but *Pakistanis.* However, Sind and Frontier provincialism, and feudal tribalism remained forces as strong as, if not stronger than, the sense of national Islamic brotherhood that had created Pakistan, that had brought speakers of diverse languages and ethnic origins within a single nation's borders a quarter century ago.

The language bill would not "bury" Urdu, Zulfi insisted, because Urdu would remain Pakistan's national language; it would only elevate Sindhi to official-provincial-language status. That Punjab, the North-West Frontier, and Baluchistan accepted Urdu as their official provincial as well as national language was irrelevant, Bhutto argued; none of those provinces had ever had a language whose literature was as old and rich as Sindhi. Yahya Khan had called him "a provincialist," but Yahya himself "could not read Sindhi. Sir, there are names in Sindhi which you cannot write in any other language." Teaching Sindhi to every child in Sind was a "legitimate aspiration." Bhutto angrily denounced those who had put up posters saying "We want a Pakistani President and not a Sindhi President." "Have I usurped the President's chair?" he asked. "Have I come here by a military *coup d'état* to be the President of Pakistan?"[2]

There were no shouts of "No!" or "Jiye Bhutto!" to reassure him inside the sorely divided assembly, so he became more strident: "You can throw me out of the chair of the President of Pakistan . . . I am not a coward. Do you think I am going to be intimidated by these demonstrations? . . . [T]here is no cause for this agitation. . . . They must have something to exploit the poor people, the poor Urdu-speaking people."[3] *They* were stirring this Muhajir-Urdu revolt against him because he was named Bhutto.

> How can I assure you that I am a Pakistani? What else can I do to show that I am a Pakistani? What more do you want me to do? From where [do] you want me to get a certificate that I am a Pakistani—from God? From the age of 15 . . . I have given my blood and sweat for this country. . . . How else can I show that I am a Pakistani? . . . By killing all the people in

> Sind? By obliterating the culture of Sind? . . . We have given our lands; we have given our homes; we have given our lives . . . to people from all parts, to the Pathans, Punjabis, to the Muhajirs living in Sind. . . . What else can we do to show our loyalty, our love and our respect for Pakistan and for our Muhajir brothers?[4]

But many who heard him remained skeptical, suspecting this "Raja of Larkana"—whom some now called "Raja Dahir"—of special pleading on behalf of his provincial constituency. Punjabi popular folk "wisdom" still holds that "if you meet a Sindhi and a snake on the road, and you have only one bullet in your gun, shoot the Sindhi!"[5]

The Teaching, Promotion, and Use of Sindhi Language Bill, which had been enacted by Sind's provincial legislature in early July 1972, was signed by Governor Talpur and soon operationalized. Sindhi regained its ancient place of cultural distinction and official support. Thereafter it was harder for Bhutto to convince many Punjabis and Pathans that despite his international training and national experience he was not a "Sindhi-firster" after all.

Blood continued to be spilled daily in the streets of Karachi, and Zulfi took to the airwaves in vain. "Fellow citizens," he remonstrated, "have we struggled these past so many years for the restoration of democracy to seek recourse now in street-fighting and fratricidal killing? Have we struggled for democracy to see it thus disgraced? . . . I say no, I have made every effort, strained every nerve, to see a flowering of our democratic institutions. I shall not be an idle spectator witnessing their destruction."[6]

But Sind went on burning. From Lyari to Larkana, from Jacobabad to Latifabad, from Hala to Sanghar, Sindhis fought Muhajirs. As the daily death toll mounted, Bhutto broadcast nightly urgent appeals. "We have much to lose," he cried on 16 July 1972. "When I hear these words, these harrowing words like 'Revenge' or 'Shaheed Minar' [Martyr's Monument], I shudder. I did not come to office to build 'Shaheed Minars'. I came to office to build a monument of peace, a monument of friendship, of brotherhood, not to ignite the forces of bitterness, of extremism, so please believe in me."[7] Only fewer believed now. Instead of streets lined with cheering multitudes on his every visit to Karachi, Zulfi saw only lines of solemn police behind barricades or burned-out vehicles, broken bottles, brickbats, and rubble. Dismayed, saddened, and disappointed, Zulfi was frustrated by his provincial government's impotence in the face of growing turmoil. "This is not my individual . . . popularity which is at stake; it is a whole future we have to build. . . . You have been unfair to me. . . . I have tried hard, tried very hard to find a solution. . . . You can see the results. The cancer spread and spread and spread. Have you ever considered how many decisions I have to take? . . . [The] constitution, Martial Law, democracy, autonomy, Kashmir, the Biharis, India, Bangladesh. . . . All these problems have fallen into my lap. I do not want to run away from these problems. I can never run away from them. I am not the one to run away."[8]

"At times he used to say, 'I feel like giving up everything and going away,'" Husna recalled. "But I used to say, 'You'll never be happy, because your life is

Pakistan, and I know that after a while you'll be hungry for it and want to come back. . . .' Unfortunately, he wouldn't delegate, because he wanted to do everything perfectly, himself. And I used to say, 'God forbid, suppose you collapse, what would happen to the Party?' "[9] Rafi Raza cautioned him in much the same way, especially when awakened at three or four in the morning by his president's manic voice, questioning him, as Zulfi did many of his close advisers in the middle of the night, about some trivial matter. But Zulfi rarely rested, except for respites at Husna's hideaway in Karachi when they watched an old movie or two. "He loved Ava Gardner . . . Elizabeth Taylor . . . Raquel Welch. . . . He loved *Casablanca*. He liked Ingrid Bergman also," Husna remembered. And he identified with Bogart's role in that film classic, which they watched often. Perhaps Zulfi felt that Karachi was becoming a bit like Casablanca had been during World War II: both a haven and a trap, a place many refugees fled to, but from which few escaped.

The level of Muhajir frustration and anger against Bhutto and his party could be gauged from the number of voices now calling upon the army to bring back Ayub Khan. Zulfi termed it a "conspiracy"; Ayub Khan had "sucked the blood of his country." More Pakistanis, particularly in Sind, now feared for their lives, their property, their progeny, however, than feared Ayub or the army. Bhutto and his party had promised them a safer, fairer, happier world, but precious little had changed for the better. Economic indicators kept falling, production dropping steeply by midsummer of the first year of what Bhutto had promised would be a people's revolution. He appealed to Karachi workers for "more time" and again promised not to fail them. To Pakistan People's Party workers, he laid the fault at the feet of "the agents of big business and capitalism," the "anti-people forces." He blamed them for the arson that had reduced no fewer than 120 PPP offices in Karachi to rubble overnight. "But let them burn our offices," Zulfi shouted. "We have installed the PPP offices in the hearts of the people."[10]

Zulfi Bhutto's prestige was also charred. Many Pakistanis had believed his rhetoric and ascribed to him superhuman powers. He would do away with suffering and misery, bring instant prosperity to what remained of Pakistan. The prestigious Civil Service of Pakistan (CSP)—whose British-trained members still ran the bureaucracy, though possibly feeling less secure than they had before Bhutto took over—knew better, of course, but there were only some 350 of them at the top of a government work force that numbered more than a million. The Roedad Khans and Ghulam Ishaq Khans of the CSP watched quietly, waiting for this political hurricane to blow itself out.

To the seniormost officials of Sind that July 1972, President Bhutto said, "I want you to be happy. I also want you to reconcile yourselves to our existence . . . we have to come to an understanding, establish an equation, a working equation. There should be no ill-will. We all have a job to do. We will take the decisions and you as civil servants will implement them. You must understand that the decisions which we make are not personal decisions. We have no scores to settle."[11] Those officials of Sind understood *Siasat,* of course. "Let us relax," he told them, smiling.

> Let us all work together for a common purpose. I have told the Ministers that they must hear you, must seek your advice. . . . But once a decision has been taken and even if you have been overruled, you should implement it without reservations. . . . You are serving in Sind. It is a great opportunity of service . . . a very rich province. It has everything. . . . Let us today decide to clear the slums, both mental and physical.[12]

A month later Zulfi complained to Mumtaz about how "filthy" Karachi was, how "shocking" the profusion of disease-breeding flies. "I thought we were going to make a determined and resolute bid to clean the metropolis of Karachi. . . . I am also told that a little bit of rain in Karachi has again dislocated the traffic on account of the damage to the roads. . . . Is this not a great pity?"[13] Chief Minister Mumtaz Bhutto pleaded severe shortage of funds; there was no money for new urban beautification projects, barely enough to pay those already employed. But Zulfi hoped for a "first-class" police force to cope with the city's riots, fearing it would otherwise be "necessary to call upon the armed forces to intervene."[14] To call out the army would be a public admission of failure, yet not doing so would leave Karachi and other cities of Sind, including Larkana, open hunting grounds for every variety of criminal and prove equally catastrophic to his party's power.

In early September Zulfi returned to Karachi to find "everyone . . . preparing to strike again," as he warned Mumtaz. Most bureaucrats were "disgusted," workers depressed, business leaders fed up, police useless, students lethargic, angry, disillusioned, frustrated. Even members of Mumtaz's cabinet, with whom Zulfi lunched the day before he wrote his "brother," were sullen and resented the chief minister's high-handedness. "It is true that the leader must exercise control over his ministers," Zulfi conceded, trying to mollify Mumtaz even as he cautioned him against "this growing resentment and intrigue," adding a warning he had begun to repeat more often, "Precious time is running out."[15]

From Karachi Zulfi flew to Lahore, where he met with Maudoodi and other Maulanas of the Jamaat-i-Islami, anxious to win their support for the constitution to be drafted by a committee of the National Assembly, a constitution that he had promised the nation would be ready in March 1973 or in April at the latest. He was also keen to recognize Bangladesh, a prerequisite to the return of ninety-three thousand Pakistani prisoners—their incarceration remained a bitter, constantly mentioned failure of policy. Zulfi was eager to talk with Mujib again, and tried to phone him from Lahore to arrange an early meeting in London. Sheikh Mujibur Rahman was not, however, ready to risk antagonizing his Awami League lieutenants by agreeing even to speak with the man still deeply hated throughout Bangladesh, as much as was Yahya Khan. Many Bengalis blamed Zulfi Bhutto more than Yahya, in fact, who was considered a dupe of his much shrewder political manipulator. Maudoodi and other orthodox opposition leaders denounced any proposal to "recognize" Bangladesh as intolerable; they said that it would be seen as rewarding "aggression" and would tempt India to grab another piece of Pakistan. Still, Zulfi hoped that he and Mujib could agree upon a "confederation" of Pakistan and Bangladesh, bolstering the economy of both nations. Zulfi and Mujib were, indeed, faced with many similar domestic problems—unable to deliver half of what they had promised, increasingly mistrusted by their mullahs and military, and swiftly losing the support of student radicals as well as hungry workers. Had they agreed to work together before the war, not only would all those killed and wounded in 1971 have been spared but Zulfi and Mujib themselves might have lived at least a decade longer.

"What is built on hypocrisy and deceit must finally crumble," President Bhutto warned his nation on 11 September 1972, the anniversary of Quaid-i-Azam Jinnah's death. He repeated his cry of the previous year, "Oh My Quaid, when will the long night of repression end?" as though he were not now in control. "But how shall God bless any efforts unless you, the people, are with me, by my side, giving me strength and power, giving me confidence and hope."[16]

If Sind was on fire, the Frontier kept smoldering, and even Punjab had begun to heat up. Zulfi's political support was weakest along the Frontier, both in Baluchistan and in the North-West, but he wisely reached agreement with the most popular and powerful leaders of those two provinces, allowing him to carry on at the center without facing constant attacks from Frontier members in the National Assembly. In late April of 1972 he signed a tripartite political agreement with Khan Abdul Wali Khan, leader of the National Awami Party (NAP), and with Maulana Mufti Mahmood, leader of the Jamiatul-Ulemi-i-Islam (JUI), the two most powerful Frontier parties.

The agreement promised NAP and JUI full control of the Frontier provinces as long as Bhutto had their political support in the National Assembly. Arbab Sikandar Khan was appointed governor of the NWFP by Bhutto, and Mir Ghaus Bakhsh Khan Bizenjo, another important member of NAP, was appointed governor of Baluchistan. A few days later Sardar Ataullah Khan Mengal of NAP was elected to lead Baluchistan's Provincial Assembly as chief minister, and JUI's Maulana Mufti Mahmood was elected chief minister of the NWFP. Bhutto wrote to congratulate them on 1 May. A week later Mengal replied, appreciating the central government's economic support of Baluchistan's development, especially Bhutto's approval of that backward, impoverished region's request to use its own Sui gas. Mufti Mahmood was less pleased, however, for Zulfi's closest Frontier friend, the People's Party opposition leader in the NWFP Assembly, Hayat Mohammad Khan Sherpao, was put at the head of the central government's powerful Water and Power Development Agency (WAPDA), and refused to release funds for three major NWFP irrigation projects. Sherpao, Wali Khan, and Mufti Mahmood had long been bitter political rivals in the Frontier, a struggle for provincial power in which Zulfi gave whatever support he could to Sherpao. As the uppermost user of the Indus River, the NWFP considered itself cheated of its "natural water rights" by WAPDA's funding of irrigation and power projects in Punjab and Sind. The NWFP was, in fact, annually obliged to import wheat from the Punjab. Except for Sherpao, all other members of WAPDA at this time were Punjabis.

Bhutto's second-most-important Frontier adviser was his Minister of home affairs and frontier regions, crusty old Abdul Qaiyum Khan, leader of the Frontier's Muslim League, Wali Khan's most hated enemy. Qaiyum Khan was Zulfi's most conservative cabinet colleague, considered by many a spent political force now, yet still respected for having worked so closely with Quaid-i-Azam Jinnah in helping him to organize the Muslim League along the Frontier. Qaiyum Khan, of course, had access to all police and intelligence reports, and kept his president fully informed of everything Wali Khan and his "treacherous" NAP supporters (all of whom he considered "pro-Indian" and "pro-Russian") did, both at home and abroad. Wali Khan's venerable father, Khan Abdul Ghaffar Khan, had been a friend and "disciple" of Mahatma Gandhi, a leader of the Frontier's "Red Shirts," and a leader of the Afghan-supported separatist Pakhtunistan (Land of the Pakhtuns) movement, and had long since gone off to live in exile in Kabul. Qaiyum Khan now warned Zulfi, however, that old Abdul Ghaffar Khan was "thinking" of coming back to the Frontier, that "Indian agents" were very "active" in Baluchistan, and that the "dream" of a separate Greater Baluchistan remained a serious potential threat to Pakistan's integrity. He therefore urged "upgrading" of the staff of Pakistan's Central Intelligence Bureau, to which Bhutto acceded immediately. The home minister also advised raising "mixed formations" of new "scout platoons" to be deployed around Baluchistan and along the Frontier: "We have learnt a bitter lesson in East Pakistan of purely Bengali Regiments."[17] Zulfi noted next to that suggestion, "agreed."

Several Punjabi members of Bhutto's People's Party had begun publicly to attack Bhutto from the floor of the National Assembly. The most arrogantly outspoken was 35-year-old Ahmad Raza Kasuri, who charged that the president was hiring "murderers and criminals" to "kill democracy."[18] Everyone who knew young Kasuri, however, knew he was "like that," usually shouting something outrageous. In public school he had asked questions till his teachers lost patience. He was "rusticated" for an entire year from Lahore's Law College in 1964. Classmates described him as a "big mouth," "loud talker," "noisy, impossible fellow." He was impulsive, bright, articulate, impatient, often angry, ambitious enough to finish law school and begin practice in 1967. Kasuri's insults made some people hate or fear him, yet he impressed enough voters in Lahore to win a PPP seat in the National Assembly. Some people said Kasuri was very much like Zulfi Bhutto, in fact. He too had a highly respected well-to-do father, Nawab Mohammad Ahmad Khan, who had been honored with the title Khan Bahadur (Great Khan) by the British. Like Sir Shah Nawaz he too was proud of his son, wanting him to enjoy "the best of everything"; some might say he had been overindulgent.

Kasuri had been so excited by Bhutto's fiery rhetoric the very first time he heard him speak in 1966 that he resolved to devote himself to the PPP and also to launch a career in politics. He admired Bhutto so much that he resolved to be just like him, a "revolutionary" leader of the people in Pakistan's struggle to restore democracy—people's rule—to his militarily repressed fatherland. Like young Zulfikar, young Ahmad Raza was an idealist. He became so active, in fact, in helping to organize the People's Party that he was appointed to its Central Working Committee. Like Zulfi he hated Ayub, and his own passionate speeches against that evil "dictator" echoed virtually everything Zulfi said, adding stronger language of his own at times. Yet now, less than a year after coming to power, Bhutto seemed in many ways just as bad as Ayub had been, perhaps a bit worse. Ahmad Raza Kasuri felt betrayed, as did many other young, sincere, hard-working, radical idealists who had devoted themselves to the PPP, believing all the promises made by its Quaid-i-Awam. Most of the others were afraid to utter a word in public against Bhutto, for they knew of his temper and his feudal sensitivity to all criticism. Kasuri, however, was audacious, "self-destructively" bold, some declared. He had always been "itching for trouble," his friends said, and some feared that he would soon "get it."

"We have planted a very small and delicate plant of democracy," Bhutto told a Western journalist that October of 1972. "But suddenly, now, after fifteen years of sealed lips . . . people once again have gone on a verbal rampage without any regard for each other's . . . feelings. They justify this as freedom of expression . . . disregard all laws, from treason to perjury, and call it freedom."[19] Was it Kasuri's latest "verbal rampage" he had in mind? The head of Intelligence at this time, Saied Ahmad Khan, had warned Bhutto in a "top secret" note of a "rift" among the People's Party members of the National Assembly from Punjab; many of them "feel the President is too busy with International Affairs." He urged that greater "care" be extended to the dissidents, suggesting that the governor of Punjab, Mustafa Khar, might do more by way of "entertaining" those disaffected members because "Punjab has to be kept on the right side of the Government at all cost."[20] Zulfi wrote to and spoke with Khar and Punjab's Chief Minister Malik Meraj Khalid about the urgent problem, and both agreed to pay more attention to troublemakers like Kasuri. But Kasuri continued his "bombasts against me," Zulfi wrote Malik Meraj and Khar that September, "surprised" they had not as yet "ostracized" him for such outrageous behavior. Zulfi even started to "suspect" his own chief minister, who often sat next to Kasuri in the National Assembly, of perhaps agreeing with him, as he "cast subtle aspersions on my leadership." Zulfi feared "intrigues" were "brewing" all over the Punjab, and "a kind of cut-throat mentality" was "taking shape" there.

Devout older Muslims as well as disillusioned young idealists were now losing faith in Bhutto, some because of what they saw or heard about the unofficial doings of Governor Khar. Many people wrote to warn Bhutto about the private life of his most powerful governor, but he always staunchly supported Khar because Khar was his closest friend, his alter ego, in many ways a more youthful, more virile, handsomer reflection of his true self. They had driven together through volleys of fire, had been in and out of the same jails, tasted the same tear gas, knew the same women, faced the same gangs of goondas targeting them both with bullets and stones, and now they were tarred with the same "rumor-mongering" tales of lasciviousness and lechery. "Our country is unfortunately a country of liars and a country of rumours," Zulfi replied to one such letter. "Rumours spread like wildfire and people tell lies by swearing on the

Holy Quran. Only recently it was on the lips of everybody that I had got married again. Neither I can help this kind of rumour-mongering nor can any of my colleagues. The Jamaat-e-Islami has paid workers whose duty it is to go from door to door telling lies and to sit in restaurants fabricating stories to damage the name of the Government. . . . I am disgusted with the conduct of some of the parliamentarians . . . casting aspersions on my closest comrades-in-arms will not be taken lightly."[21]

Like Zulfi Bhutto, "Sher-i-Punjab" (Lion of the Punjab) Mustafa Khar was a large landowner, a stirring speaker, and a great lover. He too spiced his Urdu with sharp, racy language, saying whatever he felt, rarely beating around the bush when he got angry. At the end of July 1972 in Lahore he threatened to "pull out the tongues" of those who "abused" Bhutto or talked of "breaking the country." Soon after Zulfi called Ahmad Raza Kasuri's "bombast" to his attention, Khar called a press conference at his house in Lahore and warned all "legislators" to remember that "but for the leadership of Mr. Bhutto they would not have been returned" to so much as a local school board. At Islamabad in mid-October he said much the same thing about himself, confessing that his own "success" was "due entirely" to Bhutto's "popularity." A month later he attributed the prompt solutions of "problems half a century old" to the president's "glorious leadership," "immensely benefitting the masses." Some of Khar's sycophantic supporters had started to hail him as "Quaid-i-Punjab" (Leader of the Punjab), but he called upon them to desist: there was "only one" *great* PPP leader, "Quaid-i-Awam Bhutto." Khar knew how to keep Bhutto happy, and Zulfi always enjoyed spending time as often as possible at the Lahore house of his handsome, convivial comrade-in-arms.

Ahmad Raza was not the only Kasuri to bother Bhutto during the first autumn of his presidency. Law Minister Mian Mahmud Ali Kasuri (no relative of Ahmad Raza; *Kasuri* simply means "town-dweller"), who also was minister for parliamentary affairs, thanks to which he chaired the National Assembly, and had been elected by his colleagues there to chair the important constitution-drafting committee, resigned from the cabinet in early October. Mahmud Ali Kasuri was one of Pakistan's most respected liberal lawyers, having served the International Association of Democratic Lawyers, Lord Russell's War Crimes Tribunal, and the World Habeas Corpus Commission. As the most courageous member of the cabinet, Mahmud Ali had often argued with Bhutto over the earlier retention of martial law and excessive "emergency" regulations. He had first resigned in September when he saw that Bhutto was determined to introduce a centralized presidential system rather than the more democratic parliamentary federalism to which the PPP had long been committed—and which had initially attracted Mahmud Ali to join the party. Bhutto feared that the resignation would "inevitably delay" framing a constitution, and at first persuaded Mahmud Ali to retract it, promising to support his constitution-drafting labors. Mahmud Ali, whose assembly seat was also from Lahore, was warned by Governor Khar to toe the strict line, but then found Bhutto turning a deaf ear to his appeals for Bhutto's public affirmation of recently promised "support." On 4 October he resigned again, sending a copy of his letter to the press. The following day Bhutto publicly accepted it "with immediate effect" and urged him to resign his National Assembly seat. Bhutto then moved his old friend and lawyer Abdul Hafeez Pirzada into both vacant cabinet posts, knowing that he had to act fast to prevent further "desertions" or weakening of public confidence in the promised constitution.

Every political challenge, such as that posed by Mahmud Ali Kasuri's public resignation, galvanized Bhutto into action with dazzling speed. Attacks that might have intimidated or paralyzed less adroit political leaders served only to send a flood of adrenalin to Zulfi Bhutto's brain. He convened the All-Parties Conference in Rawalpindi less than two weeks after Kasuri resigned, and four days later he had cajoled and convinced, pleaded with and browbeat every leader of the nine major parties represented in Pakistan's National Assembly and of all four provinces to reach an unprecedented unanimous accord on a new constitutional formula. It was the most scintillating political waltz any Pakistani leader had ever performed. Deft maneuvering, personal charm, awareness of the strengths and weaknesses of each of the invited players achieved in four days what none of his predecessors, from Liaquat Ali Khan to Yahya Khan had been able to do in almost a quarter century of futile constitutional wrangling. Now that Mahmud Ali was off his team, Zulfi approved a "parliamentary federal" formula as the keystone of his October Twentieth Constitutional Accord. He was always ready to discard any baggage that threatened to sink his political bark, and hence tossed overboard without a second thought the more centralized French presidential system, of which he had long seemed enamored. He would be happy to serve as prime minister rather than president. To insure Pakistan's "stability," however, Zulfi insisted on raising the Westminister simple majority formula for any vote of "no confidence," to no less than two-thirds of the National Assembly, at least for the first fifteen years of the new constitution's existence. That would assure him three consecutive full terms in premier office, which on the eve of his forty-fifth birthday seemed enough to Zulfi, his dark hair now a fringe of fast-receding gray, his waistline bulging from all the late official banquets at which he ate and drank much more than was good for his deteriorating health.

Bhutto met with Maulana Maudoodi on the eve of his Pindi conference, bringing that leading Muslim conservative and his opposition Jamaat around to accepting the accord by agreeing to make Pakistan an "Islamic Republic," whose president and prime minister would always have to be Muslims, and by promising that "no law" would be passed under his constitution against any "teachings and requirements of Islam" as articulated in the Holy Quran and Sunnah (teachings of the Prophet). Islam was to be the "State religion of Pakistan," and Zulfi satisfied Maudoodi that the "Islamic Socialism" he had long promised when first campaigning against Ayub had nothing in common with communism, and would in no way violate any laws of Islam. Frontier Mullahs of JUI had, of course, earlier joined NAP leaders in allying themselves with the PPP, and now were satisfied with the new constitutional formula, which promised a federation, "wherein the units will be autonomous with such boundaries and limitations on their powers and authority as may be prescribed." Zulfi's reluctance to agree to that fundamental federal formula had been one of the major causes of Kasuri's resignation. Some ten members of Bhutto's own PPP from the Punjab, including the irrepressible Ahmad Raza Kasuri, had associated themselves with Mahmud Ali Kasuri's liberal–left-wing opposition to their president's "Bonapartism," so the constitutional accord also promised "preservation of democracy" against any and all "oppression and tyranny," and the "elimination of all forms of exploitation." To Zulfi Bhutto's mind, politics was not only the "art of the possible" but the promise of reconciliation and the integration of opposites and everyone's favorite desire.

Bhutto launched a whirlwind tour of the North-West Frontier Province in mid-November 1972, visiting all of its nine districts and six tribal political agencies; addressing twenty-four large meetings in eleven days; and covering some fifteen hundred miles with his large entourage of party faithful, media reporters, People's Guards, and members of his new paramilitary elite Federal Security Force (FSF). Everywhere he spoke friendly crowds were bussed in, PPP flags and huge pictures of his face flew overhead, and he announced time and again, "We have laid the foundations for democracy." He repeatedly asked for "more time," however, to achieve all he had promised, all he was still promising to do. The tour ended with a mammoth meeting at Jinnah Park in Peshawar, where he urged that the moment had come to recognize the "reality" of Bangladesh, pointing out that it would be good for Pakistan's economy and lead to the return of all Pakistani POWs. He also promised that now that the "principles" of a "democratic and federal" constitution, "ensuring provincial autonomy" had been agreed to by all parties in the National Assembly, that the permanent constitution would assure Pakistan a true "Awami Raj" (People's Rule) starting in the new year.

The next day Zulfi went to Lyallpur and announced his decision to release all 617 Indian prisoners of war, held there since the 1971 War had begun. It was a humane, wise, and bold unilateral initiative, Bhutto diplomacy at its best. It cast a revealing light on both India and Bangladesh for continuing to detain Pakistani prisoners long after the war had ended. Zulfi knew that PPP hopes raised in the aftermath of the Simla summit had been sadly deflated by now, and he was anxious to wean Bangladesh back toward its Muslim "Brother"—Pakistan. He had received many reports of how "disillusioned" Bangladeshis were becoming with West Bengal's tough Hindu Marwari bankers and business tycoons, yet how desperately Mujib needed financial help of every kind. He also understood the anger and frustration of the relatives of the more than 90,000 Pakistani prisoners over his government's impotence in regard to their freedom. New Delhi remained adamant, however; recognition of Bangladesh preceded prisoner repatriation.

Bhutto flew to Karachi from Lyallpur to inaugurate Pakistan's first nuclear power plant there on 28 November, promising his "fullest support" to Professor Salam in his top-secret plutonium-production labors. The president personally chaired Pakistan's Atomic Energy Commission and the Ministry of Science and Technology he had established earlier in the year, to which he gave the "highest priority." Zulfi never deviated from his early resolve to produce atomic weapons that would allow Pakistan to restore its military balance with India, or possibly even tip that balance in favor of the much smaller nation. "We are to wage a relentless struggle against all evils that are retarding our progress," Zulfi told his cheering audience in Karachi. "I want to achieve results in the shortest possible time. We are, therefore, utilizing all those means which can hasten our progress."[22]

The fifth anniversary convention to celebrate the founding of Zulfi's PPP in 1967 was held in the army's football and hockey stadium in Rawalpindi on the last day of November and the first day of December 1972. Thousands of delegates—1,700 from Punjab, 650 from Sind, and 1,000 from the Frontier provinces and Azad Kashmir—poured into Pindi's train and bus terminals for several days before the great show started. They camped under canvas on the spacious grounds surrounding the stadium, a political army ready to hail its leader. Bhutto's photograph was set beside that of Quaid-i-Azam Jinnah's on a green banner some 140 feet long and almost 20 feet high that stretched across one section of the stadium and faced the fortunate delegates. Hundreds of People's Guards were on duty at every entrance to the stadium, and the prayers, sermons, speeches, and robustly repeated slogans were all amplified to often deafening magnitude and ear-piercing pitch. Mumtaz and Khar shared overall responsibility for the show.

Zulfi flew in for the concluding grand ceremony, bringing all the delegates to their feet to cheer their Quaid-i-Awam. The gold-embroidered tight collar on the shimmering green silk achkan he wore and his thinning crown of silver hair certainly showed how much Zulfikar Ali Bhutto had changed since his fiery days as the open-collared, two-fisted leader of a people's crusade to oust the old military dictator with self-aggrandizing ways. Zulfi called again for recognition of Bangladesh and urged his followers to remain "confident" in his leadership of the nation and the party, promising that he would "never compromise on principles."

Strikes continued to spread in Sind, and right-wing Jamaat-i-Islami student protest marches in Lahore and elsewhere in Punjab against the recognition of Bangladesh led to hundreds of arrests before year's end. Baluchistan's Quetta and much of its impoverished rural hinterland also rocked with riots, armed Bugti tribesmen riding against weaker provincial government forces, and many Sindhis forced by Baluchi armed antagonists to flee from Lasbela to Karachi. The Tehrik-i-Istiqlal (Freedom Party) opposition party of retired Air Marshal Asghar Khan had been strengthened of late by half a dozen Punjabi defectors from the PPP, led by Mahmud Ali Kasuri and Ahmad Raza Kasuri. Zulfi suspected a "conspiracy" paid for with "foreign money." He appeared on television on 20 December, just one year after he'd taken over as president, and warned his opponents to be careful. "They are responsible for the debacle and now they try to be sanctimonious. . . . Now that we have the authority of the Government, you think we cannot tackle them? . . . We know how to deal with them . . . they opposed Pakistan, tooth and nail; they abused Quaid-i-Azam. I do not know why people have such a short memory."[23]

Pakistan's industrialists and money men had longer memories than Zulfi Bhutto wished. Capital was swiftly fleeing the country as inflation soared, defense expenditures were climbing despite the loss of half the army and national population, and the cost of the new elite internal security forces was rising precipitously even as urban and rural security collapsed. Bhutto called the expatriation of wealth "sabotage" and declared that he had no intention of allowing the economy "to collapse," though the stock market had been doing just that. Then he warned all black marketeers and smugglers throughout Pakistan, borrowing a favorite turn of phrase from his good friend President Richard Nixon, who had just invited him to the White House, "I must make it abundantly clear, crystal clear that there is a limit to our patience."[24]

The opposition to Zulfi Bhutto grew as the new year began. Threats and harsh words were shouted from the floor of the National Assembly on 2 January 1973, as several opposition leaders denied having agreed to any constitutional accord. On 3 January, Zulfi hoped to replay the previous year's triumphant speech in Karachi's Nishtar Park, but less than a fifth of the hundred thousand who had hailed his victorious return to Karachi just one year before could be lured or driven to listen to him. This time security was much tighter and more evident; barbed wire and steel bars surrounded the isolated stage from which he looked across a clearing of some ninety feet to the first VIP section of seats for his loyal Sindhi followers, who clapped each time he paused and periodically chanted, "Jiye Bhutto!" Others in the park were less responsive, either listening in silence or raising cries of opposition, especially when he asked for a "People's vote" on recognition of Bangladesh. "*Na Manzoor!*" (Not acceptable), they shouted, much the way so many thousands of young men had done all across Punjab for most of the previous month. Bhutto was visibly shaken, enough to shout in echoing agreement, "Na Manzoor, Na Manzoor, Na Manzoor!" That won the crowd to his side, of course, as Zulfi knew it would. The formerly sullen and angry listeners jumped to their feet to applaud, crying out, "Jiye Bhutto!" and "Pakistan Zindabad!"

"By all means, don't recognise Bangladesh if you don't want to," Zulfi told them. "But also, don't blame me when the prices of sugar and wheat flour register an increase. . . . With heavy expenditure on defence and an annual burden of Rs. 90 crore [c. $90 million] on account of Bangladesh."[25] Until Pakistan recognized Bangladesh, Bhutto explained, it would have to go on paying the full annual international debt of both nations, and that, he asserted, was why they all suffered economically. Nor would any POWs be returned prior to recognition, for that was India's way of adding pressure—though Zulfi clenched his fist and shouted that "neither India nor Russia nor the United States" could "pressurise Pakistan into taking this decision." That too brought forth a popular approving echo; Bhutto was still skilled at eliciting positive feedback from a crowd. And when audience apathy settled in again, he resorted to a once-familiar tactic: he bared his chest and thundered at his "opponents" out there, wherever they might be lurking and conspiring, "Who is it that wants to shoot me? Let him come and take a shot. I am not scared."[26] But soon after that he was spirited away by a phalanx of FSF guards, who moved swiftly to keep the crowd from crossing the no-man's strip till he was back inside his bullet-proof limousine.

The next day Zulfi flew to Larkana to celebrate his forty-fifth birthday with his royal guest, the sheikh of Abu Dhabi, and his family in his newly fenced and security-gated Al-Murtaza ancestral home. In proper regal fashion he distributed land deeds to 105 thitherto landless peasants from every part of Sind, at Naodero on the actual day of his birth. Many of the peasants cried as they kissed the deeds, bowing to kiss Sind's red dust from their president's shoes and then dancing with happiness. To such haris of Sind, he was, indeed, *Raja* (king) not just of Larkana but of the entire land. He had always been good to his peasants—too good, his opponents complained; the land he distributed had been taken from absentee landlords, who had in recent years come to Sind from Punjab, without compensation. Zulfi always remained especially fond of celebrations such as this one, and later he and the sheikh enjoyed a few good days' shooting. President Nicolae Ceausescu of Romania came to Pakistan on 8 January, and Bhutto had to be back in Karachi to drink many toasts to his health and that of Madame Ceausescu that evening, and to "ever-expanding collaboration" between their two "Socialist Republics."

Zulfi had invited Shah Reza Pahlavi of Iran to Larkana for his birthday, but the latter could not come until shortly after mid-January. He spent several days shooting game and talking business with Zulfi, who desperately needed military supplies from the shah, primarily to deal with growing Baluchistani opposition to his central government. The U.S. embargo on the shipment of arms and spare parts to Pakistan, imposed by Congress after the Bangladesh War had started, made it impossible to get the materiel needed to strengthen the air force for use against tribal "insurgents." Zulfi's dependence on the Shah and the shah's aloof acceptance of the cordial and gracious Bhutto hospitality left unpleasant memories of "Pahlavi" in Zulfi's mind after his guest had gone home.

"There was an uncomfortable perversity about him," Zulfi later wrote, reflecting on this singularly wealthy king of kings. "He could be jealous and mean in small things . . . unrelentingly ruthless and disparaging about personalities . . . in their absence. . . . He could tell a big lie without blinking. He spoke disparagingly about almost all his neighbouring countries and their leaders. . . . He had a complex towards me. He respected and feared my capabilities." Zulfi noted that the shah was "intensely envious" of him and—even more revealingly, perhaps, of his own fall—that "his grandiose designs and fanciful ambitions . . . contributed in no small measure to his ruin. . . . He lost all touch with reality."[27]

When Zulfi visited Nixon in the White House in September 1973, he later recollected, much to his chagrin: "In good faith I *hinted* to him [Nixon] that the Shah was not all that stable as they thought and that they should look beyond Iran and lift the arms embargo on Pakistan. It was one of the arguments I used to get the ten year old arms embargo lifted but it was my honest view as well. Kissinger was present in the meeting. Either Kissinger or Nixon or both passed on my views to the Shah. The Shah was mad as hell with me. He was infuriated. He sulked and spoke against me." Zulfi felt obliged to fly to Teheran soon after his faux pas to apologize and to try to mend fences with his sorely offended mighty neighbor. But the Shah never forgave him.

Before the end of January 1973, federal troops were ordered into Baluchistan. Lasbela had become a hotbed for "miscreants," and Akbar Bugti let it be known to Bhutto that he suspected Governor Bizenjo was preparing for the "secession" of his troubled state. In Baghdad, moreover, a "Baluch Liberation Front" office had opened, and hidden transmitters kept beaming broadcasts toward Pakistan from Iraq in four regional languages. Karachi was a beehive of rumors about an imminent "Russian-backed invasion" and Indo-Soviet "sabotage" aimed at turning Baluchistan into another Bangladesh. Wali Khan's 83-year-old father, Khan Abdul Ghaffar Khan, had come to Peshawar from Kabul after eight years in exile to a royal hero's welcome unprecedented in living memory; every Pakhtun tribesman in the NWFP was there. The reception made Zulfi as nervous about the return's heralding a NAP-led Frontier separatist movement as he was envious of Abdul Ghaffar Khan's popularity. Home Minister Qaiyum Khan had advised against "allowing" the old "troublemaker" into the country, but Bhutto wisely realized that any attempt to stop him at the Khyber Pass would have been as impossible as it had been for the British raj to subdue all the armed Pakhtun tribals, whose rugged homeland spanned that and every other NWFP pass.

On 14 February 1973 Bhutto wrote to Mir Ghaus Bakhsh Bizenjo, "terminating" his appointment as governor of Baluchistan "with immediate effect"; "little purpose would be served" by going into "details" about his reasons for doing so. He appointed Nawab Akbar Bugti as the new governor, and the following day he issued a proclamation removing Baluchistan's Council of Ministers: Sardar Ataullah Mengal's ministry had "failed to take effective measures to check the large scale disturbances . . . resulting in loss of life and property on a massive scale."[28] A week earlier Chief Minister Mengal had warned at a press conference that if his government were to be toppled by the center "the future of democracy in the country is in danger." Zulfi insisted, however, that Baluchistan under Bugti was as "democratic" as it had ever been, and characterized what others called the current "crisis" a mere "storm in a tea cup."

Bhutto changed the governors of Sind and the NWFP at this time as well, thereby reminding "alarmist" critics that the naming of provincial governors was one prerogative of the president. He replaced the NAP governor of the NWFP with a Pathan, Mohammad Aslam Khattak, and though willing to leave JUI's Mufti Mahmood at the helm of his ministry there, Wali Khan insisted he resign and the Mufti agreed. The NAP-JUI "accord" with the PPP thus died in less than a year. Both Frontier provinces were now placed under President's rule, and Tikka Khan's troops kept pouring into Baluchistan, fanning out to surround "guerrilla camps" reportedly discovered in Marri tribal country. Still President Bhutto said that "no shot" had been fired, "not a single person killed" as yet by the army. But he warned his outraged opposition in the National Assembly that "if anyone resorted to violence" it would "recoil on them and finish them off forever." He wanted that to be "clearly understood."[29] He insisted, moreover, that "democracy" was the "main aim" of his PPP. Wali Khan, however, called him "Adolf Bhutto" and told his Frontier followers that it would be impossible for them to enjoy "undiluted democracy" until Bhutto was "removed" from office.

President Nixon called for the lifting of the 1971 embargoes on all sales of arms to South Asia on 14 March 1973. India and Bangladesh immediately denounced the move as one designed to escalate tensions and diminish prospects for peace between India and Pakistan. Bhutto was relieved and delighted. He would no longer have to rely upon the doubtful dependability of the Shah of Iran for the arms and spare parts he desperately needed, not simply to keep India from moving west but to hold restive Marri and Mengal tribesmen of Baluchistan within their barren region. "Unafraid of any adversary without and watchful over the enemy within," Bhutto confidently assured his nation that Pakistan Day (23 March), "we will persevere in the long, hard task of building a Pakistan free from exploitation and at peace with all those neighbours who wish it well. . . . A rejuvenated Pakistan will be a source of strength to the Muslim world and an asset to the Asian-African community."[30] Later that very day he took the salute from the joint services in their Rawalpindi parade, assuring them that "by the grace of Allah" "the dark clouds are about to disappear." Pakistan's future now looked "bright" to Bhutto, who focused all his political wits and wisdom on insuring the enactment of a new constitution.

The secular North-West Frontier NAP leader Wali Khan joined forces with the Sindhi Hur leader, Pir of Pagaro, and Islamic fundamentalists of Maulana Maudoodi's Jamaat-i-Islami—unlikely as that combination sounded to all who knew them—in the United Democratic Front (UDF) against Zulfi Bhutto's government. The Pir of Pagaro, with some forty-thousand armed Hurs at his command, had long been a rival of Bhutto's, and with former Governor Talpur alienated now as well, brought a substantial Sindhi force into the Frontier camps of Wali Khan and former Baluchi Governor Bizenjo, all blessed by Maudoodi and other maulanas in their grand opposition alliance to the PPP. Youthfully charismatic Syed Sikandar Ali Shah, the Pir of Pagaro, was elected president of the UDF, and he called upon his followers to boycott the National Assembly barely more than a month before Bhutto promised the nation a new constitution.

Bhutto was swift to check his opponents in every political move they dared to make against him. A UDF rally in Rawalpindi on Pakistan Day, 23 March, turned into a pitched battle: Wali Khan's tough frontiersmen and the Pir of Pagaro's Hurs fighting PPP Punjabi troops trucked in from Sialkot and Campbellpur to wage "Operation Sugar Candy" against the Pathans bussed in from Peshawar. When the gunfire stopped and the smoke cleared over Pindi's Liaquat Park, ten were dead and almost a hundred wounded. None of the UDF leaders had even had a chance to speak. A week later Bhutto solemnly informed the nation over Radio Pakistan that the UDF had launched a "unilateral declaration of illegality" against the motherland.

On 2 April Zulfi met with many leaders of the opposition, excluding Wali

Khan and the Pir of Pagaro and several others, of course, and generously conceded many constitutional demands that he had firmly rejected over the past six months. He was especially conciliatory to the religious parties, weaning Jamaat-i-Islami as well as Jamiatul Ulemi-i-Islam and Jamiat Ulma-i-Pakistan away from Wali Khan and his secular-minded "socialist comrades" by agreeing that the prime minister as well as the president must be Muslims and be required to take "a solemn oath" to that effect; that special facilities would be set aside throughout Pakistan for the study of Arabic; and that "correct and exact" versions of the Holy Quran would be printed and distributed. The Islamic Ideology Council, earlier accepted in the October Twentieth Constitutional Accord, would now be "strengthened."

Nimble political dancer that he was, Zulfi also reaffirmed his complete commitment to his People's Party Manifesto, insisting that it was the prime duty of the state to "provide work" for all of its people, as well as "food, clothing, housing, education and medical relief." In an aide-memoire he circulated to the opposition leaders on 4 April 1973, moreover, Bhutto promised that "given time to fulfil our mandate, we will, Inshallah, not fail the people of Pakistan."[31] Finally, though he had most tenaciously argued the previous October in favor of the nondemocratic two-thirds majority requirement stipulated in the accord for passage of any no-confidence motion tabled in the National Assembly during the first fifteen years of the constitution's life, Zulfi now offered to reduce that interval to ten years. He liked to be magnanimous, aristocratically gracious, making such offers of his own volition, insisting, "This is proof enough to demonstrate that there is no desire of the majority party or any person to perpetuate one party or one-man rule." But in return he demanded an end to the opposition's boycott of the National Assembly; insisting that all members get back to work on 7 April and "stay there till the Constitution is framed."[32]

The Pir of Pagaro promptly rejected Bhutto's new offers and demand, but virtually all his UDF "followers" abandoned him. They returned to work and passed the new Constitution, almost unanimously—some, who underestimated Zulfi's political astuteness, said "miraculously"—on 10 April 1973. Assembly Speaker Fazal Elahi Chaudhry formally presented the gold-embossed leather-bound authenticated Constitution to President Bhutto at the President's House on 12 April. Thus the Islamic Republic of Pakistan was reborn under its 1973 Constitution "in the name of Allah, the Beneficent, the Merciful." The preamble began by noting that "Sovereignty over the entire Universe belongs to Almighty Allah alone, and the authority to be exercised by the people of Pakistan within the limits prescribed by Him is a sacred trust." Zulfi fondly hoped that his most impressive political achievement to date would long outlive his own tenure in high office, ten more years seeming to him at this time perhaps all that remained for his enjoyment. He would not be vouchsafed as much.

11

Foreign Triumphs, Domestic Tragedies

(April 1973–1974)

Having badgered and cajoled, bullied and begged enough members of the National Assembly to pass his constitution with enthusiastic acclaim, Zulfi Bhutto knew he could rest easy on his democratic laurels for at least a year or two. Earlier opposition demands for fresh elections and a caretaker government to run Pakistan as soon as a constitution was enacted died down. Wherever Bhutto went he was hailed and lauded with patriotic praise for his noble national political initiative. Wali Khan's dark pre-Constitution warning that "Bhutto and Pakistan cannot co-exist," was now attributed to "sour grapes," as were other negative remarks of former opposition leaders, such as that the almost universally hailed document was "neither Islamic nor Federal nor Democratic." All Pakistanis love a good fighter, and doubly appreciate a leader who can beat opponents at their own game by whatever means, fair or foul. Bhutto had again proved his claim to the title Quaid-i-Awam, and found his stock rising wherever he went to speak in the aftermath of early April's constitutional euphoria.

On 21 April 1973, Iqbal Day, Zulfi spoke of the "re-awakening" of South Asia's Muslims "symbolised by the Islamic Republic of Pakistan."[1] That same day he told the cadets of Pakistan's Military Academy at Kakul, "You are not playthings to be used and exploited for selfish ambitions. You are the custodians of our frontiers . . . the sword-arm of our defence. We cannot permit the insult or humiliation of any citizen of Pakistan."[2] Over the new World Service he inaugurated that day for the Pakistan Broadcasting Corporation, Bhutto proudly announced that "we are blessed with a Constitution unanimously acclaimed as democratic, federal and Islamic." He spoke of "My Government" as "your Government," thus sharing with every Pakistani abroad who heard him the "vibrant hope" he felt and majestically conveyed. His enthusiasm proved contagious. Few could resist it, other than such "spoilsports" as Ahmad Raza Kasuri, who spewed angry epithets whenever he heard Bhutto's voice or name. But to Pakistan's silent majority Bhutto was unrivaled, unmatched—his political bat as powerful as the bat of any cricket star, his tongue as sharp and potent as the rapier of any field marshal.

Bhutto was euphoric not only about the Constitution but about his upcoming state visit to Washington, scheduled for 16–22 July 1973. It had been sixteen months since his last meeting with President Nixon, on the eve of a long flight home via Rome and Teheran, when dark clouds of humiliating defeat in Bangladesh shrouded Pakistan, and only he was deemed fit to take up Yahya Khan's burden. Nixon's superpower support had given him the foreign backing he needed to convince the army and civil service that he could, in fact, lead Pakistan out of the deep depression into which it had fallen by December 1971. Nixon's support was now aimed at lifting the 1971 embargo on arms, and Zulfi knew that his friend would help him in any and every possible way, despite Congress's opposition. Nixon and Bhutto appreciated each other, each seeing in the other perhaps the almost perfect reflection of his own insatiable political appetite; each hearing from the other his own oratory and slick rhetoric. They were a uniquely compatible pair, despite polar differences in their backgrounds, and shared many instincts, aspirations, passions, and dreams for themselves and their people, each within the limits of his own national environment.

A week before he was to leave for Washington, Bhutto addressed the National Assembly on foreign affairs, requesting prior authorization to "recognize" Bangladesh at his own discretion. Zulfi planned to go to New York after meeting with Nixon, and hoped to conclude ongoing negotiations there with Indian as well as Bangladeshi representatives. He promised that "repatriation" of all Pakistani prisoners of war would remain "the prerequisite to the re-establishment of normal relations."[3] A year had passed since the much-acclaimed Simla Accord, yet nothing had as yet been accomplished toward achieving a more durable peace in South Asia. Zulfi knew that he was blamed by New Delhi as well as Dacca for the prolonged stalemate and wanted to be sure he could act quickly once he reached the United States. The Vietnam War was over and the United States had recognized China, and Bhutto was sharp enough to understand that this was "an age of détente and reconciliation, not of belligerency and strife. It is an age when the new generation all over the world has rejected the philosophies of hate. . . . I say to my people as I would say to the people of India: do not be sworn to eternal hostility against each other; if you do, only your comon enemy, which is squalor and poverty, will triumph."[4] Here was Zulfi at his best, but frustrated opposition members of the assembly had staged a walkout before he had even uttered those words of wisdom. Wali Khan and his NAP backers, and men like the former air marshal Ashgar Khan and Ahmad Raza Kasuri would not sit and listen.

"If I were sitting with them I could have made speeches, inflammatory, volatile, convincing," Zulfi shouted in reaction to the insult, "because there is much to be said on both sides of the issue. This is what democracy means . . . Mr. Speaker, Sir, politics is a great art. But it involves ethics. What is the ethics in their politics? What is the morality of their thoughts?"[5] Zulfi had, of course, argued the opposite side of this issue, as he had of most political positions he took, and as he would continue to do. He was always ready to switch at a moment's notice; it was one of the keys to his singular success not only in climbing to the top of the slippery pole of Pakistani power but in clinging to the precarious perch. "We are a product of the people," Zulfi cried out to the chamber, most of whose opposition seats were now offensively empty. "We have come here to guard the ramparts of Pakistan and not to destroy and decompose what remains of Pakistan."[6] He meant it. He believed every word. To Bhutto, politics and patriotism were inextricably joined, each a bulwark of the other.

Even though the hour was late and the seats only half filled, and it was just a week before his long and carefully planned trip, Bhutto did not stop talking but warmed to his theme. "They know that each step that we have taken is in the highest interest of the people of Pakistan and that we will never take any step, never contemplate any measure that will be inimical or detrimental to the greater glory of Pakistan and its ultimate interests. That is why they have bolted out. . . . These are runaway people. These are people who do not want to face the truth. These are people who would like to misguide the youth of this country . . . They would like to see the blood of the people shed for their own selfish interests. [T]hey walked out with a purpose. . . . [T]hey want to create bloodshed. They want the warm blood of this warm country to be shed on streets, not for a national purpose, but for their own selfish interests. They want to provoke the people."[7] But Zulfi promised not to be provoked.

The resolution on Bangladesh passed shortly after 3:00 A.M. on 10 July 1973.

That same day Bhutto flew to Karachi, where he spoke the next morning to the press. Heavy rainfall had caused severe flood damage from Dera Ismail Khan to Karachi since 7 July, when more than four inches of rain had fallen on Karachi in less than four hours. Zulfi was eager to leave before the next monsoon downpour. He had less than a week, after all, before he was due in Washington, and had planned layovers in Rome, Geneva, Paris, and London en route. Begum Nusrat had flown to Cambridge the month before to attend Benazir's graduation from Harvard, and would remain in the United States until she and Zulfi faced the cameras together in the White House and at subsequent state functions. Zulfi's presidential plane carried an entourage of handpicked companions and the PIA's loveliest hostesses. After the past few months of political haggling and infighting, he badly needed a holiday and was looking forward to this chance to unwind and relax. The "insurgency" in Baluchistan kept escalating, though Tikka Khan poured his best regular army troops into that impoverished, underpopulated border province, whose tribal links with Afghanistan and Iraq as well as Iran had long raised the specter of a "Greater Baluchistan" separatist movement. Less than a week earlier Bhutto had decided to appoint a secret committee of his most trusted advisers—Ghulam Mustafa Khar, Ghulam Mustafa Jatoi, and Rafi Raza—to "find a satisfactory long-term solution to the problem of Baluchistan." Too many "ghastly mistakes" had been made there. "One Government after another . . . foiled and fumbled." He charged his trusted troika to "find an honourable and lasting solution without being cornered. . . . [W]e must always keep in mind that the antagonists of today might get together tomorrow to pull the rug under our feet."[8]

The Marri and Mengal tribes continued to fight, daily gaining support as well as sympathy for their struggle against imported Punjabi and Pathan troops, viewed by most Baluchis as an "army of occupation." Former Chief Minister Sardar Mengal reported that the people were being "starved" by the government's "blockade," and it was sheer "desperation" that drove them to guerrilla warfare. Former Governor Bizenjo called Governor Bugti "the most hated man in Baluchistan." Bhutto had alerted Governor Bugti in February to reports he had received of an "NAP-sponsored plot" to "assassinate" him. In early May, moreover, Zulfi reported to Bugti that the Pir of Pagaro and other leaders of the UDF opposition had met in Quetta at the Khan of Kalat's house to "discuss my assassination" and to draw up "tentative plans for creating simultaneous trouble in NWFP, Baluchistan and Sind."[9] Bugti was by then hard-pressed by Sardar Mengal, Nawab Khair Bakhsh Marri, and other opposition leaders in his Provincial Assembly, who had no confidence in Bugti's appointed chief minister, Jam Mir Ghulam Qadir Khan, and demanded that the governor convene a session that June to vote on a motion to that effect. Bugti himself was well aware of the crisis of confidence over the Jam of Las Bela. Zulfi suspected that his powerful governor had chosen this "meek" chief minister in good measure because of the latter's reputed willingness to "take orders." Zulfi never trusted Bugti as he did Khar or Mumtaz, for they were his "brothers," after all, and belonged to his own party, which Bugti had never joined. Zulfi used Bugti as his most effective foil in Baluchistan against Bizenjo, Mengal, and Marri. He hoped Bugti could help him and Tikka stop the Baluch insurgency before it became a full-scale war, but always worried about what might occur if Bugti joined his "rebel" Baluch brothers or invited external support for their "separatist" movement from Baghdad, Kabul, New Delhi, or Moscow.

"Perhaps Khair Bakhsh Marri is trying to play a miniature Bangladesh," Zulfi had written to Bugti in early June. "Perhaps he wants to get some Marris across the border . . . to internationalise the problem with the help of Kabul Radio and Kabul propaganda. . . . We must therefore bottle up the migration even if it is a handful of Marris. You know the situation on the ground better. I therefore want your views."[10] Without admitting they were his personal views, Bugti sent Bhutto several "anonymous" reports, detailing the situation in Baluchistan, one of which noted that the central government's

> Counter Insurgency forces are in . . . disarray. Bhutto . . . has not yet realised the situation. He is still dreaming his dreams of one party government. His love for whole and sole power has blinded him to the danger of not sharing power. . . . Bhutto is convinced that sooner or later Bugti will turn against him. . . . Bugti knows this and so a situation of mistrust prevails. . . . The national leadership does not support Bugti. . . . The people's party press gives him no build up. . . . [So] Bugti is unable to give the people what they want. . . . If affairs continue to move as they are moving today, Baluchistan is lost to Pakistan.[11]

Flying above the dark clouds that hovered over Sind and Baluchistan, however, Zulfi Bhutto could relax, sipping cool drinks, tasting a succulent dish or two, lighting up a Havana. He was on the way to meet again with the world's most powerful leader, and thanks to his mastery of the artful dance of politics, he had achieved little less than a political miracle: Pakistan had been brought from darkest depths of defeat and despair to such peaks as the Simla summit and adoption of a new constitution. Now he would soon bring home the last prisoners of war and could then cut a deal with Mujib, if that self-deluded Bengali knew what was good for him. And he would bring home hard-currency funding and arms from Washington.

Zulfi knew he was scheduled to meet with Nixon just a few days before the Shah, for he had visited Teheran in May and learned many things there. Nixon would be interested in hearing what he had learned about the Shah. Zulfi hoped that by playing his cards right, he would be able to leap Pakistan over Iran as Washington's favorite and closest ally in the Gulf region. He could hardly believe how many new U.S. jets and tanks had been lined up around airports in Iran. Pakistanis were better soldiers than Iranians, and he was certainly a better bet than the tottering Shah for Washington largesse. He hoped for nothing less from this flight than to turn Pakistan into Washington's premier sword-arm of Asia.

By the time he reached Rome, Zulfi felt intoxicated by such thoughts and by all he had imbibed. At Rome's airport, however, he was brought down to earth with a disconcerting message from the White House: the state visit would have to be "postponed" because of Nixon's "sudden illness," which required that he be "hospitalized." The "illness," of course, was Watergate. Nixon and Kissinger decided that the Shah was too important and too sensitive to put off, but that Bhutto could be kept on hold for a few months without jeopardizing global "security." The only major damage might be to Zulfi's ego, which White House and State Department pundits knew was strong enough to survive virtually any blow.

Zulfi informed his ambassador to Switzerland that he would stay a bit longer in Rome, and a few more nights in Geneva when he reached there. He then wired Paris and London the same top-priority messages, deciding that his reply to Washington could wait a bit. He recalled how Kissinger had used the same poor-health cover to fly directly from Pindi to Beijing several years earlier, when Yahya helped the secretary of state arrange that first prerecognition visit with Mao and Chou. Zulfi simmered but remained outwardly cool, too astute to fire off the message that came to mind. Some day all of them in the White House and Foggy Bottom would wish they had been more considerate of the sensitivities of Zulfikar Ali Bhutto! Though his Washington appointments were cancelled, it never even occurred to Bhutto to fly directly home to flooded Karachi or drenched Islamabad-Pindi. A two-week holiday in Europe proved, indeed, much more salubrious than rushing on across the Atlantic might have been. Khar and Mumtaz flew immediately to Geneva to join him, and by the time they reached London's Heathrow on 23 July it looked as if half the Pakistanis in the U.K. were at the airport, waving green and white flags and shouting, "Bhutto Zindabad!" and "Jiye Bhutto!" as he stepped out of the plane. Zulfi had an opportunity briefly to meet there with the Shah, who was on his way to Washington. Nixon had "recovered" just in time for that state visit.

Prime Minister Edward Heath hosted a formal dinner for Bhutto at the Savoy the evening of 24 July, and Zulfi's spirits were clearly high by the time he rose to respond to the welcoming speech: "The Prime Minister has bowled us over with his charm and . . . I am speechless. . . . We have settled all the problems between ourselves. We have pledged that in future there would be no underground nuclear tests between Pakistan and Great Britain. We have agreed to retain our forces in Europe and . . . in passing we have also discussed the exasperating situation in the subcontinent." Zulfi was rather sobered by the lack of laughter that greeted his tasteless remarks, and concluded without further attempts at humor that it had been "a rewarding visit" and he was most thankful for the "opportunity to discuss with your leaders candidly a number of problems that affect us." He raised his glass to toast the prime minister and "the people of Britain,"[12] but ignored the queen, even as she had ignored his visit.

On 17 July 1973, while Bhutto was touring Europe, Sardar Mohammad Daud Khan, cousin and brother-in-law of Zahir Shah, who had also just flown to Rome, staged a palace coup in Kabul, proclaiming Afghanistan a "republic." President Daud, as he would be officially called till the bloody attack that killed him and most of his relatives and guards on 27 April 1978, permitted the king's wife and child to fly to peaceful exile with him outside Rome. Daud, who had warmly supported Pakistan's Frontier "Pakhtunistan" separatist movement more than a decade earlier, during his premiership, had long enjoyed cordial relations with Moscow as well as New Delhi, both of whom immediately extended diplomatic recognition to his new republic. Bhutto waited until he reached home before offering Pakistan's recognition—which he did less than a month after Daud took control, for soon it became clear that the coup was no threat to Pakistan but had been launched only to prevent a tougher imminent revolt by the much younger Communist "Turks" within Kabul's Assembly, Comrades Nur Mohammad Taraki and Babrak Karmal.

On 14 August 1973, after completing twenty months as president, Zulfi Bhutto changed his title to prime minister and addressed the people on National Day over radio and television, under his now formally operative constitution. "Today we bid good-bye," he said with premature optimism, "finally and for all time, to the palace revolutions and military coups which plagued Pakistan for nearly two decades."[13] He encouraged all voices to be "raised in defence of all view-points" but warned against "pandemonium." Democracy could work only if there was "discipline." He cautioned against "separatist tendencies," and urged the opposition to rest "content" and "wait until future elections." He reminded his listeners that "we have pledged to Allah" to permit no "intrigues subversive of Government." He spoke at some length about Pakistan's external relations, and sounded very much like Nixon discussing Watergate when he said, "Let me make it clear that, when I talk of the Opposition, I am mindful of the fact that the Pakistan People's Party may itself be in opposition some day. . . . I am no great Khan that I should entertain dreams of ruling

Pakistan in perpetuity."[14] He had as he spoke already ordered Governor Bugti to arrest former Governor Bizenjo, former Chief Minister Mengal, and Sardar Marri, but made no mention of those "insurgents."

The heaviest, deadliest rains in more than a decade poured from the skies over Pakistan that night. Floods washed away more than ten thousand villages, ravaged crops, and claimed countless peasant lives as well as precious livestock. Torrents of mud mixed with deadly detritus raced across millions of acres of scarred Punjabi and Sindhi soil. The twenty-one-gun salute to the birth of the new Constitution was muffled by the downpour, and the festive outdoor celebration earlier planned to hail the new prime minister and president, former Assembly Speaker Fazal Elahi Chaudhry, was canceled. Pakistan's economic losses from the unprecedented floods would prove to be greater than the total cost of the 1965 War.

In late August Zulfi sent Minister of State for Defense and Foreign Affairs Aziz Ahmad to Delhi to conclude negotiations of an agreement with Indira Gandhi's special representative, P. N. Haksar—signed on 28 August 1973—that started the repatriation of the ninety-three thousand Pakistani prisoners of war. Bengalis still in Pakistan were to be sent home to Bangladesh, and the prime ministers of Pakistan and Bangladesh would at some future date "meet to decide what additional . . . persons who may wish to migrate to Pakistan may be permitted to do so"—more than fifty thousand considered themselves Pakistanis. Two decades later, however, those unfortunates had not yet been accepted by the nation they considered their own; most Sindhis as well as Punjabis believed that Pakistan had taken in more Muhajirs than either Karachi or Lahore had room or work for. Mujib repeatedly threatened to put 195 Pakistani prisoners of war on trial for "war crimes" committed in Bangladesh. India offered to take custody of the latter prisoners till final arrangements could be made concerning their treatment after Pakistan recognized Bangladesh—which Bhutto now had the assembly's permission to do at his own discretion. With this Delhi agreement concluded just two weeks before his rescheduled summit with Nixon, Bhutto had fresh reason to feel proud of his diplomatic waltzmanship.

"Citizens," Premier Bhutto addressed his nation on the eve of his flight west that September,

> mindful of what actually happened in 1971, we cannot but have reason to be satisfied that we secured the withdrawal of the invading armies from our territories and have now obtained the assurance of the release of our prisoners of war. . . . [I]n the stately environment of the nineteenth century Europe, a Metternich would have achieved more . . . but we need not be apologetic about it.[15]

Pakistan's modern Metternich was welcomed warmly by Nixon on 18 September 1973. The president called preserving "the integrity and independence of Pakistan" the "cornerstone" of U.S. policy. Rain forced the reception, which was to have been held in the Rose Garden, indoors. "I am sorry that the ceremonies were somewhat marred by the rains," Bhutto told Nixon, "but nothing can mar the eternal friendship and warmth between our two peoples." Vice President Spiro Agnew hosted lunch that afternoon, and Zulfi toasted him. Dinner, of course, was hosted by Nixon, and Henry Kissinger was present. In response to Nixon's after-dinner remarks, Bhutto commented on their earlier secret discussions in Nixon's office:

> We discussed three matters, economic matters, cultural matters and military matters. I don't know how, but cultural and military matters got intertwined, perhaps because Dr. Kissinger was there. (Laughter). . . . We were told that since military and cultural matters are interrelated, we must know that Jill St. John is booked for the Soviet Union—(Laughter)—and Raquel Welch is earmarked for China—(Laughter)—as far as our old friend Pakistan is concerned, Tallulah Bankhead is there available for Pakistan. . . . So we told our friends candidly that what we are interested in is not obsolete spare parts, but in red hot weapons. (Laughter).[16]

Zulfi considered Nixon "a great and lofty president," and "a world statesman."

The next day Bhutto addressed members of the National Press Club, informing them that the "new Pakistan" was "determined to nurture democratic institutions." The new Constitution "ensures the maximum autonomy" to all federal units, while it "safeguards national solidarity." Socially and economically, Pakistan was "forward-looking" and soon would be "no longer dependent on the assistance of its friends." In a joint statement issued on the eve of Bhutto's departure from Washington on 20 September, Nixon promised "strong U.S. support" to Pakistan, and Bhutto promised to try better "to control narcotics traffic" and "to eliminate poppy cultivation."[17]

That night Bhutto addressed the 28th session of the General Assembly of the United Nations in New York. Pakistan would not recognize Bangladesh, he reiterated, until all 195 of the Pakistani prisoners still held for possible trial were released and repatriated. He no longer hoped to win Mujib over to his earlier "commonwealth" concept of confederation but still dreamed of resurrecting pan-Islamic ties, of restoring "fraternal sentiments," with Bangladesh that might help erase the traumatic memories of the darkest days of 1971.[18]

Turning to India, Bhutto argued that "but for the unresolved dispute over the State of Jammu and Kashmir," Pakistan had "no territorial ambitions":

> If we support the cause of Jammu and Kashmir, as we must, it is because the right of the people of Kashmir to decide their future has been recognized by India and expressly pledged to them by the United Nations. The resolutions of the United Nations on Kashmir have not been implemented and the dispute has festered and led to constant strife and major conflicts in the last twenty-five years. To remind the world community of its responsibility, to ask for the fulfillment of the pledges solemnly given and a hundred times repeated, is not to seek confrontation but rather the only way to avert conflict. . . . [T]he important issue of self-determination for the people of Jammu and Kashmir, to which the United Nations and both

India and Pakistan are committed, will have to be faced and honourably resolved for the good of all of us.[19]

On Kashmir Zulfi Bhutto had always been a passionate yet principled statesman and patriot, one of whom every Pakistani could be proud, as all of them were who heard him eloquently articulate his nation's plea "for sanity and justice in the relations between Pakistan and India." With long-divided Germany and Korea moving toward peaceful settlement of their disputes, and with the Vietnam War finally over, Bhutto wisely noted that "peoples of the world, the youth at least, are turning away from the narrow appeal of national or doctrinal prejudice." Why should not India and Pakistan and Bangladesh join that promising parade toward a more rational future? Even the superpowers were moving inexorably in the direction of détente. "Who is there, indeed, that would not applaud the relaxation of tensions that cast their sombre shadows over the entire globe?"[20]

If it was "a long road," Zulfi presciently warned, from "confrontation between the great powers to conciliation," it might prove an "even longer one" to the "ideal of co-operation" worldwide. He pledged Pakistan's support to "all genuine movements for the emancipation of peoples from alien domination." This was "the cause of the Third World. It is the struggle of Asia and Africa. It engages us emotionally and enlists our whole-hearted sympathy." Though Pakistan had to date been "excluded" from the Conference of Non-Aligned Countries because of its commitments to SEATO and CENTO, Zulfi hoped not only to join the conference but to lead the Third World, that "major segment of humanity," which felt "the stirrings of a revolution" that could "never be suppressed." Bhutto waxed ever more eloquent, proudly proclaiming, "We are all the children of revolution, a revolution whose onward march may sometimes be arrested by our own weaknesses and contradictions but cannot be stopped by any force in the world. . . . But we in the Third World should not be swayed by conspiratorial theories and forget the fact that the fault lies on the ground and within ourselves. To quote Meredith, we are betrayed by what is false within."[21] Alas! Bhutto often understood the Zulfi side of himself far better than he managed to control it.

Bhutto also spoke to the Foreign Policy Association and the Asia Society while in New York, and two days later was the guest speaker on the popular television program "Meet the Press" on the eve of his flight home. When asked if he feared that "big-power détente" posed "serious dangers" to other parts of the world, Zulfi responded that "the great ones" should realize that "the romance of revolution prevails over the machinations of great or small powers." He was reminded that as a student in California he had worked for Helen Gahagan Douglas in her congressional campaign against Richard Nixon, and was asked if he thought Nixon had used tactics in that campaign similar to those he had used in Watergate. "I don't remember that campaign," Zulfi replied. "I wasn't intensely . . . involved . . . I was a foreign student and I thought that she should be supported because she was a woman."[22]

He was welcomed back to Pindi's international airport by a noisy crowd of supporters and admirers, some of whom were less than attentive during the playing of Pakistan's anthem. "I noticed just now that somebody was playing a record," Zulfi scolded, "while some others were raising slogans. Playing of the national anthem is a very solemn occasion when everyone should remain silent. Similarly, when the Holy Quran was being recited, some of my friends were raising slogans. . . . It is my duty to tell you that this is wrong. . . ." Nixon's verbal construction seems to have caught the prime minister's ear, but then he must have noticed how his greeters reacted to his pedagogic scolding and softened his tone: "Since I am . . . talking to you not as Prime Minister but as your brother, I am telling you these things."[23] Then he reported on a successful trip: "I could have done no more."

The following day—28 September—the first batch of Pakistani prisoners of war returned from India. Prime Minister Bhutto welcomed them to "a new Pakistan, a vibrant Pakistan imbued with the urge and determination to forge ahead and carve out its rightful destiny."[24]

Much though Bhutto spoke of the "new Pakistan," it was Khar and Mumtaz, Jatoi and Pirzada who continued to stand or sit beside him while all the old intractable problems of poverty and feudal inequities, provincial conflicts, and tribal disorders plagued and confounded their heroic attempts to cope or solve, even as nature heaped floods on top of prolonged drought. Export earnings prior to the floods had been so promising—cotton leading the way toward what looked like a major economic turning point—that Zulfi felt optimistic about full recovery. After the floods of August, however, high surtaxes had been added to virtually all imports, stimulating urban discontent, last-quarter strikes, protest marches, and riots. Zulfi's costly new elite Federal Security Force (FSF) provided him ample protection and information but were never otherwise deployed.

Haq Nawaz Tiwana, first director-general of the FSF was soon replaced by the much shrewder, more experienced Masood Mahmood, who had been introduced to Zulfi Bhutto in 1958 at the bar of their club by Police Inspector-General Ikram Khan. Bhutto had probably consumed "half a dozen martinis," Mahmood estimated, recalling that meeting, and had muttered, "I know *Masood,*" waving aside Ikram Khan's more formal introduction. "You'd better call me Zulfi," he told the man who was later to become his closest confidant, and then the "approver," whose evidence would help send Bhutto to the gallows, "or I'll slap you!"[25] They hit it off well from the start. Masood liked Bhutto's genial, convivial manner and his sense of humor, though, as Masood recalled, "he was the worst lecher!" Having entered the Indian Police Service in 1945, after graduating from Lahore College and studying law for two years at Lincoln's Inn in London, Masood was a man of wide-ranging experience and sophistication. He was an attentive listener, moreover, and still enjoyed a martini or two himself, "*very* dry!"

On 28 September 1973, a fourth attempt was made on the life of Wali Khan, as he was driving in his Jeep from Mardan to Swat. The outspoken opposition NAP leader's driver and bodyguard were both killed, but Wali Khan walked away unscathed. Bhutto's tour of the North-West Frontier, scheduled to start the following week, was immediately canceled. Zulfi flew instead on 2 October from Lahore to Karachi, where he received a regal welcome. Though the month-long Ramadan fast had begun and Karachi was sweltering in 100-degree heat, the airport reception crowd was impressive, thousands of schoolchildren waved Pakistani flags as he drove from the airport and sweating workers, given time off on this "special holiday," turned out to greet their Quaid-i-Awam. From Karachi Zulfi flew to Larkana. He was still relaxing in his newly air-conditioned home there at the start of the Ramadan (Yom Kippur) War in Israel that October 1973.

Zulfi could hardly contain his delight as word reached him of the two-pronged Arab attack against Israel, launched without warning on the Day of Atonement and fasting, by Egyptian and Syrian forces. He quickly wired UN Secretary-General Kurt Waldheim, urging fulfillment of Security Council Resolution 242. Zulfi told his service chiefs, with whom he met in Pindi on 11 October, "[a] cease-fire will not do. . . . It is essential that Arab territories held by Israel—illegally by Israel—should be vacated."[26]

Zulfi now launched a historic initiative toward a second Islamic summit, inviting the heads of all Arab and Islamic nations to meet in Lahore the following February. "It was his most important innovation," his daughter Benazir reflected during her own later brief tenure as prime minister. "He gave Pakistan this linkage to the countries of the Gulf. . . . [O]n defense lines, on economic lines, on foreign policy lines, he carved out this bloc of Islamic countries . . . uniting the countries of the Muslim world, which gave birth not only to the Islamic Conference, but also to a newfound assertiveness. . . . [T]o have unified action . . . he sent soldiers abroad, he sent labor abroad. . . . [A]fter [my father's] death Zia tried to say that he [Zia] was a 'Soldier of Islam,' but he [Zia] was a soldier of the devil."[27]

The first Islamic summit at Rabat in 1969 had been attended by only twenty-four of the thirty invited states, including India, much to Pakistan's frustration and dismay. The Lahore summit, however, thanks to Zulfi's strenuous personal efforts and singular energy, would lure no fewer than thirty-eight Muslim states to Pakistan, and prove to be Bhutto's greatest diplomatic triumph. It attracted every major Muslim head of state except the Shah of Iran, who was smarting over Zulfi's "unstable" report about him to Nixon. Bhutto flew to Teheran in mid-October 1973 to try to reassure the Shah and mend fences he had destroyed in what he thought was an off-the-record remark in the White House. The Shah was cool to Zulfi's apology and rejected his invitation to the summit. Zulfi felt convinced, he later wrote, that the Shah "had a complex towards me . . . He was intensely envious. . . . His grandiose designs and fanciful ambitions of being the modern Cyrus the Great if not Greater contributed in no small measure to his ruin. . . . He lost all touch with reality."[28]

Zulfi wired messages of support and sympathy on 16 October to President Hafez Al-Assad of Syria and Hasan Al-Bakar of Iraq in their struggle on behalf of the "indomitable Arab nation" against "the Israeli aggressor." The same day he also wired King Faisal of Saudi Arabia and King Hussein of Jordan that "the

Government and people of Pakistan have always fully supported the cause of the Arab people and will continue to do so."[29]

At a press conference in Karachi four days later, Bhutto said that in the current Middle East conflict "the whole of the Muslim world is on trial. . . . Pakistan is not a non-aligned state. Pakistan is finally aligned to the principles of justice and international law and to a durable structure of international peace."[30] Pakistan was doing "everything within our power and capacity" to help the "Arab position," and to give "tangible support" to the Arab states in their struggle. "It must be remembered that a war is being waged, every minute is precious and every action has to be taken deliberately," Zulfi announced. "It is not possible for me to make full revelation of all our efforts. . . . We know the people of Pakistan have faith in their Government and in the successful execution of this Jehad."

While Bhutto busied himself with pan-Islamic affairs and secret plans for amassing a nuclear arsenal for Pakistan, the internal situation deteriorated in Sind as well as in Baluchistan and all along the North-West Frontier. The former governor of Sind, Mir Ali Ahmad Talpur, found the

> country drifting towards strife, chaos and disintegration. The present Pakistan Government seems to be intent on pushing it over the precipice. Its actions in Baluchistan are nothing short of anti-national; in Frontier it seems to be pushing straight for a showdown with democratic forces; the Punjab is being ruled through simple gangsterism; and Sind is . . . deplorable. . . . people dying in encounters with police. . . . As for Mr. Bhutto . . . if he stays, the country cannot survive. He is really a show boy. . . . The reality behind his rule is something very ugly.[31]

Governor Bugti resigned on the last day of October 1973, writing to say that he had earlier accepted responsibility for the administration of his province "with the thought and intention of being able to do justice for the downtrodden masses of Baluchistan . . . but . . . in spite of my best efforts, things don't seem to get done as I would want them done, or are not allowed to be got done!"[32] Zulfi suspected that Bugti only wanted more power and might rest content with "elevation," under the new Constitution, to the chief ministership of Baluchistan, which was the democratic job former Governor Khar had now taken for himself in Punjab. Khar was sent on a secret mission in early November, therefore, to try to talk Bugti into staying on, for the "insurrection" in Baluchistan kept getting harder to handle. Zulfi may also have hoped that his big three prisoners, Bizenjo, Mengal, and Marri, might actually agree to form a new, even more popular Baluchistan cabinet. But though he had air-conditioned their Sihala Rest House prison, none of those redoubtable NAP leaders would even begin to consider any political deal while they remained under armed guard.

Zulfi's most trusted Baluchi supporter was conservative Sardar Ghaus Bakhsh Raisani, who had been governor before Bizenjo, and would soon be reappointed to that job by his prime minister. Raisani now still served as Zulfi's minister for food, agriculture and rural development and regularly sent

"*Secret*" reports to Bhutto about how badly Bugti and his comrades misbehaved. Bugti himself made no secret of his frustrations with the central government, and whenever interviewed by a foreign correspondent confessed that the "area of insurgency" in Baluchistan was "increasing." As was the death toll of Pakistani soldiers as well as Marri and Mengal tribesmen, thousands of whom would be killed in the next year of that secret bloody civil war.

U.S. Secretary of State Kissinger, en route to China, stopped at Pindi for dinner with Prime Minister Bhutto in early November to discuss the ever-troublesome problems of the Middle East. "We have a vital stake in the region," Zulfi told him. "We have taken a position which we believe is based on principles. . . . We shall remain in touch with our Arab friends. And whatever modest contribution Pakistan can make for a just settlement . . . we shall always be available. I want you to remember that. . . . In Pakistan we have had our problems. . . . We have broken the back of our basic problems."[33] Less than a month later, French Secretary of State for Foreign Affairs Jean de Lipkowski flew in for top-secret meetings with Bhutto, but Kissinger was kept informed of them by cypher messages transmitted from the sealed telecommunications chamber in the new U.S. embassy compound in the diplomatic quarter of Islamabad.

Bhutto sought to cement Pakistani ties with Turkey by ordering erection of the Kamal Ataturk Memorial in his hometown of Larkana to commemorate the fiftieth anniversary of the founding of the Turkish Republic on 30 November. "I make no secret of the pleasure I feel at the building of this memorial at a place where in my boyhood, I read about Ataturk and marvelled at his courage, his lucidity and his determined fight against reactionism and obscurantism."[34] The Turkish minister of state for land reforms flew in for the gala ceremony, and stayed with the Bhuttos at Al-Murtaza—outside of which a monument to commemorate the achievements of President Sukarno of Indonesia was being erected. In several respects both men had been political role models for Bhutto: Ataturk igniting his passion for Pan-Islamic unity and Muslim modernism; Sukarno stimulating other passions, including his resolve to be free of any superpower control.

The twentieth of December 1973 marked the second anniversary of Zulfi's rise to presidential-now-prime-ministerial-power. His adept dance at the top had rendered all his angry critics and formidable adversaries helpless. Though the political opposition still sniped at him daily, from the left as well as the right wings of his own party and every other tribe or group, nothing brought him down. Indeed, nothing seemed to touch or faze him. He moved smiling or scowling through the worst weather, the harshest accusations, the loudest epithets, as though coated invisibly with impervious plastic—fearless, death defying, power affirming. The crowds continued to love him, cheer him, laugh with him, for he always told them a spicy joke or twenty, and all of his gestures were fun to watch, as was his remarkably expressive face. His voice, moreover, was a miracle of primal energy, its potent power rumbling, echoing, uplifting. He raised the spirits of the saddest, poorest haris and ragged tribesmen and of the oldest as well as the youngest women, many of whom walked miles just for a glimpse of him or to hear the titillating cadences of his honeyed tongue. He was the living image, the three-dimensional vital incarnation of many Pakistanis' wishful dream of themselves: tough yet tender, a practical idealist, smart as a cat-o'-nine-tails whip, swifter on the draw than a Texas cowboy, and better in bed or as the head of state than any Indian, Russian, or Englishman. He was Quaid-i-Awam Zulfikar Ali Bhutto! What did it matter if "Roti, Kapra aur Makan" all cost so much more nowadays? Election slogans were forgotten the world over, and inflation was a universal problem, was it not?

"You know how it is," Zulfi admitted to an Indian journalist, who had flown in that December to interview him, "people are charlatans. . . . They want some pecuniary benefits. They put it in the garb of some noble ideal and say that I have to fight in election; I need funds; I have to buy cars; I have to pay workers; this is needed to buy this and that is needed to buy that. After all, there is inflation in the world. People find it difficult to work out a living and so they enchant people that I am a guerilla leader and I can do this and I can do that."[35]

Enchanting people was what Zulfi did best. He had always been enchanting, from his earliest years in Larkana, through his "salad days" in Bombay and Karachi and California. He had the gift, the eyes, the voice, the power to enchant, to mesmerize, bewitch, enthrall. He had fingers so remarkably tender and dexterous that his mere "touch" could be "felt," remembered for more than a decade by at least one Pakistani beauty, whose cherubic face had smiled up at him as he entered a ballroom, "filling it" instantly, that young star recalled, with his "magical" presence.[36]

Prime Minister Bhutto laid the first brick of Karachi's long-awaited steel mill at Pipri on 30 December 1973. The mill would be constructed with Soviet aid, marking the dawn of a "new era" of Pakistani "technical maturity" as well as "international collaboration in this particular field." It would "help organize the defence potential of Pakistan on a scientific basis," Zulfi noted, and Port Qasim, currently under construction next to Karachi's older port and shipyards, would be used to expedite the mill's completion. He had flown to Karachi at year's end not only to launch the new steel mill but also to address the annual Sind convention of his People's Party.

The gala two-day gathering of Sind's political elite was held in Karachi's new National Stadium's hall, decked outside and in with enormous hardboard pictures of Chairman Bhutto, looking a bit like Chairman Mao, his achkan collar open, an enigmatic smile on his pudgy-chinned face, the fast-thinning hair gone white, a fitting portrait of this people's prime minister. A few days before the convention began Zulfi had replaced Mumtaz Bhutto with Ghulam Mustafa Jatoi as chief minister of Sind. Too many reports about Mumtaz's inability to keep Karachi "clean" and "safe," and to control his cabinet and his own nocturnal "wild boar hunting" had reached Zulfi's ears. He continued, of course, to trust and repose full faith in his "brother," but even Mumtaz had complained about having "worked myself into a dangerous state of health," and insisted that for his own "branch of the family . . . monetarily our position has become disastrous."[37] So Zulfi decided to give Mumtaz a well-deserved rest, and to elevate Jatoi from the Ministry of Railways he had only grudgingly agreed to head to the job he had set his heart and sights on long before Zulfi Bhutto came to power.

As central minister of railways Jatoi had merely been able to order a new railway station built in his wadero-capital of Nawabshah, but as chief minister of Sind he would virtually be able to bring Nawabshah to Karachi, and vice versa. Robust Chief Minister Jatoi entered the Karachi convention hall shortly before noon on that last day of 1973, and was pleased to note that Mumtaz had absented himself. Jatoi acknowledged the cheers of his Nawabshah delegation, raising one hand above his heavy head, barely smiling, his cheeks too full and his mouth too tight to smile because he had put on so much weight. Zulfi arrived late, but as he entered, the crowd of more than a thousand jumped to their feet, chanting, "Jiya Bhutto!" and "Quaid-i-Awam Zindabad!"

The tallest of Masood Mahmood's FSF guards surrounded Bhutto, covering him with catlike care and precision wherever he went because there had been many "rumors" of "assassins" moving down from Baluchistan or Afghanistan and the Frontier, or possibly heading west out of Delhi or Bombay—"hired guns" who would try to kill this great new leader not only of Pakistan but of the entire Islamic world, for he would soon host the summit conference in Lahore that would bring to his side the king of Arabia and the presidents of Libya and Syria. So there was more security and less "pressing of the flesh" as Zulfikar Ali Bhutto climbed to the dais and started to speak. But the roar kept him waving and smiling, for these were his people, his party family, the party he had conjured up out of his own enchanting mind and spirit, which he alone led. Others like Ministers J. A. Rahim and Dr. Mubashir Hasan had helped him organize and launch it, yet where would any of them be now without their Quaid-i-Awam?

Bhutto had brought Pakistan's first steel mill into being and was gathering weapons of science and secret knowledge that would soon bring the nation more power than any of his mediocre military predecessors had even dreamed possible. With those devastating floods of August, and the quadrupling of crude oil prices since the Arab-Israeli October War, however, his major problem now was to find money enough to do all the things he still had to do in order to achieve his destiny, to reach those golden goals he had set for himself and his people, his Pakistan. Ample money was here in Karachi, but the private bankers and greedy capitalists, the Habibs and Haroons, all of whom were "jealous" of his brilliance and greatness, just as Ayub had been, would never give him the financial support he so desperately needed. None of them understood the urgency he felt, the pressure of racing time, the imperative of fast action, and none of them truly trusted him. Zulfi knew that, which was why at the formal dinner that New Year's Eve, when he briefly addressed Karachi's industrial and banking leaders, Pakistan's capitalist elite, he reassured them that they had "nothing to worry about." The "wild rumors" that he was about to nationalize all banks were "not so." He calmly waved aside their concern, raising his glass for one last bubbly drink before he left for a more pressing engagement.

On 1 January 1974, as a New Year's "present" to the "People of Pakistan," Bhutto announced the immediate nationalization of all banks. The doors and windows through which thousands of millions in hard currency had fled the country during the past two years were closed. The banks were secured by Bhutto's FSF guards; they too were part of his "present" to the "People of Pakistan."

12

Prime Minister Bhutto at the Peak of His Power

(1974)

Zulfi celebrated his forty-sixth birthday that January of 1974 with a most successful wild boar hunt, by day as well as at night in his compound in Larkana. He was joined there by his old friend Piloo Mody, then president of the Swatantra Party of India, who hoped that together he and "Zulfi, My Friend," would bring peace to South Asia. Piloo's American-born wife, Vina, had flown with him to Larkana. "I always liked Zulfi," she recalled. "Of course, he was a womanizer, but I think he tried to do some good, and he was a generous friend and host."[1]

Messages of acceptance to the forthcoming Islamic summit kept arriving from more than thirty heads of state, and Zulfi knew it would be his most important opportunity to show the world how great a diplomat he was and to treat his powerful and noble guests with the generous hospitality that every Pakistani aristocrat was taught from childhood. Nothing tasty or succulant would be missing from the conference in Lahore, nothing tempting or delicious unavailable to sate the appetites of his guests, by day or at night. Before the summit began, however, there were mundane matters of administration to handle.

Nawab Bugti's resignation as governor of Baluchistan was finally accepted by Prime Minister Bhutto in early January, and Mir Ahmad Yar Khan of Kalat was named to replace him. This did nothing to stop the civil war raging in the rugged and remote regions of that vast province, for the Mengals and the Marris and Bizenjo tribesmen knew that as long as their sardars remained prisoners of the central government, and Tikka Khan's troops kept firing, there would be no peace for Baluchistan, only the sword of continuing conflict. Bhutto nonetheless insisted wherever he went in Sind and spoke that busy month that as long as his "People's Party was strong, Pakistan was strong and stable." And he truly believed that so long as he remained his people's leader, the party could not help but be strong and reflect the "will" of the people because his was a mystic "union" with them, an instinctive understanding, intuitive empathy, which guided his footsteps along the otherwise obscure path to the watering place of wisdom.

Wherever he stopped on his frenetic tours he collected the written appeals of those who came to see and hear him—half-naked and barefoot haris or town laborers, the impoverished and powerless, women and children—all were encouraged to toss their folded paper messages, scribbled prayers and hopes onto the platform where the prime minister stood. And each of them prayed that he or she would be the lucky one in ten thousand whose plea might be answered someday, land returned, taxes reduced, fines or prison sentences negated by the great leader of the people, who could by a word or the stroke of a pen, save or destroy a life, a dream—the whole world of countless haris, whose hollow-cheeked faces and imploring eyes looked up to him as they would to a Sufi saint or God Almighty. Like every ancient raja of Sind before him, Bhutto made it his practice during his birthday month to tour his region of Larkana. This year he toured his neighbor Jatoi's region of Nawabshah as well, accompanied by the new chief minister, meeting their people, meting out "justice."

Pakistan's leading jurists and advocates met for two days in early February 1974 in Karachi to discuss their ideas of justice and the many injustices that remained under the Constitution of 1973, whose fine section on "fundamental rights" had been suspended the day after that document was almost unanimously adopted. Draconian Defense of Pakistan Regulations (DPR) continued instead to proliferate, as they had during the darkest days of military rule, and Bhutto's own attorney general, Frontier barrister Yahya Bakhtiar, called for their termination, advocating "full and unfettered" restoration of that elusive "rule of law" that barrister Bhutto so often bragged of when he traveled abroad. Bhutto, who addressed the learned gathering on 9 February, insisted that he was "bound to view the issue of fundamental rights in a framework which lays greater stress on the fundamental rights of the nation as a whole to preserve its territorial integrity, to combat subversion . . . to retain a sense of security. . . . Once the situation changes—and the change can best be evaluated by the Government which alone has all the relevant data at its disposal—the state of emergency will be discontinued."[2]

Senior Supreme Court Advocate A. K. Brohi strongly disagreed with the prime minister's attempt to defend his rule by regulations for reasons of "security." Brohi cautioned that "all too often, alas, the doctrine of the rule of law and the principle of democracy are invoked by the custodians of political power not so much to render justice as to work injustice." It was not the last time that Brohi and Bhutto would confront each other at opposite sides of a legal dispute. "A vast amount of legislation today consists of ordinances . . . treated as having the force of law by means of a polite legal fiction," Brohi reminded his colleagues of Pakistan's bench and bar. "Whether humanity will take the path that leads to freedom or the road to serfdom is an issue which is still an open one. . . . The highest ideal for man is to let his conduct be regulated by the forces of reason, motivated unconditionally by the pursuit of the ideal of justice,"[3] concluded the soft-spoken advocate, who would serve as the "federation's" lawyer three years later, when he spoke with equal passion in defense of the martial coup launched by Zia ul-Haq against Zulfikar Ali Bhutto.

Senior officials of most of the thirty-eight Islamic countries (the PLO delegation was led by Yasser Arafat) attending the Lahore Conference met on 18 February 1974, and for three days the Muslim foreign ministers hammered out draft resolutions that would be considered by the conference, convened in Lahore on 22 February. Until the very eve of the formal convening, however, whether Prime Minister Mujibur Rahman of Bangladesh would attend remained unclear. But that evening Prime Minister Bhutto dramatically appeared before a battery of television cameras and radio microphones to announce in solemn, almost lugubrious tones his momentous decision to "recognize" Bangladesh. Though Mujib had not as yet promised to drop his charges against the 195 Pakistani prisoners still held in India for possible trial as war criminals, Bhutto's bold initiative cleared the way for Mujib's attendance at the conference, thus at least symbolically reuniting the war-severed wings of old Pakistan. It was Zulfi's first summit coup but would not be his only diplomatic victory during the most jubilant and spectacular week in the recent history of Lahore.

"In the name of Allah, most gracious, most merciful," Prime Minister Bhutto began in welcoming his "Majesties, Royal Highnesses, Excellencies," and "dear brothers in Islam," to Pakistan's premier Islamic capital, whose old Mughal Fort, great Masjid, the largest mosque in all of South Asia, and beautiful Shalimar gardens provided the proper backdrop for this singular gathering of the Muslim leaders of more than one-fifth of humankind. As the home of Iqbal, moreover, and the venue for adoption of the Muslim League's famous 1940 resolution, known to history as the Pakistan Resolution, Lahore remained particularly significant in Pakistan's own history. Bhutto's address and the conference itself focused on the Arab-Israeli conflict, and how to resolve it. "The tragedy of Palestine has agitated Muslim minds for half a century," Bhutto argued. "Israel has gorged and fattened on the West's sympathies, nurtured itself on violence and expanded through aggression. . . . [s]ituations arise in which there is no choice but war against the usurper. Such a situation was created for the Arab peoples."[4]

Zulfi never directly mentioned Kashmir during the conference but made a number of oblique references to it, including, "Your host country . . . has been a victim of international conspiracies and is concerned with an intense question in which, . . . its stand is based on nothing but justice and concern for Muslim rights." He also pointedly warned in reference to the Arab-Israeli conflict that "disengagement . . . is not peace." No Indian who heard or read that phrase had any doubt but that he was thinking of Tashkent and Simla. Not that Indira Gandhi or any of her representatives had been invited to the summit. But she and Swaran Singh and Kewal Singh listened carefully, and read and reread Bhutto's official, and most of his unofficial, statements, for he was no undergraduate declaiming his passionate pan-Islamic message to an unsophisticated coed in Southern California. His attentive audience included King Faisal and Presidents Sadaat, Boumedienne, and Assad, the charismatic Colonel Qaddafi and Yasser Arafat. "Among the Arab territories occupied by Israel, Al-Quds

[Jerusalem] holds a special place in Muslim hearts. . . . Except for an interval during the Crusades, it has been a Muslim city—I repeat, a Muslim City. . . . Let me make it clear from this platform that any agreement, any protocol, any understanding which postulates the continuance of Israeli occupation of the Holy City or the transfer of the Holy City to any non-Muslim or non-Arab sovereignty will not be worth the paper it is written on." Then he coyly added, "This is not a threat." Yet, "In this respect, there is a fire in our hearts which no prevarication, no skillful evasions on the part of others, will ever be able to quench."[5]

It was not only as their most eloquent Muslim host and brother, however, that Zulfi Bhutto addressed this stellar audience in Lahore but as a Third World leader, whose new economic strategy could topple the global dominance of both superpowers and the entire industrialized West.

> The Third World has emphasized, time and again, that poverty and affluence cannot co-exist. But . . . we ourselves have not fully realized the nature and value of economic power nor grasped the urgent need of developing science and technology for our progress, indeed for our very survival. . . . The war of last October has, however, precipitated a chain of events and created an environment in which the developing countries can at last hope to secure the establishment of a more equitable economic order . . . by the demonstrated ability of the oil-producing countries to concert their policies and determine the price of their resources. This may well be a watershed in history . . . an unprecedented shift will occur in the global monetary and financial balance of power. The Third World can now participate in the economic and financial councils of the world on an equal footing with the developed countries.[6]

Master of realpolitik strategy that he was, Bhutto always thought ahead; he could see himself at the helm of a Third World coalition, including China, someday, whose economic resources, as well as its vast population, would not only equal but even exceed the strength of either the First or Second World of Western and Eastern powers. With Qaddafi and Sadaat seated on one side of him and Faisal and Mujib on the other, he already felt potentially much stronger than India; it was an overwhelming transformation, less than three years after the disastrous Bangladesh War. "The Muslim countries are now so placed as to be able to play a most constructive and rewarding role for cooperation among themselves. . . . It is time that we translate the sentiments of Islamic unity into concrete measures of cooperation and mutual benefit. It will bring us strength in spirit and substance." In contrast to the "materialistic West" and "spiritual East," Islamic nations have been called the "People of the Middle," Zulfi noted; Pakistan, as "the midmost nation of the People of the Middle," was charged with the mission of "mediating conflicts." Zulfi was good at that, born diplomat that he was. He found it as easy to hunt or speak with any Arabian sheikh as he did to drink or dine with any North African colonel or radical revolutionary. Yet Bhutto alone could never have brought so many disparate leaders together, without the common enemy of Israel and its "imperialist supporters" to assist him.

> As I survey this splendid gathering, I recall that as a young student twenty-six years ago, I was asked to address a student body of a University, almost wholly non-Muslim, on the Islamic Heritage. . . . I spoke of Muslim unity against exploitation and of Muslim revival and sketched a plan for a Muslim commonwealth. I ventured to predict that a movement in this direction would take shape in the next twenty years. . . . [L]ike all of us, I have been assailed by doubts whether this vision of mine would be fulfilled. Today, despite all difficulties in our path, I bow my head in gratitude to Allah for making me witness to a scene which should dispel those doubts.[7]

Not since the heyday of Mughal imperial glory had Lahore seen so many resplendent Muslim monarchs parade in its Shalimar gardens, where exultant cries of "Bhutto-Mujib *bhai-bhai!*" (Bhutto-Mujib are brothers!) alternated with "Bhutto Zindabad!" Few Pakistanis feared that any of the 195 prisoners would ever face a Bangladesh or Indian court, for Mujib openly said that he hoped the "matter" of the remaining prisoners would soon be "resolved." All who saw the warm smiles and brotherly treatment Bhutto lavished on his "old friend" Mujib suspected that a satisfactory deal had already been cut. A tripartite South Asian summit would soon be arranged.

If Zulfi was the impresario hero-host of the Lahore Conference, Qaddafi was its visiting superstar. The handsome young Libyan colonel captured the hearts, if not the minds, of virtually every Punjabi he smiled at or waved to, and nearly a hundred thousand turned out when he spoke to them at Lahore's new Qaddafi Stadium. He hailed Pakistan as a "citadel of Islam in Asia," and promised that Libya stood "ready to sacrifice its blood" if Pakistan were ever threatened. "Our resources are your resources!" he shouted. The captivated young crowd cheered almost loud enough to be heard without transmitters or monitors in New Delhi. Qaddafi enjoyed Pakistan so much that he lingered there four days after the conference ended and the others, less close to Bhutto, flew home. Zulfi and Qaddafi resonated warmly to each other, and the Libyan firebrand promised his older "brother" much more than simply the oil and monetary support Bhutto needed. Libya was the first Arab nation to ship its oil at the cost of production to Karachi that year.

The day after Qaddafi left, President Kenneth Kaunda of Zambia flew into Pindi and vigorously affirmed how "non-aligned" Pakistan was, adding significant non-Muslim support to Bhutto's dream of hosting an even grander Third World summit someday. Malaysian Prime Minister Tun Abdur Razak attended the banquet Bhutto gave for Kaunda, and he and the prime minister of Lebanon both congratulated their host on the "tremendous impact" of his Islamic summit. The following Saturday, 9 March, Sheikh Zaid bin Sultan al-Nahiyan, president of the United Arab Emirates (UAE), flew back to Pindi for his own state visit. The sheikh of Abu Dhabi loved to hunt wild boar with Bhutto, and they flew together to Larkana, where he laid the foundation stone for a women's hospital that remains one of the best such institutions in Pakistan. The sheikh considered the Lahore Conference "a landmark in Islamic history and the beginning of a new era of Muslim unity and brotherhood."[8] A PLO delegation and a much larger Egyptian parliamentary deputation also flew in and out of Pakistan in March, as did the chief of the Turkish General Staff, and special Soviet Envoy Boris Podtserov, and Vice-President Kang Riang Uk of the Democratic People's Republic of [North] Korea. A Saudi Army team was scheduled to arrive in early April. Zulfi Bhutto's brilliant diplomatic initiative had put sleepy Pakistan high on everyone's list of important places to visit. The Lahore Conference paid handsome immediate dividends, bolstering Pakistan's no longer ailing economy.

Zulfi slept at the official residence of his close friend Chief Minister Mustafa Khar throughout the Lahore Conference, but a week later Khar submitted his resignation, and four days later Bhutto appointed Hanif Ramay to the post. Khar had of late become a political liability to Bhutto. His notorious private life, discussed in grotesque detail by his former mistress and fifth or sixth wife, Tehmina Durrani,[9] did not offend Zulfi, who shared most of his younger friend's personal tastes and social habits. What did bother Bhutto, however, was Khar's increasing alienation of his own Punjab Assembly supporters in the PPP. For the usual reasons of arrogance, impatience, lack of interest or concern in the "petty" problems of others because of his total preoccupation with more "important" problems of his own, Mustafa lost the backing of more than forty of his legislative assembly colleagues. Former Punjab Chief Minister Malik Meraj Khalid, a most astute political leader and also a friend of Bhutto's, had been displaced by his former governor without a word of thanks or apology. Cabinet Minister Khurshid Hasan Meer, another of Bhutto's earliest and strongest PPP lieutenants, had also recently fallen out with Khar—especially after some of Khar's sycophants insisted on calling him "*Quaid-i-Punjab*" (Leader of Punjab). Meer and others in the party high command felt there should be only two *Quaids* in Pakistan, *Quaid-i-Azam* Jinnah and *Quaid-i-Awam* Bhutto. Perhaps it was simply something Khar said during those several nights of the Lahore Conference when Bhutto stayed with him—tone of voice or a crooked smile—that made Bhutto advise him to "take some time off." It was not a permanent break as yet. Exactly one year later Zulfi would reappoint Mustafa governor of Punjab, though not for long.

Chief Minister Mohammad Hanif Ramay was six years older than Khar, three years younger than Bhutto, but in some ways seemed centuries remote from those feudal landlords, for his was a middle-class urban mentality, self-effacing, artistically cultivated, forbearing. He had admired Bhutto's idealism and courage in attacking the military regimes of Yahya and Ayub Khan, whose once powerful heart was fast failing in Islamabad. Ramay was a poet and a painter who saw the People's Party as a "revolution of hope" that would transform Pakistan into a "true democracy."[10] He worked hard to win popular support for the party's manifesto, believing in it. Though at first he was allied to Khar, he soon came to distrust him and to believe that Khar was incapable of leading Punjab's majority province toward any progressive or "better future" for the workers and peasants. So when Bhutto replaced Khar with Ramay, it almost seemed an important change, not the sort of change in name only that had occurred when Jatoi moved to Mumtaz's chair. Yet everyone knew that Ramay could do precious little without Bhutto's support, indeed, he even waited for the prime minister's approval before naming his own cabinet, for especially in the year of the Islamic summit Bhutto's word was law, in Lahore as well as Karachi, Abu Dhabi, and Tripoli.

Minister of State Aziz Ahmad flew to Delhi on 5 April to meet with the foreign ministers of India and Bangladesh, seeking to decide the fate of the 195 prisoners still held by India. The tripartite negotiations opened with cool and almost hostile exchanges, and the negotiations that were to have taken no more than two days dragged on for three without much apparent progress. Prime Minister Bhutto flew to Paris to continue his own top-secret negotiations with the French government for aid in his highest-priority effort to develop a nuclear arms capability. In a terse statement to the French press, Bhutto warned that unless all Pakistani prisoners were released "quickly," the entire "process of normalisation" of relations in South Asia could be dangerously "reversed." Next day the tripartite agreement was signed in New Delhi, and Bhutto flew home to another hero's welcome from his entire cabinet and party. He seemed incapable of making a wrong move, diplomatically at least.

Bhutto next focused more creative problem-solving attention on Baluchistan, holding a series of meetings with Tikka Khan and his top military advisers in Pindi. He decided to cease military operations in that rugged province from mid-May onward and offer "amnesty" to all Baluchi "rebels" who turned in their arms before mid-October. It was the sort of bold gesture that appealed to Bhutto's astute political mind, for he was never afraid to reverse himself overnight, testing his opponents' resolve and often outwitting them with his ingenuity.

Wali Khan did not accept Bhutto's overture; nor would Bizenjo, Mengal, or Marri speak to his peace-offering emissaries as long as they remained behind bars. Hence, both Frontier provinces remained beyond the Sind-Punjab limits of Bhutto's political reach, defying him it seemed, challenging his claim to represent democratically, indeed, to embody the popular will, to epitomize the aspirations of all Pakistanis. A less egotistical political leader might have made his peace with that reality graciously, granting that it accurately reflected the tribal complexity of Pakistan's provincial pluralism and was no mortal attack against himself. But Zulfi could never accept political opposition with equanimity. He believed it his destiny to lead all of Pakistan, and he wanted every Pakistani to love and support him in that task, that leadership labor for which he was born and had so uniquely educated and trained himself. It was an insult to his *Izzat*, his personal feudal "honor," to be challenged, mistrusted, openly attacked in provincial assemblies and press conferences.

"I had come from Kabul only after I was assured that there was democracy in the country," venerable Abdul Ghaffar Khan, Wali Khan's 85-year-old father now under house arrest in Peshawar, told Yusaf Lodi that April. "But what I saw here was the worst ever dictatorship, it was virtually a one man rule."[11]

Frontier Governor Aslam Khattak and his chief minister, Inayatullah Khan Gandapur, managed just barely to control their almost equally divided Provincial Assembly, thus keeping Wali Khan's NAP out of power in Peshawar. Zulfi suspected that both Khattak and Gandapur were secretly dealing with Wali Khan, and urged his young Frontier cabinet minister Hayat Mohammad Khan Sherpao to return to Peshawar, where he would take control of the PPP block in the Provincial Assembly and then displace Gandapur as chief minister. Zulfi liked Sherpao and trusted him, but Khattak feared that if he appointed this charismatic young man chief minister, he would no longer be in control of the NWFP. In May Khattak was thrown out by Bhutto, and shortly after that Sherpao moved in to take charge of the North-West Frontier Province. But he would not long enjoy that much-coveted provincial power before being assassinated.

Prime Minister Bhutto flew to China on 11 May 1974, and was soon closeted for an hour and a half with Mao Tse-tung and Chou En-lai, both of whom promised him military support, including aid in developing Pakistan's nuclear capability. Vice-Premier Teng Hsiao-ping hosted a banquet for Bhutto the next night, and spoke effusively about the Lahore Conference, hailing it as a "new contribution to the Third World countries' cause of unity against hegemonism." The latter term, of course, was China's way of referring to the Soviet Union. Before the end of his welcoming speech, Teng added a similarly modern Chinese reference to India: "Our Pakistani friends may rest assured that, come what may, the Chinese Government and people will, as always, firmly support Pakistan in defence of national independence, state sovereignty and territorial integrity and against hegemonism and expansionism, and firmly support the people of Kashmir in their struggle for the right to self-determination."[12]

At the mention of Kashmir, the Indian ambassador left the banquet. "I am really baffled and puzzled," Zulfi commented on the departure in his response to the vice-premier's speech, "that when we are promoting good relations, friendship . . . that the representative of India should choose to leave this banquet. Does India want conflict and confrontation instead of cooperation and friendly relations? If India wants that, then I can tell you that Pakistan is prepared for it."[13]

Less than two weeks later, on 18 May 1974, India answered Bhutto's Beijing reaffirmation of Kashmir's right to a plebiscite by exploding a plutonium device in a deep salt cave in Rajasthan, close enough to Sind seismically to jolt that province, shaking the revived self-confidence of many Pakistanis in their so recently enhanced defenses. "Let me make it clear that we are determined not to be intimidated by this threat," Bhutto assured his nation the next day: that "We will never let Pakistan be a victim of nuclear blackmail. . . . In concrete terms, we will not compromise the right of self-determination of the people of Jammu and Kashmir. Nor will we accept Indian hegemony. . . . From the day I assumed office, I have been conscious of the dire necessity of our having a coherent nuclear programme."[14]

Bhutto was at his best in times of crisis. Whether it was a superfluity of adrenalin or his instinctive pugnacity in the face of any threat, especially physical, he always responded with heroic defiance. He seems so long to have anticipated early death that he appeared to show no fear when faced with its prospect, baring his chest rather than ducking, redoubling his speed in the race to achieve nuclear self-sufficiency rather than agreeing to sign the "no-war pact" India once again offered him or the nuclear nonproliferation treaty. He argued now that India had "dynamited and shattered to pieces" the NPT, and that "a no-war pact in the face of a nuclear threat would amount to capitulation."

Mujib had flown to New Delhi to meet with Indira Gandhi while Zulfi was in Beijing. Despite seeming cordiality, no agreement was reached at this mini-summit on the major issue of Bangladesh's strong objection to India's erection of a huge dam (Farakka Barrage) on the river Ganga near Calcutta, which could divert so much of Bangladesh's water as potentially to turn many lush districts into deserts. Bhutto had considered postponing his scheduled visit to Bangladesh in June, but when he read the reports of Indo-Bangladesh conflict over Farakka, decided it might be the ideal moment to fly into Dacca with his entourage and seek to sweep Mujib back to a much closer relationship with himself and Pakistan. The visit proved a remarkable success, considering how recently he had been reviled as "worse than Yahya" by so many Bangladeshis.

"Let us forget the enmity and bitterness of the past and inaugurate a new chapter of hope and prosperity for our peoples," Sheikh Mujib said, welcoming Bhutto to his first Dacca banquet.[15] "We have come here with our gaze firmly fixed on the future," Zulfi warmly responded. "[T]he people of Pakistan cherish abiding good will and affection for the people of Bangla Desh. They wish . . . relations between Pakistan and Bangla Desh, Mr. Prime Minister, will be founded firmly on the principles of sovereign equality and mutual respect for each other's political independence and territorial integrity."[16] He toasted Mujib's "good health and long life"—a life that would be brought to a bloody end on 15 August 1975.

Even as Prime Minister Bhutto tried in Dacca to recapture the friendship, if not allegiance, of some of the Bengali Muslims whose faith in Pakistan was violently shattered in 1971, religious riots were rocking Punjab. The small cosmopolitan Qadian Ahmadi community, one of whose centers was in Rabwah, Punjab, came into the most violent confrontation with neighboring Sunni fundamentalists since the 1953–54 anti-Ahmadi riots launched by Maulana Maudoodi's followers in Lahore. Disciples of Qadian Ahmadis worshipped their "Prophet" Mirza Ghulam Ahmad, and were, therefore, considered "heretics" by orthodox Sunnis. The high level of education that prevailed among Qadianis, however, helped them attain positions of disproportionate power and prominence in Pakistan's civil, military, and diplomatic services. Their wealth and power aroused the envy of poorer Pakistanis, and most fundamentalist mullahs and maulanas reviled them as "non-Muslims" or "traitors" to Islamic law.

When Bhutto returned from Bangladesh and found the Punjab in so volatile a state, he moved swiftly to defuse the problem. He appealed directly over radio and television for immediate restoration of "order" in and around Rabwah, alerting the army and announcing his appointment of Justice Samadani of Lahore's High Court to serve as a one-man committee of inquiry into the Rabwah "incident." The vexed question of the religious status of the Qadiani community was also brought before the National Assembly by Bhutto's Law Minister Hafeez Pirzada. On 10 September the assembly officially declared the Ahmedis a minority, thereby barring individual members from attaining either the prime ministership or presidency of Pakistan under the Constitution. The action was viewed as a fatal blow to the political aspirations of former Air Marshal Asghar Khan, and as a major victory for Prime Minister Bhutto because many thitherto anti-PPP fundamentalists crossed over the assembly lobbies in Punjab and Islamabad to join what they now considered the more "orthodox" People's Party. Maulana Kausar Niazi, Bhutto's leading religious affairs adviser and minister of information in the central cabinet, was now in a position of ascending power. "Maulana Whiskey," as he was called by some of his secular opponents in the cabinet, was so close to Bhutto personally that he later asserted that he had performed the orthodox "Islamic ceremony" that joined Zulfi and Husna in "marriage"—which Bhutto himself repeatedly denied. Husna Sheikh, on the other hand, "absolutely" considered herself "his wife . . . in the eyes of God and the witnesses. . . . [W]e were, as far as we were concerned, married."[17]

Although Islamic fundamentalists now had gained greater power inside both Bhutto's government and party, several of his closest early comrades in starting the "People's Revolution" dropped out. J. A. Rahim was replaced as minister for production and commerce by Rafi Raza in July, Finance Minister Mubashir Hasan was let go in October, and Minister of Health Khurshid Hasan Meer was out of the cabinet before year's end. If any trio represented the early revolutionary ideals of the PPP, it was Rahim, Hasan, and Meer. Rahim was the theoretician who had masterminded the party's manifesto; Hasan, the socialist planner for nationalizing Pakistan's economy and redistributing its wealth; and Meer, the activist advocate, and organizer of the mass meetings held in Pindi and Lahore. Meer still remained deputy secretary-general of the party but now saw his influence over his Quaid-i-Awam "slipping" as Bhutto turned his face from the left to right wings of his kitchen cabinet.

The most brutal and tragic fall from Zulfi Bhutto's inner circle was that of Rahim. Until 3 July 1974 he had been one of a handful of confidants, advisers, even "friends"—though some insisted that Zulfi had no friends, only followers, servants, or family. Yet since 1966 at least Rahim had been among his closest followers, and had served Bhutto not only as a senior minister in his cabinet but as secretary-general of the party, which Bhutto continued to chair as prime minister. It was hardly surprising, therefore, that J. A. was one of a select group of cabinet ministers and important political advisers invited to dinner at the prime minister's house on 2 July. The handsomely embossed invitation said 8:00 P.M., and though J. A. knew Zulfi well enough to understand that he rarely dined so early, he also knew that Zulfi liked his guests to be punctual. Rahim liked a drink or two before dinner, so he showed up on time, as did all the others. But Bhutto was nowhere to be seen at eight, nor did he appear even so much as to show himself from the room "upstairs," where he remained closeted on some "more important business," at nine, ten, or eleven o'clock. All of the waiting coterie knew their leader and his habits well enough, of course, to appreciate the fact that he rarely slept at night, and often sat down to start his dinner at midnight. There was plenty to drink and various tasty morsels to eat, so most of the guests sat patiently. What else should they do, after all? Most of them felt honored to have been chosen thus to serve.

Rahim was getting tired by midnight; perhaps he had too much to drink on an empty stomach, or had twiddled this thumbs long enough. Shortly before 2 July turned into 3 July, he put down his empty tumbler and said in an irritated, clearly audible voice, "You bloody flunkies can wait as long as you like for the Maharaja of Larkana, I'm going home!" His words elicited no response, only shocked, anxious, knowing glances from one courtier to another. All of them could anticipate Bhutto's angry response to the patent insult from a Bengali who had been so well trained as a diplomat, and till now had exhibited singularly sound political judgment.

> On reaching home I went to bed," J. A. recalled. "About 1.0 a.m. I was woken by my servant who said that there was a crowd of people before the house and someone at the door was demanding that it should be opened. . . . I then armed myself with my revolver. . . . Some men of the FSF were climbing up the front balcony for the purpose of entering my bedroom. . . . I went to the front door downstairs. . . . Said Ahmed Khan, Chief of the Prime Minister's Security, who was at the head of that mob of armed FSF thugs, answered that he had come to deliver a message from the Prime Minister. . . . I then ordered my bearer to open the front door. . . . As the door opened, Said Ahmed Khan and several others rushed in . . . armed with rifles or sub-machine guns. No letter was delivered, not even a verbal message. . . . Besides being beaten by fists I was hit by rifle butts. I was thrown to the ground and hit while prostrate. . . . My son [Sikander] tried to intervene to protect me and was himself assaulted by FSF men. I lost consciousness. . . . S. A. Khan ordered his men to seize my son . . . and then to lead us away. . . . I was hardly able to stand, much less to walk, though I recovered consciousness. I was dragged out by my legs, then thrown into a Jeep . . . bleeding profusely from a wound in the nose, the left nostril having been ripped open. No medical attention or first aid was offered. . . . After nearly a couple of hours Mr. Rafi Raza arrived at the police station. . . . He drove us in his car to my house shortly after 3.00 a.m. . . . On the 20th July I was flown to Karachi where no one was allowed by the police to see me at home until I left for Europe.[18]

"What I cannot get over is Rahim's diatribe and invective against Larkana," Zulfi later wrote from his own prison cell, reflecting on his life as he awaited death. "What harm did poor Larkana or the poorer people of Larkana do to Rahim? He was royally entertained at Larkana for days on end. . . . If I was Rahim I would never forget the hospitality of Larkana."[19] But Zulfi Bhutto was not Rahim, of course, though he may have thought at times, and in certain perverse ways, that he was. "Rahim's private life has not been very complimentary," he wrote in that dark and stinking death cell. "It is a story of mistresses and misbehaviour. . . . Why then did I get friendly with him and make him the Secretary-General of the PPP and Senior Minister and all the rest? A very good question," he mused. Many "good questions" kept invading his mind now, as they must so often have done through those long sleepless nights and sunless days, while he awaited final judgment.

How many times did he think of his impulsively foolish explosion from the floor of the National Assembly exactly one month before Rahim was so savagely beaten? That 3 June 1974 he was, of course, intolerably provoked by the treacherous Ahmad Raza Kasuri, who kept needling him, criticizing him for the pettiest, stupidest things, just to try and show how big he was, when everyone knew Kasuri was nobody. "Nobody" at least compared to Zulfikar Ali Bhutto, convener of the Islamic summit, friend of Qaddafi, Nixon, and Mao, who almost alone had toppled the field marshal and by now had outlived him. "You keep quiet," Prime Minister Bhutto impulsively shouted to the rude, noisy nobody-Kasuri. "I have had enough of you: absolute poison. I will not tolerate your nuisance."[20] By then he thought he had learned to control such outbursts, but still they came like lightning, cloudbursts over a torrentially swollen river. He never knew when such an episode might happen, and after it was over, he sometimes barely understood why he had said what he did, what had incited him to intemperate speech. Yet there it was, before the entire assembly. No more than the truth, of course, but so imprudently spoken.

"Strange, how unpredictable he was," recalled his old hunting friend, former Air Marshal Zulfikar Ali Khan, Pakistan's ambassador to the United States during Benazir Bhutto's premiership. "He could be charming to one person, and for no reason, just turn on another person who was close to him. It was very strange how he did that! No rational explanation. I never understood that."[21] Hanif Ramay suggested that the strange "split" or "division" in Zulfi Bhutto's personality was perhaps attributable to the enormous "cultural differences" that divided his mother from his father. "He was two different people really, you see."[22] Rafi Raza and Admiral S. M. Ahsan also remarked on that split personality, which partitioned Zulfi into a rational, intelligent, responsible Bhutto on one side, and an irrational, impulsive, uncontrollable Bhutto on the other. Whether he was, in fact, one or twenty-one Zulfi Bhuttos may never now be known, but he was a complex personality.

In August of 1974 Prime Minister Bhutto appeared before a joint session of Parliament to urge extension of the "state of Emergency" under which his government ruled, and for which he required parliamentary approval every six months. Opposition leader Wali Khan insisted, however, that the constant "state of crises" and "Emergency" in which Pakistan found itself was Bhutto's own "fault," his "creation" to keep everyone in perpetual "fear" and "terror," for then he could announce what must be done to "save" the nation. When Bhutto rose to address the legislators, Wali Khan, his NAP followers, and other members of the opposition, including Ahmad Raza Kasuri, walked out, with fists raised in defiance. Bhutto's response was a long and scathing critique in which he forcefully argued that every crisis along the Frontier from Baluchistan to Peshawar was directly attributable to Wali Khan himself, whose close ties to Kabul and Delhi were also noted.

"Well, Mr. Wali Khan thinks that we are so naive and we are so innocent that we are not aware of his activities. Does he not know that we have smashed in Peshawar the plot of those people who were throwing bombs and hand-grenades, and things like that?

"[Wali Khan] gave credit to Yahya Khan for holding the first general election in the history of Pakistan. . . . We know what is the implication of that statement. We understand that fully well," Bhutto said, sensitive to the opposition's repeated demands that he call fresh elections yet determined not to be pushed prematurely into what he knew might prove to be a trap, for having ousted or alienated Mumtaz, Talpur, Khar, Rahim, and several Kasuris, he had lost support from both wings of his own party. But Wali Khan's praise for Yahya Khan as a relative "paragon" of democratic freedoms compared to Zulfi really angered Zulfi: "I will not give him that credit, but . . . if ever Yahya Khan, in his whole tumultuous life did some right thing by mistake that was the time when he banned the NAP."[23] Zulfi alluded to Wali Khan as an "agent" of Afghanistan and India but stopped short of accusing—that he would do next year.

Sardar Daud moved closer to both the Soviet Union and India for support of his "revolutionary" regime in Kabul, and also appealed more openly to Wali Khan for the tribal Pakhtun support that *Khan* controlled along the rugged Frontier with Pakistan, where Wali Khan and his venerable father were indeed the "kings" of the Khyber and Jalalabad. Wali Khan visited Afghanistan that summer, en route to London and when returning from "medical treatment" there. Each time he met with Daud in Kabul, and was hailed wherever he went in Afghanistan with shouts of "Pakhtunistan Zindabad!" His personal power and unique popularity all along the Frontier was secret to no one who knew the Pathans and their lifelong devotion to old "*Badshah*" (Emperor) Abdul Ghaffar Khan and his regal son, Wali Khan. Zulfi was well aware of his leading rival's popularity.

"Pakistan has a comprehensive and vigorous nuclear programme," Bhutto admitted to NBC's Jack Reynolds in Pindi that September, "but it will be directed for peaceful uses and when I say peaceful uses, I mean it."[24] Bhutto also insisted that in Baluchistan he was fighting "against feudalism and tribalism," that "Baluchistan is the last bastion of the worst form of feudalism and tribalism in these parts." In the aftermath, moreover, of the 1971 "dismemberment of Pakistan," as he called it, those tribal "chiefs" he had imprisoned launched a "mini-rebellion" against "the federal government." He now estimated that there were "at the most" only four hundred tribal "rebels" involved. "Theirs is a struggle for their chiefs who are behind the bars." He vowed that he was "always ready" to negotiate a "political settlement" but insisted on the precedential need to modernize the region, first building more roads, more hospitals and schools, and finally, "the writ of the federal government must be respected there." When asked why he kept so many of the Baluchi leaders in prison or under house arrest if he truly believed in "democracy" and "political settlements," Zulfi replied, "Anyone who tries to preach secession or says that he will break down the edifice of the state or says that he is going to see to it that there is bloodshed . . . these are people who bring in little toys—little bombs—from Afghanistan. They plant them here and they throw hand grenades there. . . . What do you expect me to do? Invite them to a tea party?"[25] Three weeks later, Bhutto "gratefully" informed "the nation" that the "insurgency against the state" in Baluchistan "has come to an end."[26] About five thousand Marri tribals had "presented themselves with their arms to the authorities," he declared, and he had decided to "grant them a full amnesty." So the "insurgency" was "over," but the fighting continued.

Bhutto flew to Moscow on 22 October for a state visit that lasted three days. At the farewell banquet he toasted the "health and happiness" of Leonid Brezhnev, Nicolay Podgorny, and Kosygin. He was "happy" to recall his first visit to Moscow thirteen years earlier, "when we made the first substantial move towards constructive cooperation between our two countries . . . for oil exploration."[27] The joint communiqué that concluded the visit noted that a new trade agreement would be negotiated between the USSR and Pakistan, both of whom desired to "develop mutually beneficial trade and economic cooperation." But Moscow was much colder to Zulfi on this visit than it had been thirteen years before, for the USSR's Treaty of Peace and Friendship with India was still in its early years of bright promise, and Soviet ties with President Daud seemed much more friendly than they would all too soon prove to be. So Zulfi flew home swiftly to welcome his good friend Henry Kissinger back to Pakistan.

Bhutto cast aside his written text to say a few words "from my heart," not just to Kissinger but to his lovely new bride, whom Zulfi toasted first. "I remember fondly waltzing Matilda but we prefer tilting Kissinger," he improvised. "But they say that Dr. Kissinger doesn't tilt any more—but why did he get married?" In a slightly more sober tone he lifted his glass again, focusing on Kissinger: "We, who are your admirers, would like you to be considered as a modern Metternich. But the difference is that Metternich's nation had lost the war and Metternich came after a Napoleon." He asked Kissinger, who was headed for Kabul and Teheran, to "Say hello to Daud for me and tell him that we would like to be friends with them and when you go after that to Iran please give our warmest regards and affection and respect to His Majesty, the Shahinshah of Iran."[28] Neither Kissinger nor the Shah ever forgot that Zulfi had told Nixon the Shah was "unstable." Zulfi turned to smile again at Mrs. Kissinger, to whom he raised his glass a bit higher. Kissinger had brought a cordial invitation from newly inaugurated President Gerald Ford for another state visit to Washington early the next year. Zulfi was delighted to accept.

November of 1974 would not be as pleasant a month as October had been for Prime Minister Bhutto, who always enjoyed foreign trips and jet-set visitors, brilliant banquets, and tall beauties. The month started with another noxious, insulting speech in the National Assembly by Wali Khan, who repeated his "stale" old phrase "Bhutto and Pakistan cannot coexist!" As if *he* were the patriot! Ahmad Raza Kasuri applauded loudest, of course. Pure "poison" that one! Another Intelligence report had just reached Bhutto of Kasuri's having said on 27 October that "the Prime Minister . . . had broken Pakistan into pieces to install himself into power and that he was a traitor and son of a traitor."[29] And the shooting continued in Baluchistan, as if those Marris and Mengals had not heard him say the "insurgency" was "over." His chief minister of Baluchistan explained to the press that "incidents of bomb blasts" and other "subversive activities" in the region were all carried out with the "connivance" of "insurgents" from across the Durand Line border.

The explanation did nothing to make the news any less disturbing, less embarrassing. So two days later, on 9 November, at a public meeting in Sadiqabad, Zulfi gave notice to Afghanistan that it was "inviting trouble" if it continued to "interfere" in Pakistan's Frontier affairs. The next day at another public meeting, he tried to explain why the economy was so bad and there was so little wheat: oil prices had skyrocketed, which was good, after all, for brother Islamic nations—the reason he had not opposed the OPEC price hike.

That same night Ahmad Raza Kasuri was driving his aged father, mother, and aunt home from a wedding party they had attended when his car was ambushed by automatic gunfire from both sides of the dark road. One of the bullets knocked out the car's lights as he sped away, thinking they had escaped harm because the shots had missed him. But old Nawab Muhammad Ahmad Khan Kasuri, who was seated next to his son, had been fatally hit, "his shirt and seat soaked with blood."[30]

Ahmad Raza Kasuri managed to drive his father to the nearest hospital in Lahore, but the operation failed to resuscitate him. Police soon arrived to ask what had occurred, for they had "heard shots" and "wondered" about it. In his First Information Report (FIR) Ahmad Raza told them exactly what he could remember, every detail of the drive, the road, and the ambush, and when they asked if he had any idea of who the perpetrator might be, he answered without a moment's hesitation, "Zulfikar Ali Bhutto!" That was the name he repeated, and insisted upon having recorded in the FIR. Several police officers tried to dissuade Ahmad Raza from naming the prime minister, but Kasuri had never been an easy man to talk out of anything once he had it fixed in his mind. So no amount of pressure, advice, subtle warnings, or reasoning could move him to retract one jot of the FIR accusation.

"When Kasuri's father was killed," Bhutto's FSF Director-General Masood Mahmood recalled, "the Prime Minister phoned me up. 'Mian [Abbas] has done balls! Instead of killing Kasuri, he's got his father!' On the *open phone,*" Masood added, still shocked at his chief's audacity more than a decade and a half later. "Not even to use the sacraphone," he added, shaking his white head in disbelief.[31]

Masood Mahmood's subsequent testimony, as the first of two "approvers" (prosecution witnesses) in Bhutto's last trial, was crucial in convincing the judges of Lahore's High Court that Bhutto was guilty of conspiring to "murder" Kasuri. Zulfi, however, insisted on the third day of his own trial testimony that

> every material statement that Masood Mahmood has made in the High Court, almost every one, is an improvement or a new statement. . . .

> Masood Mahmood has been in custody, a questionnaire has been given to him . . . he has been told to say whatever he wants to in his forty days in Martial Law custody. . . . His confession, or his statement as an approver, is not a full and complete disclosure and falls on that defect. . . . Ghulam Hussain [the second approver] admits that not only did he not make a complete disclosure but that it was a falsehood. . . . So both approver No. 1 and approver No. 2 stand self-condemned. They are liars. They have admitted that they are liars.[32]

The National Assembly was recessed at the time of the elder Kasuri's murder, but shortly after it reconvened, on 20 November 1974, Ahmad Raza entered, and "Rising to a privilege motion he held up a bottle with his father's blood and a blood-drenched shirt," People's Party Member Salmaan Taseer of Lahore recalled. "He dramatically announced his father's death to the Parliament and accused Bhutto's regime of murderous attacks against members of Parliament."[33]

On the third anniversary of his return from the United States to "pick up the pieces" of Yahya's tattered and fallen power over Pakistan, Prime Minister Bhutto told the Foreign Press Association,

> Every democracy must have an effective leadership. . . . The objective situation in Baluchistan is that whereas only three years ago when the People's Party Government was formed, it was simply not possible for any citizen of Pakistan . . . to go into Dera Bugti without the concurrence and consent of Sardar Akbar Bugti . . . that is not the position today, we have got the writ of government firmly established in all these places. Now there is full government, an established government. A day will come when the people of this country will say that I played a pioneering role for the real consolidation of Pakistan in breaking these barriers and obstacles to civilization and progress. You know, a new situation has arisen. . . . Now the police have gone there and the agencies have been established. . . . [But] [w]e will not get the ideal peace even after a permanent settlement. . . . [T]ribesmen have a long record of inter-tribal quarrels. This results in a little bit of shooting here and there. . . . [I]n the world of today, you have tension, you have problems . . . someone taking a shot at a bus—that has gone on for years and will continue . . . till we reach very high standards, till all the roads are opened up, till all the people get educated. . . . There are people who indulge in vendetta, who go by the law of revenge. Those things are there.[34]

Siasat!

13

From "Leader of Pakistan's People" to "Leader of the Third World"?

(1975)

On the eve of the new year, the worst earthquake in recent memory rocked Pakistan's north-west frontier districts of Swat and Hazara, killing more than five thousand people, seriously injuring four times that number. The intensity of the quake made some people fear that an atom bomb had exploded; others said it was God's "warning" to Pakistan's "evil" and "faithless" people. Prime Minister Bhutto flew into the most devastated region and promised to mobilize all government resources swiftly. He urgently appealed to the world for aid, which came with heartening speed from the United States and Saudi Arabia, the United Kingdom, Australia, and Canada. Foreign Minister Ahmad Khalifa al-Suweidi of the United Arab Emirates flew to Larkana with his amir's special gifts, then remained to help celebrate Bhutto's forty-seventh birthday.

Zulfi's second state visit to Washington was set for early February, and he tried to be certain that no adverse publicity from the Kasuri ambush cast a pall over that important first meeting with President Ford. Given the importance of the murdered Nawab Khan Kasuri and the greater fame of the man charged by the son with having "ordered" the murder, the ambush was investigated by a special tribunal, which in January of 1975 examined no fewer than twenty witnesses, including Ahmad Raza Kasuri himself. The rapid-fire 7.62 mm guns used to murder Nawab Kasuri were a new type of weapon recently imported from China for the elite FSF. Several months before that fatal November ambush, in fact, five shells from the same caliber gun had been recovered after an earlier unsuccessful attempt on the life of Ahmad Raza Kasuri from a fast-moving Jeep in Islamabad on 24 August. Ahmad Raza had not mentioned Bhutto by name in his FIR at that time, however, for he had recently antagonized another political rival as well. The tribunal investigating the fatal Shah Jamal-Shadman Chowk ambush concluded, however, that the "motive behind this occurrence is political," and that "the perpetrators of the crime are well organised, well equipped, resourceful and persistently after Mr. Ahmad Raza Khan's life."[1]

Faced as Bhutto now was with more defections from his mistrustful left wing in Punjab, and finding it impossible to be himself with mild-mannered, soft-spoken Hanif Ramay, Zulfi decided to bring his closest Punjabi friend, Mustafa Khar, back as governor of Punjab. Khar was one of the few people with whom he felt totally comfortable at any time of day or night, for they were tuned to the same wavelength, whether in politics or partying, with populist rhetoric, wine, or women, they saw most things in the same garish feudal-modern light, dreamed the same dreams of limitless glory. And though Khar had started his political career as Jatoi's disciple, he had swiftly shifted to abject adulation and total emulation of Bhutto. Zulfi became his true role model, and among all the innermost circle of confidants, Ghulam Mustafa Khar came closest to being a clone of Zulfikar Ali Bhutto.

Khar came to Larkana for the birthday party and hunt, and stayed on for nights of plotting and partying. Whenever Mustafa was with Bhutto, work was always mixed with play of every variety. Each stimulated the other's sensual appetites as well as thoughts. Like locker-room athletes or adolescent fraternity brothers they matched wits, admiring each other's potent powers, stimulated by such competitive challenges to outdo themselves, surpass past performance, aspire to greater heights of conquest and glory. Drink and age and what he called his "colon problem" (probably liver or prostate) had started to take its toll of Bhutto's prowess, but Mustafa's youthful vigor helped to revitalize him.

"I never wished to put you in a corner in order to extract an abject kind of confession from you," Zulfi wrote Mustafa that month, confirming his decision to bring his buddy back to the sumptuous Governor's House in Lahore. "I have very few personal relationships and I prize them highly. . . . I abhor the capricious attitude, which makes one overlook the good and concentrate on the bad. . . . even though I have been receiving persistent advice from impartial . . . persons to take a very cold-blooded attitude towards you. No stone has been left unturned to prejudice me against you. However, where long-term relationships are involved, much advice, sound or otherwise, cannot deflect me from a more human approach."[2]

Intelligence Bureau reports of Khar's public gross behavior and statements and his more "private" conversations, in which he expressed "true feelings" of contempt for Bhutto and many "devious" schemes for his own return not simply as "Quaid-i-Punjab" (leader of Punjab) but as "Prime Minister of Pakistan" were almost daily brought to Bhutto's desk. What Mustafa Khar "really wants," Bhutto was warned, was nothing less than the very pinnacle of Pakistani power, which everyone knew could be occupied only by one leader, one quaid. Zulfi knew all that, had always known it about Mustafa, for Khar was precisely like himself. Had Khar been the older one, Zulfi would have schemed those very same schemes to dethrone him. But none of that changed the personal "relationship" that drew them inexorably to each other; that equation transcended political power plots or fear of treachery because power plots and treachery were inherent in the very blood, the DNA, so to speak, of both men. Mustafa would not have been Mustafa had he not aspired to be king, no more than Zulfi would have been Zulfi, but how could that "deflect" them from the "more human approach" that made them virtual twins?

The major problem regarding Khar's gubernatorial functioning was the Constitution. According to that document the chief minister, not the governor, was to run each state. The drawback to being chief minister was that one had to preside over a cabinet supposed to represent the majority of the state assembly, and had to have a seat oneself in that elected body, with all the trying, time-consuming problems of managing a parliamentary democratic system. It was easier to govern Punjab as governor, just as it was so much simpler to preside over Pakistan as president, which was why Zulfi had delayed his own transition to prime minister for as long as possible. But before Zulfi and Mustafa could iron out all the details of the latter's exact new job, other more urgent matters intervened.

Bangladesh's deteriorating economy and the growing dissatisfaction with Mujib in every strata of society, including the army, made him opt for a more autocratic presidential form of government by 25 January 1975. He suspended the Constitution, proclaiming his Awami League the "National Party" of Bangladesh, relying more completely on Soviet support for his now less popular administration. Bhutto had been funneling secret "discretionary" funds to several anti-Mujib parties during the past two years, and before the end of August 1975 that investment would pay off handsomely. Orthodox Islamic as well as Marxist Communist Bengalis on both wings of the cluttered spectrum of Bangladeshi politics now combined in an opposition chorus to Mujib's inept, tottering regime. Abdul Huq, general secretary of Bangladesh's Marxist-Leninist Communist Party, had written on 16 December 1974 to "My dear Prime Minister" Bhutto, with "much pain and anguish" to appeal "for funds, arms and wireless instruments" to use against the "puppet Mujib clique . . . today totally divorced from the people." That "TOP SECRET / MOST IMMEDIATE" letter reached Zulfi on 16 January 1975, when he minuted on its margin "Important," authorizing "help" for this "honest man," whom Bhutto rated as "fairly effective."[3]

Abdul Malek, another one of Zulfi's agents in destabilizing Bangladesh, traveled to Saudi Arabia for support in the promised "liberation of 65 million Muslims [of Bangladesh], who are anxiously waiting for your guidance and leadership."[4] Bhutto sent Maulana Niazi to Saudi Arabia and the Emirates as well, hoping to marshal enough diplomatic and financial support—as well as weapons—to pressure Mujib or his successors to amend the Bangladesh Constitution to call that state the *Islamic* Republic of Bangladesh, and to establish an "Advisory Council" of "Muslim-minded political leaders" similar to that which Maulana Niazi ran for Bhutto's government in Islamabad.

Zulfi flew to Rome on 3 February—breaking his journey to Washington there overnight—and enjoyed a dinner meeting with Italy's President Aldo Moro. He had also planned a brief secret trip to Romania on his way back from Washington, where he was scheduled to meet with President Nicolae Ceausescu to discuss Pakistan's "heavy water plant" and its need for nuclear technology, in which Romania had "acquired a great deal of experience." That trip would have to be postponed until October.

Bhutto met with President Ford and Vice President Nelson Rockefeller in

Washington 5–7 February, and with Secretary Kissinger and World Bank President Robert McNamara. The ten-year-old U.S. embargo on shipment of new arms to Pakistan was officially lifted, though spare parts and ammunition for all previously shipped U.S.-made equipment had been pouring into Pakistan for the past few years, as Nixon had promised Zulfi on his first state visit. U.S. arms shipped to Iran had, of course, also earlier been flown in to bolster the army's ongoing battle against the Baluchi "insurgency." Ford also released some 300,000 tons of U.S. wheat for immediate shipment to Pakistan, in addition to the 100,000 tons that had been shipped a few months earlier. U.S. aid to victims of the earthquake was increased, moreover, with many private as well as public foundations generously pledging funds, food, and clothing for the hard-hit Frontier region. Zulfi and Nusrat enjoyed a "festive lunch" with David Rockefeller at the top of New York's Chase Manhattan Bank building, and Bhutto immediately sent a beautiful silk Pakistani rug to Rockefeller as "nothing more than a souvenir" to commemorate that "pleasant event."[5]

The next day, 8 February 1975, a bomb exploded in Peshawar, instantly killing Zulfi's closest Frontier PPP lieutenant, Senior Minister Hayat Mohammad Khan Sherpao. Wired word of that assassination canceled the rest of Bhutto's sojourn abroad, and he returned home without further festivity or delay en route. He had no doubt, moreover, as to the identity of the murderers, though no party or individual would ever claim "credit" for killing Sherpao. Zulfi felt as certain that Wali Khan was the man responsible for Sherpao's death as Ahmad Raza was about who "killed" his father. Hayat Mohammad Khan Sherpao had, after all, served Zulfi Bhutto in the NWFP much the way Mustafa Khar did in Punjab, and as Mumtaz had in Sind. He was the one Frontier politician, in fact, of whom Bhutto had no fears, no doubts, feeling total fraternal trust in him. Sherpao served, indeed, as his eyes and ears in Peshawar, as Wali Khan and his Badshah father well knew, and like them, of course, Hayat Sherpao also was a "king" *(Khan)* among the Pathans, though of a smaller tribe.

The very day he reached Pakistan, Bhutto ordered the arrest of Wali Khan and his closest associates in the NAP, declared that party "dissolved," confiscated all of its properties, sealed all of its Frontier offices, and seized all funds and other movable property belonging to NAP leaders. It was not quite what the white-hot fury he felt demanded, indeed, required of him under hoary rules of *Siasat*'s vengeful game, but he now was, after all, Pakistan's prime minister. He had just been in the Oval Office talking softly with President Ford, and in Rockefeller's Chase Manhattan Bank building. He could not simply redress the murder of his friend the way his great-grandfather would have done it, though he must have been sorely tempted to call in Masood Mahmood to give him Frontier marching orders. Yet from the moment he heard that dark news of Hayat Mohammad Khan's murder, Zulfi knew that his own days were numbered. He had always expected early death, of course, only now the memory of that Bombay astrologer's refusal to say a word about his "life" past fifty returned to haunt his sleepless nights. The banning of the NAP and the arrest of its leaders triggered a walkout by all other members of the United Democratic Front opposition from the National Assembly.

Sherpao's murder made Zulfi pause again in his decision to reappoint Khar as governor of Punjab. He met twice with Khar during his first week home, once over dinner with Rafi Raza, when Khar promised that he "would strictly comply with the constitutional position" of listening to the chief minister, should any "clash . . . inevitably arise" between himself as governor and Hanif Ramay, who was to stay on a while longer as chief minister. The promise sufficed to satisfy Zulfi, though Rafi Raza and the director-general of the powerful Inter-Services Intelligence (ISI) agency remained skeptical about Khar's reappointment, for the agency kept a weighty file on Mustafa, his brothers, and their friends, part of which was a detailed report of their "corruption" submitted by the director of the Anti-Corruption Establishment, an independent investigative body, to the Punjab government.

ISI ultimately agreed to Zulfi's expressed desire to reappoint Mustafa as governor but refused to allow him to remove Ramay, who was persuaded to stay on as chief minister as long as possible. Ramay finally quit in mid-July to take a promised seat in the Senate.

Zulfi reappointed Mustafa Khar governor of Punjab on 14 March 1975. Some nights of partying followed in the grand old British Governor's House in Lahore, but the second honeymoon proved briefer than Zulfi had hoped it would be. The "Lion of Punjab" roared as arrogantly as he had before, and even Zulfi could not put up with that for more than four and a half stormy months. Khar was dismissed in late July. He tried then to get Zulfi's permission to run for Hanif Ramay's vacated Lahore seat in the National Assembly, but Zulfi turned him down, indeed fearing by now that Mustafa wanted nothing less than Zulfi's own prime minister's throne. Khar tried to run without the PPP ticket, proclaiming himself an independent, but the party machine crushed him so easily that he turned to the Pir of Pagaro. He joined the Pakistan Muslim League group headed by that leader of the Hurs before year's end. Khar then openly denounced his erstwhile friend and former leader as a "cheat and a liar," using stronger Urdu words as well. Their divorce was final. Zulfi believed that one more of those "enemies" by whom he was "surrounded" was his potential Brutus: Ghulam Mustafa Khar, who recently remarked with a mischievously Mephistophelian gleam in his dark eye, "I know things about him [Bhutto] that no one else does!"[6]

Sind provincial politics now became almost as troublesome to Prime Minister Bhutto as the Punjab, and even Mumtaz, Zulfi was warned by Jam Sadiq Ali, Chief Minister Jatoi's minister for housing, town-planning and local government, was politically "flirting" with the Pir of Pagaro, with whom he had "all the links."[7] Jam Sadiq had himself been quite close to Sind's most powerful Pir as the scion of the wealthiest family, and elected PPP member, of central Sind's Sanghar district. When Jatoi became chief minister, he brought Jam Sadiq to Karachi's Government House as his senior cabinet minister, knowing that Jam Sadiq and Mumtaz had never seen eye to eye, though the Jam was a supporter of Zulfikar Ali Bhutto. "I would submit that Mr. Mumtaz Ali is doing the same things [in Sind] as Mr. Bugti did in Baluchistan," Jam Sadiq reported to Zulfi. "He has all the benefits and patronage of the Government, which is ultimately going to be utilized for achieving his future political ambitions."[8]

With Mumtaz in his cabinet as minister for communications, Zulfi saw more of him now than he had a year before, and received almost daily written reports from this closest of his colleagues—the one brother he continued completely to trust. Mumtaz, of course, viewed Jatoi as his arch rival, and knew that whatever poison Jam Sadiq poured into Zulfi's ears about him was first cleared with the chief minister himself. Mumtaz was outraged and asked,

> Have Mr. Jatoi and others received a green light against me while I am under the Prime Minister's orders not to even defend myself? Are my intentions, sincerity and membership of the Party in doubt and that of Mr. Jatoi beyond it? The Prime Minister will recall that I am one of those who repeatedly pleaded with Mr. Jatoi to join the Party. . . . In Nawabshah all those who were elected against the Party ticket . . . enjoy full patronage of the Sind Government, whereas Zardari who was elected on the Party ticket and has remained loyal to it and Reza Mohammad Dahri . . . are being crushed. Their crime being that they have turned down many approaches to join the Jatoi group.[9]

Siasat!

Though Mumtaz was accused by his rivals of being so "pro-Sindhi" that he would hire no one born elsewhere, he urged Zulfi to appoint a "Punjabi" minister of railways "because the Railway personnel is mainly Punjabi." And Mumtaz noted so many "mischief-makers" in the higher ranks of Pakistan's railway bureaucracy that he argued for a "purge" of them as a prerequisite to improving railway services and wasting less money. Mumtaz himself had already fired a number of those top railway officers, whom he described to Zulfi as "complete failures" and "intolerably corrupt" as well as "disgustingly immoral,"[10] but so many remained to be weeded out "if the Railways are to be put right," that he despaired of finding time for it himself. Hence his recommendation for a separate Punjabi minister, "because of the vastness of the task which cannot possibly receive the attention it deserves if the Minister has other portfolios. . . . The colossal development programme, which has not even got off the ground, needs to be watched and pushed round the clock if it is to be implemented in time and people are to be prevented from swallowing up all the money." Zulfi barely found time to read such letters. He merely ordered his secretary to "file" it with the thousand other letters he received that week.

Wherever he journeyed around Sind, now that he was no longer chief minister, Mumtaz wrote his cousin-nephew-brother, he found what he termed the new provincial government's "time servers," who had launched "a deep conspiracy against the Party. Party workers and the loyal students are being hunted ruthlessly, their homes razed and their families harassed when on the other hand opposition elements . . . enjoy almost complete immunity and sponsorship of the Provincial Government."[11] Later in the year, on his way from Karachi to Larkana, Mumtaz met with PPP workers in Dadu and found them "really up in arms. . . . Even the flood relief work continues to be dishonest, benefitting not those for whom it is intended but the man in between."[12] Jatoi and Sind's PPP President Abdul Waheed Katpar vehemently denied all such charges, insisting that Mumtaz "misrepresented" the facts. "We are not doing such things," Katpar wrote his "Most respected Sir" leader. "Then why is it, that conspiracy is attributed to us? We are doing our humble best for the Party and for Qaid."[13]

With provincial and party affairs in disarray, Zulfi usually found satisfaction in focusing on foreign policy. His visit to Washington had been most successful, and as he reported to the press and over television, soon more new planes and tanks and guns would arrive to beef up the army, and wheat enough to help feed some victims of disaster, and those "in between." Very disconcerting, however, was an accord just reached between Indira Gandhi and Sheikh Mohammad Abdullah of Kashmir, who, after spending more than a decade behind Indian bars, now decided to accept political control of the State of Jammu and Kashmir under Article 370 of the Indian Constitution, which gave Kashmiris special privileges and status within the Indian Union. Zulfi swiftly denounced the agreement as "illusory," one that could in no way resolve the Kashmir conflict, or change the "inherent nature of alien occupation."[14]

Bhutto called upon the people of Jammu and Kashmir to strike in "protest against the agreement" between Indira Gandhi and Sheikh Abdullah. He urged all Kashmiris to stop work on the last Friday in February to demonstrate their "faith and national unity," and "unshakeable resolve to uphold their right of self-determination." Because most Muslims observe Fridays as their weekly day of rest, the "response" to Zulfi's call was lauded by him as a "total and unblemished success." He insisted that the Gandhi-Abdullah agreement was in violation of the Simla Agreement he had earlier reached with Indira Gandhi, calling it "a setback" to the establishment of "durable peace in the South Asian region." It was for "the people of Kashmir" to "decide whether they want to join Pakistan or India."[15]

He had written Mrs. Gandhi to "protest" her agreement with Abdullah, but the Kashmir Action Committee led by Maulana Farooq had called for observance of "Self-Determination Day" on 17 March. Zulfi cautioned against "escalating this problem," for he knew how passionately most Pakistanis felt about Kashmir and how easily violence might erupt. To the west, Zulfi explained, "Our relations with Afghanistan are sad and bad. . . . [W]e want to give Afghanistan a Simla-type agreement, a treaty of non-aggression, but if in spite of that Afghanistan commits aggression, Afghanistan will not be the winner. . . . We will not be browbeaten."[16]

News of the assassination of Saudi Arabia's King Feisal on 25 March 1975 by one of his own nephews was received with "profound sorrow and extreme shock" by Prime Minister Bhutto, who proclaimed ten days of national mourning for "one of the greatest statesmen of our times." Zulfi flew to Riyadh that evening to attend the funeral summit there, expressing his personal condolences to the former crown prince, now King Khalid. A day later he flew home and addressed a specially convened session of the National Assembly in Islamabad. "It is no secret to my countrymen that Pakistan had a special place in the King's heart," Bhutto recalled. "Whether it was in the war of 1965 or that of 1971, he gave us his full and steadfast support as victims of aggression. . . . [W]ho amongst us can forget his decisive support to the holding of the Second Islamic Summit in Pakistan a year ago?"[17]

When the ten days of mourning ended, Zulfi embarked on one of his singularly popular meet-the-people tours of Punjab, speaking first at a mammoth rally at Jinnah Park in Sialkot on 3 April. Though these carefully planned and well-controlled public meetings, which attracted multitudes of people—most of whom were bussed in from neighboring towns or villages—were not supposed to be part of any political campaign, Bhutto's opponents all knew how effective he was at using the power of office to his own party's political advantage. Wearing his Mao cap and open-necked peasant shirt, Zulfi was Pakistan's most popular and effective speaker. He dazzled and excited all who gathered to hear his eloquent, sonorous voice, and to watch his often wild and histrionic gestures. He had always been a great showman, a brilliant entertainer. The crowds kept growing, more than 300,000 at Sheikhupura, closer to 400,000 at Gujranwala, and more than half a million by 6 April in Lahore, where he promised the immediate release of funds for development and the construction of two new hospitals, and a more than doubling of the size of the medical colleges there. Mustafa Khar was with him, of course, and the nights they spent together at the Governor's House delightfully capped that week's otherwise arduous tour of the Lahore Division.

Bhutto returned to Islamabad to address the National Assembly, now consisting only of his own supporters, informing them that his government's decision to raise the prices of many edibles, including wheat and sugar, was not only "unavoidable" but "in the best interests of the country." It was at any rate in the best interests of many landowners who grew both crops. He then flew to Karachi, and a day later received his Rumanian counterpart President Nicolae Ceausescu, who had flown in for a two-day summit—which Bhutto promised to reciprocate in October. "The dedication of the Rumanian leadership to the cause of progress and its vision and courage in upholding the rights of medium and small nations have earned deep respect and admiration from all peace-loving nations," Zulfi told his dictator friend at the banquet for him. "India's nuclear explosion in May 1974 cannot but be viewed with grave concern by all peace-loving states. . . . We are living in times which are both exciting and grave. Joy is mixed with agony."[18]

On 20 April, Bhutto hosted a banquet for Vice-Premier Li Hsien-nien of China, who had come nonstop from Beijing on a six-day state visit. Bhutto and Li exchanged promises of "steadfast support" for each other's country against "hegemonism and expansionism." They drank many toasts during those six days in April, and discussed in detail ways in which China could help Pakistan develop its newly drafted program for building at least a dozen nuclear power plants over the next quarter century. At Chasma in Punjab a large nuclear plant was well started, which Bhutto hoped to see operational by 1979 but would not live to launch. A smaller nuclear plant was also started in Karachi. "Pakistani friends can rest assured that the people of China will always remain their trustworthy friends," Li promised. "In the future joint struggle, we will always stand by the Pakistani people."[19]

"China is a nuclear power and a genuine nuclear power in contradistinction to phoney nuclear powers," Zulfi told David Frost three weeks later in Pindi, explaining that "India is a phoney nuclear power. . . . [I]t does not have the economic infrastructure to really support a nuclear programme. People are starving there, they have to really suck the blood of the people . . . the Indians have to go right deep down to the bowels of their people to extract the money to do it. So, it is a giant with feet of clay."[20] Like India, Pakistan had refused to sign the Nuclear Non-Proliferation Treaty, which Zulfi felt was doomed to "breakdown." Frost asked Bhutto which world leader had most impressed him, and he picked Chou En-lai, saying perhaps as much about how he would like to be seen by others as about Chou's personality in explaining why: "Like Napoleon who was a complete man, Chou En-lai is a complete man. He knows about music, he knows about history, he knows about military science, he knows about what is happening in the world, he would be able to analyse the hippy problem . . . [H]e takes care of everything, he looks into everything, he looks into everything minutely."[21] Zulfi knew that Chou was dying of cancer, which was why Chou had sent his deputy, Li to Pakistan.

Less than a month later, Prime Minister Indira Gandhi was convicted of electoral malpractices by Allahabad's High Court, mandatory punishment for which would have barred her from holding any elective office for six years. Instead of resigning, she appealed to India's Supreme Court, and soon after that declared a state of "emergency." All parties opposed to her rump Congress rule were banned; most of her political opponents were arrested; and she autocratically canceled India's constitutional liberties and rights. Among those arrested was Zulfi's old friend Piloo Mody, Jaya Prakash Narayan, and Morarji Desai, who would replace her as prime minister in two years. Faced with the options of surrendering power or abandoning India's hard-won freedoms, Indira Gandhi had no difficulty deciding which she preferred. India's press was first blacked out, then totally censored and controlled. More than fifty thousand political dissenters were jailed, many of them tortured, for daring to disagree with Madam Gandhi or her younger son, Sanjay. "She understands nothing this Mrs. Gandhi and acts off the bat without asking anyone's advice," Morarji Desai told Oriana Fallaci, just a few hours before his arrest. "She's not a Prime Minister, she's an Empress."[22]

Asked how he felt about Prime Minister Gandhi's conviction, Prime Minister Bhutto said, "We do not gloat over the predicaments of others. We take a serious and long-term view of such situations. . . . We have to be watchful lest Mrs. Indira Gandhi, bedevilled and bewildered by the present crisis, seeks to extricate herself from this mess by embarking upon an adventurist course against Pakistan."[23] Bhutto understood his Indian counterpart well, but even "Empress" Indira knew that there were limits beyond which her own military command would never permit her to act in pique or with impunity.

Bhutto, moreover, had troubles enough with the outlawed NAP and the rest of his angry opposition to refrain from telling the world how delighted he was at learning about the political plight of his "bedevilled and bewildered" New Delhi neighbor. And why gloat? His diplomatic restraint was more admirable. His warning against any "adventurist course" sufficed to convey the clear message he meant for Madam Gandhi. Tikka Khan put his troops on highest alert all along the Indian border. Baluchistan and NWFP, however, continued to pose more dangerously explosive problems to Islamabad-Pindi than Delhi did.

Attorney General Yahya Bakhtiar argued the central government's case against Wali Khan and the NAP before Pakistan's Supreme Court in mid-June 1975. "Prime Minister Bhutto, during his visit to the grave of Sherpao, said that he had laid down his life for the preservation of the ideology of Pakistan," Yahya Bakhtiar declared. "There is ample evidence to show that NAP has been playing into the hands of Afghanistan and serving, in a way, as the frontline organization for Afghan territorial designs on Pakistan."[24]

Yahya himself was a Pathan; his home, however, was in Quetta. Scion of a wealthy Frontier family, he had been educated in Britain, and was called to the bar from Lincoln's Inn several years before he first met fellow barrister Bhutto in Karachi. Like Bhutto, he had been a close personal friend of Hayat Khan Sherpao, and argued with persuasive passion before the Court in Rawalpindi. Wali Khan sought unsuccessfully to have the ban on his party lifted. He would eventually manage to clear himself and his party of charges of "disloyalty" to Pakistan, as well as "open advocacy of violence to subvert the rule of law," but was found guilty of "advocating a policy of subversion" and therefore was obliged to remain behind prison bars as long as Bhutto continued to rule. "Mr. Bhutto always accused his political opponents of the very actions which he intends to perform himself," Wali Khan insisted, though Yahya Bakhtiar denied that charge in his subsequent rejoinder to the Supreme Court, arguing, "This cap fits Mr. Wali Khan instead."[25]

While opposition political leaders of both India and Pakistan were being kept in jail by their autocratic prime ministers, the increasingly angry opposition to President Mujibur Rahman in Bangladesh launched a coup against its one-party leader. Two young army officers, Majors Farook Rahman and K. A. Rashid planned and led the bloody murders of most of the family and closest friends of the man hailed three years earlier as "Nation-Unifier-Father." Mujib's own first minister for water resources and power, Khandaker Moshtaque Ahmed, later minister for commerce and foreign trade, was privy to the conspiracy, and immediately took over as president after the slaughter of Mujib and his family on Dacca's blood-spattered Dhanmandi Road that mid-August morning. For a few moments the charismatic president had managed with his defiant words to paralyze the first would-be killer, who caught him on the stairs inside his home and softly requested, " 'Sir, please come.' " " 'What do you want?' Mujib asked him scornfully in Bengali. 'Have you come to kill me? Forget it! The Pakistan army couldn't do it. Who are you that you think you can?' "[26] But the second intruder did what the first could not; he fired "a burst from his sten gun. Mujib didn't have a chance. The bullets tore a huge hole in his right side."

The blood-soaked majors then went to rouse Moshtaque, informing him that he was now president. Before noon he broadcast news of the fait accompli to his startled nation. Zulfi Bhutto was the first foreign leader to recognize the new "fraternal" government that same day, and called upon all "members of the Islamic Conference" and "all countries of the Third World to do likewise."[27] He ordered the immediate dispatch of 50,000 tons of rice and 15 million yards of cloth to Dacca as a "modest contribution" to "our brothers in Bangladesh." President Khandaker Moshtaque would retain power for only eighty-three days before the next coup, led by generals rather than majors. Three weeks after the first coup, Mahmud Ali, Zulfi's special adviser on how best to reintegrate Bangladesh with Pakistan, flew to London from Islamabad with the hope of winning enough support there to announce a "confederation" of the former wings of old Pakistan. Chief of Army Staff General Ziaur Rahman would not hear of that sort of "subversion" of Bangladesh's sovereignty, however, and a brief early November coup seemed to herald a return to "Mujibism," but that too was soon crushed. After being kept under house arrest several days, General Rahman was freed by his own "sepoys" and after 7 November catapulted to supreme power over Bangladesh, which he would retain till his own assassination little more than half a decade later.

"I have always known you to be a man of tremendous courage, of unusual wisdom and remarkable foresight," Begum Akhtar ("Baby") Sulaiman, former Prime Minister Suhrawardy's daughter, wrote Zulfi that September:

> On the issue of "Bangladesh" you have surpassed all expectations. You have shown generosity towards and love for brother Muslims in a manner that is beyond praise.[28]

Begum Sulaiman was not alone in feeling this way about Prime Minister Bhutto, whose popularity throughout the Islamic world skyrocketed as Bangladesh plunged deeper into dark pools of its own once most precious blood, and as Indira Gandhi's image sank to its lowest point as well, with thousands of India's bravest sons locked behind "emergency" bars. Zulfi, of course, was delighted at this happy change of his political and diplomatic image as well as fortune.

Bhutto flew to France that October, and was asked by a reporter of *Le Figaro* what he expected from his visit to Paris, replying that he had many important "matters" to discuss with French officials, economic, educational, cultural, "and we have also collaboration and association in armaments purchases." From France he went on to Rumania, where he had talks about Pakistan's need for heavy water for its secret nuclear weapons plant near Islamabad-Pindi. Responding to the request by Bucharest's Ager Press to "comment on liquidation of the gap between the poor and the rich countries of the world," Zulfi said: "These are burning questions. Indeed, it is a furnace, a live one in which mankind is placed. Its flames are spreading . . . [T]here has to be fundamental change in the outlook of nations, the developed nations, in particular . . . to get a better economic order. The battle of the OPEC countries is not their battle alone, it is a battle for all the under-privileged and exploited nations of the world. We will maintain our solidarity with them and we will support them."[29]

The euphoria Zulfi felt in the aftermath of what had happened to Indira Gandhi and Mujib, and subsequent changes in Bangladesh carried his racing mind ahead to new visions of Third World leadership and of its potential for Pakistan—which as yet remained perforce nebulous, but with the help of France, Rumania, China, Libya, Saudi Arabia, Iraq, and Iran he hoped that Pakistan would soon become a nuclear power.

As the fourth anniversary of his assumption of power approached, Zulfi Bhutto knew that he would have to face the electorate again in the not very distant future. Because the Constitution had not been adopted till 1973, he could legally have delayed that inevitable deadline till 1978, yet why should he? As the most popular, if not as yet the most powerful, leader in South Asia, as well as the convener of the most successful Islamic summit, his diplomatic rating had never been higher. The next summit he planned to host, moreover, would embrace the entire third world, from Cuba to China, from Libya to Bangladesh and Indonesia. He would bring all of them to Islamabad-Pindi this time, and whether or not India chose to accept his invitation, Pakistan would rise high above New Delhi in the eyes of most of the world. From the humiliation and ignominious defeat of 1971 so glorious a resurrection would be acclaimed by all peoples everywhere as little less than a miracle, the work of one leader, one great leader of the people: Zulfikar Ali Bhutto. To achieve such unique glory and fulfill his destiny, Zulfi knew he first had to win the next general election, which he planned to hold in about one year, rather than two, because he was always acutely conscious of his age, and wanted to finish what he had to do before he turned fifty in 1978.

Zulfi understood better than all his combined opposition just how difficult winning the next elections could be for him. However idealistic or visionary his diplomatic dreams may have been, he was always a political realist. He had little or no hope of winning any portion of the Frontier, though he would try till the very eve of the elections to "negotiate" a settlement with Wali Khan and his Badshah father. As for his own Sind and Punjab, nature seemed determined to undermine his popularity in both states; this year's floods were almost as bad as the devastating damage done in 1973. Rains had started soon after mid-July in Punjab and continued through late September to wreak havoc all over rural Sind, including Larkana. Almost as many Pakistani soldiers were assigned to filling flood-control sandbags as to guarding the Indian border. Damage estimates were in the tens of millions of dollars, and thousands were forced to stay off their lands till October—those were the fortunates who survived. Perhaps it was touring all the flood-damaged villages, hearing the tearful tales of peasants who had lost everything that convinced him once again that the time had come to launch a major new economic reform scheme, one designed truly to help the majority of Pakistan's peasants and small landowners. Cynics said that Zulfi was obviously starting his next national election campaign a bit early.

"I am addressing tonight the toiling farmers of our land," Prime Minister Bhutto pridefully began his broadcast to the nation that 10 November. "With effect from the Rabi [spring] crop of 1975–76, small landowners owning up to

12 acres of irrigated land or 25 acres of unirrigated land shall be exempted from the payment of land revenue, local rates, development cess and all cesses related to land revenue."[30]

It was his most revolutionary move to date, affecting more than eighty per cent of Pakistan's landowners, more people than had all of his earlier industrial, bank, and cotton-mill nationalizations. Millions of Pakistani peasants would never thereafter forget that it was their Quaid-i-Awam who lifted the crushing burden of taxes from their sorely bent backs. Bhutto won more rural votes that November night, for himself and his party, than he had lost in the previous three years of alienating fired bureaucrats, dismissed colleagues, and insulted political opponents combined. He lost some votes that night as well, for the grandly canceled revenue would have to be made up by doubling the taxes of large landowners (over fifty irrigated or one hundred unirrigated acres), but the number was paltry compared to the peasant support he had secured. Many of his old wadero friends would call him a "traitor" to his own "class," of course, but Zulfi was used to harsh abuse from friends as well as foes. His political skin was tough enough to take virtually anything said. He had even stopped reacting to the abusive language of a Raza Ahmad or Wali Khan, content to leave such matters for the courts to sort out.

"My fellow citizens, tonight I feel greatly relieved and thankful to Allah Almighty that I have now fulfilled the most significant promise made to the agrarian community. . . . It should, therefore, be a proud day for them. They can make it prouder by resolving that they will work hard and produce more. May they prosper. May their children be happy. This has been my prayer and this has been my struggle."[31] Cries of "Jiye Bhutto!" and "Bhutto Zindabad!" echoed from Larkana to Lahore, from Karachi to Rawalpindi and back, as tens of thousands who heard his voice danced for joy and cheered their saintly leader of the people.

That December, on the fourth anniversary of his return from the United States to "pick up the pieces" of what remained of Pakistan, Prime Minister Bhutto again addressed his nation. He started with an assessment of Pakistan's foreign policy "In this time of flux and peril," noting how "firmly anchored" his policy during these last four turbulent years had been. Alluding to the success of the Islamic summit in Lahore, the lifting of the U.S. arms embargo, and the restoration of cordial relations with Bangladesh, as well as strong fraternal relations with all Muslim states, Zulfi could, indeed, take credit for having achieved all of his major foreign policy objectives. When he turned to the economy, however, he readily conceded that nothing had "caused us greater anxiety" because with spiraling inflation, "the honest citizen who earns his bread through his toil is hard hit." But the problem was one of "global recession combined with inflation from which Pakistan could not possibly insulate herself."[32] Nonetheless, he noted many modest gains, claiming "improved consumption levels" in foodgrains, clothing, and vegetable oils for a "majority of our people." He promised, moreover, that in the coming fiscal year production would start in new fertilizer factories currently being built in Multan, and at Mirpur Mathelo, Sadiqabad, and Hazara. The Karachi steel mill was expected to start production in 1978–79, and present planning called for the building of two new sugar mills each year over the next half decade. He recapitulated, of course, the momentous land revenue remission promised on 10 November; 7.4 million Pakistanis, out of a total population of 8.9 million, would be affected, that is, the owners of some 42 percent of all currently farmed acreage would no longer be subject to any taxes. He defended his radical administrative reforms of 1973 as consistent with his party's election manifesto to "wipe out" the crushing "legacy of colonial rule." He also stressed the "urgent need" to reform the legal system, so that "quick and speedy justice" would be available to all, without the "crippling burden" of "unbearable" litigation. And he spoke at some length about pending changes in the criminal and civil justice system, starting with the cumbersome "commitment proceedings" in a murder trial, which was at this time, and would hereafter remain, very much on his mind.[39]

Two years later, from his death cell, Zulfi wrote to Nusrat:

> I was shocked to hear that you have not engaged a first class lawyer for the tribunal matter. . . . It is a question of our honour. I do not want the people of Sind to feel let down by us. . . . You do not seem to fully understand the psychology of Sindhi mind. . . . [T]o fight every inch of the way. . . . Do not be slipshod . . . follow my instruction in such matters. Too much is at stake. All our resources have to be used and all the holes plugged. It should not be like my trial. I was left to the mercy of a *lazy* lawyer who did no homework, no research and forgot documents in his car and forgot instructions. . . . It was a fatal blunder, an unforgivable one. Do not repeat it in your case. . . . A lot of dirty things are happening behind the scenes. . . . Yahya Bakhtiar had capitulated, broken down and "joined the conspiracy". . . . But why is everyone . . . doing it . . . to dig my grave?[33]

Was there *no one* living he could trust? Truly trust? Completely? Not his own wife, his own lawyer, his own closest colleagues in the party, or in the cabinet? What had become of them all? Why had they "deserted," "abandoned," "turned" on him? Why was it that the entire world seemed by then to have "joined the conspiracy" against Zulfikar Ali Bhutto? "*Time is of the essence,*" he penned, desperately underlining those words in his dark, dirty cell. Did no one else understand that?

14

Prelude to New National Elections

(1976)

Prime Minister Bhutto placed a wreath at the marble mausoleum of Quaid-i-Azam Jinnah in Karachi on 1 January 1976, launching Jinnah's centenary anniversary, which would end during Christmas week with an international conference at Quaid-i-Azam University in Islamabad, eloquently inaugurated by Chancellor Bhutto. By thus identifying himself throughout the year with Jinnah's life and unique role in the creation of Pakistan, Zulfi Bhutto sought to merge his own identity with that of Pakistan's Great Quaid, whose popular image had never been tarnished. Political opponents said it was just one more indication that Bhutto was ready to announce fresh elections early the following year, as he would, in fact, do.

Asked that January by a foreign editor if "Roti, Kapra and Makan for everybody" would remain his party's "slogan" for the coming elections, Zulfi replied, "I do not change my commitments or convictions because of the fear that in the next elections, they may not prove to be as effective as they were in the previous ones. . . . Roti, Kapra and Makan . . . is not a slogan. It is an outlook. It is a philosophy."[1]

With Wali Khan, Bizenjo, and other popular leaders of the NAP fuming behind bars, Rasul Bakhsh Talpur now allied to the separatist "Sindhu Desh" movement, and Khar and Ramay having both recently joined Pir Pagaro's Muslim League group, Zulfi knew that the next elections would not be easy to win. It was Khar's defection that had wounded him most grievously, for he had long considered Ramay a "hypocrite, double-faced, selfish,"[2] and had kept him as chief minister of Punjab till he quit in July only because ISI would not allow his firing five months earlier. After Khar returned as governor, he had fought and argued almost daily with the chief minister, who simply tried to assert his constitutional prerogatives. But Khar had proved himself "impossible." By October, then, Khar and Ramay were both outside the PPP, looking for a new party to join, and the Pir of Pagaro lured them with what sounded like the most promising prospects for returning to power. Khar had lost the Lahore by-election for

Ramay's vacated seat as an independent. Asked now about his "opposition" by an Egyptian reporter, Bhutto said:

> My fault and crime is that I have not been as strong as some others in dealing with the Opposition. I have dealt with them philosophically, in a spirit of tolerance. . . . There was a by-election in Lahore. Do you know what a gentleman who was contesting the by-election did? . . . He phoned the American Ambassador during the election. He said: "Excellency, I am the former Governor of Punjab and I want to meet you because I want to have some help and assistance from you for my election." Of course, the American Ambassador avoided him. But he had the cheek to ask for foreign support. Then he went about very proudly saying that he was getting support from the CIA. . . . Is there any self-respecting politician from any country who would openly say that he is a CIA agent, even if he were? . . . I believe that God has given me the opportunity to not only save my country but to make my country strong.[3]

Henry Byroade was U.S. ambassador to Pakistan (1973–77) at this time, and had established cordial relations with the prime minister—who called him "Hank"—and flew with Bhutto to Larkana for wild boar and bird hunting. After his arduous tour as ambassador to the Philippines under Marcos, "I was going to retire," Byroade recalled,

> but Kissinger talked me into going to Pakistan. He said, "Look, I can't do everything by myself." Can you believe Kissinger saying *that?* He said, "We've got to lift the arms embargo against Pakistan . . . so you go out there long enough to be credible and you come back here and work on the Hill [Congress]. . . ." I came back before Bhutto's brief state visit [in 1974] and I spent about three or four weeks on the Hill, every day . . . and I gave Kissinger my report. I was only going to stay about six more months, and then we got involved in the damn nuclear reprocessing plant. . . . I liked Pakistan. I stayed two more years, and at the very last minute I extended two years. . . . [Bhutto] was easy to work with, and I was pleasantly surprised. I trusted him about 96 percent. . . . He told me a lot of things he shouldn't have told. . . . He'd say, "Okay, maybe we made a mistake with the French. . . . But as a politician I cannot get up and say that we're not going to try to go nuclear."[4]

Abu Dhabi's Sheikh Zayed returned to Larkana for Zulfi's birthday hunt that year, inaugurating the hospital named for him in Larkana, and giving petrodollars enough to build another hospital bearing his name in Lahore. In February Zulfi appointed his old friend Retired Major General Nasrullah Khan Babar governor of the NWFP, and the former nawab of Junagadh, Dilawar Khan, governor of Sind, replacing Begum Liaquat Ali Khan, who retired after having served as Jatoi's ceremonial head of state for several years. Zulfi's new governor of Punjab was Nawab Mohammad Abbas Abbasi of Bahawalpur, Khar's replacement, and Nawab Sadiq Hussain Qureshi took the seat vacated by Ramay. Bhutto now had feudal or military rulers in every province. He worked hard, nonetheless, to lure back as many of his old left-wing PPP supporters as he could. Khurshid Hasan Meer was tempted to return to the party before Bhutto launched the 1977 elections. Even Ahmad Raza Kasuri was brought back by the offer of a Punjab ticket that would "assure" him a new seat in the assembly, for that mercurial man could not long remain in the opposition ranks of the Tehrik-i-Istiqlal (Freedom Party) of former Air Marshal Asghar Khan. Ahmad Raza's return to Zulfi's fold, of course, raised fundamental questions as to the credibility of his "murder" charges against the prime minister, who was now his own party leader.

In late February 1976 Bhutto went to West Germany to meet with Chancellor Helmut Schmidt, seeking missiles and other "hot" arms, and thence to Stockholm, where he assured Prime Minister Olaf Palme that none of his Swedish aircraft had ever been used against the Baluchi "insurgents." He flew to Canada on 24 February, announcing in Ottawa that the International Atomic Energy Agency (IAEA) had "approved" the agreement Pakistan recently signed with France for the purchase of a nuclear reprocessing plant. Prime Ministers Bhutto and Trudeau met to discuss the possible sale of heavy water to Pakistan. The next day in New York Bhutto was urged by Secretary of State Kissinger to abandon his top-priority scheme to acquire atomic weapons. He refused to be dissuaded, nor could he ever be intimidated by diplomatic threats, which seemed only to stiffen his resolve to build the bomb.

Tikka Khan's tenure as chief of army staff ended that February, and Bhutto had asked his senior military adviser's opinion as to which of his lieutenant generals might best be suited to succeed him. Tikka recommended Muhammad Akbar Khan, who, much like himself, was a tough, hard-hitting soldier, thoroughly loyal to Bhutto, who took orders without any questions and implemented them with no hesitation or misgivings. Tikka's second choice was his seniormost lieutenant general, Mohammad Shariff, much more diplomatic and smoother than Akbar. Zulfi liked Shariff and appointed him to a new job that he created in the vain hope of better managing the army, without whose support he knew he could not remain in power another hour. General Shariff would chair the new Joint Chiefs of Staff Committee. For the top army job he reached past six senior lieutenant generals to pick the juniormost officer of that rank: Mohammad Zia ul-Haq, then 52, and destined to outlive his patron by almost a decade. The "pen picture" of General Zia sent to the prime minister by his private secretary was of "a capable officer who owes a lot to General Gul Hasan . . . ambitious . . . quiet and watchful. . . . A clever and reasonably good professional. . . . Doubtful reliability."[5] Tikka had not even considered mentioning Zia as his possible successor because Zia had only recently been promoted from major to lieutenant general, and Tikka knew that jumping him over his seniors would oblige them all to retire, except for those, like Shariff, lucky enough to be given new jobs that carried promotions.

Zulfi's choice of Zia proved his most fatal error of personal judgment. Many of his earlier appointments seemed to betray an apparent lack of discernment in choosing people for positions of high public trust—governor, chief minister, cabinet member, head of a federal security force or intelligence agency and the like—but chief of army staff was the one post that time and again had proved to be the solid brass springboard to political as well as military power. The recent cases of Ayub Khan and Yahya Khan were instructive in that regard, so Zulfi surely knew how risky it was to pick anyone but an officer whom he could totally trust. Yet he hardly even knew Zia.

General Chishti concluded that Zia was simply "the best sycophant to win over Mr. Bhutto."[6] Wajid Hasan, who headed the National Press Trust during Benazir Bhutto's prime ministership and was a youthful supporter of her father, suggests that Bhutto "may have chosen Zia *because* he was so junior, and by elevating him over the heads of senior generals he would lose popularity and [army] support if he ever decided to try anything."[7] General Gul Hasan called Zia "a dark horse." He knew Zia better than most of his contemporaries, for "Zia served under me," Gul Hasan reminisced, adding, "I had him promoted to Brigadier and later to Major General."[8] Benazir had heard it "rumoured" that Zia had "links" to the fundamentalist Jamaat-i-Islami, as she noted in her *Autobiography*. "He was also said to be a petty thief. . . . I remember being startled when I saw him. Unlike the childish image I carried of a soldier as tall and rugged . . . the General standing in front of me was a short, nervous, ineffectual-looking man whose pomaded hair was parted in the middle. . . . He looked more like an English cartoon villain than an inspiring military leader. And he seemed so obsequious."[9] She later spoke of him as "the devil."[10]

Zulfi soon appeared much more contemptuous of his choice to replace Tikka Khan. He often made Zia the butt of public ridicule, shouting at him from the head of the dinner table, "Where's my monkey-general? Come over here, Monkey!" He would pretend to pull Zia toward himself on an invisible string and then introduce him to a distinguished foreign guest, quickly dismissing him, even before Zia finished bowing, ever smiling. Sometimes Zulfi "joked" about how "funny" Zia's teeth looked, humiliating the man he had singled out for such high and powerful distinction. Zulfi seemed to like to mortify his "friends" in public at times making fun of the promiscuity of their wives, or mimicking their voices, or commenting on unattractive facial features, or calling attention to some physical deformity. Zia apparently never took umbrage at such "jests," always smiling, bowing, even "thanking" his prime minister for "your such kind attentions, Sir!" Once later in 1976, Gul Hasan, then ambassador to Greece, returned home and went to Bhutto's office to pay a "courtesy call." He found Zia waiting in the anteroom, with many files obviously brought from GHQ to brief Bhutto on some very important army matters. Gul Hasan had nothing special or urgent to say, he recalled. "We were just chatting." Inside with Bhutto, Ambassador Hasan reminded him, "General Zia is outside, Sir." Zulfi replied, "Fuck him!"[11]

General Zia must have known why Bhutto chose him for so high, so "honored" a post. And like Bhutto, Zia ul-Haq forgot no insult, no social slight, no attack upon his *Izzat,* no challenge to his faith, his God, or himself. But Zia's humble birth, as son of a minor civil servant in Jullundur, his lack of wealth or foreign education, his seemingly unsophisticated mind and manners had taught him to be more cautious, and much more patient than Bhutto ever was, for Bhutto was a great leader of the people, after all, and Zia ul-Haq was but a "humble *momin*" (Muslim servant of God).

"Zia . . . is frustrated," Zulfi wrote from his jail cell two years later.

> Everything is heading towards the final catastrophe. . . . We are dealing with a very strange and mercurial person, a liar and a double crosser. He is . . . desperate. . . . He goes much further than merely biting the hand that fed him. He concocts false murder cases against me, sends me to a death cell with sadistic pleasure, tries to take my life and pursues like a fanatic to employ all the powers he has usurped to bring ruination to my entire family and the barbaric liquidation of my People's Party. This is worse than the kettle calling the pot black. This is more shameless than the devil quoting the scriptures.[12]

Zulfi did not understand the general he had chosen to succeed Tikka Khan, until it was too late.

On 3 March 1976, Prime Minister Bhutto inaugurated Pakistan's first International Seerat Congress, which would commemorate the birth and study the life and teachings of the Prophet. "The life and teachings of the Holy Prophet have been the corner-stone of my government's foreign policy and our land, labour laws, education and other numerous reforms," Zulfi told the large gathering of Muslim scholars and foreign dignitaries. "It has been the governing principle of the Pakistan People's Party which I have the honour to lead," he added, the future elections never far from his consciousness and concerns. "Therefore, whatever best there is in us, and whatever good we have achieved so far, we owe to our adherence to the teachings of the Holy Prophet, particularly his concepts of equality and social justice. . . . Then why is it that we . . . are suffering from social injustice, inequality, parochialism, prejudice, hate, exploitation and all such vices? Why is it that most of us are still living under the feudal distinction of tribes and castes?" Zulfi asked. "Obviously, something has gone wrong with us, or we have failed to understand the teachings of the Prophet. Indeed, it is a very disturbing situation. . . . May Allah bless this congregation and its deliberations."[13]

He left the rest of that congress to his minister for religious affairs, Maulana Kausar Niazi, to handle. Zulfi flew off to Gilgit for a few days at cooler altitudes, returning to Pindi on 6 March to meet with visiting Soviet Deputy Foreign Minister Nikolai Firyubin and to dine with Prince Karim Aga Khan, head of the Ismaili community. The next day the Shah of Iran and his lovely wife flew into Pindi for a five-day state visit, his first formal visit since Bhutto came to power. President Fazal Elahi Chaudhry hosted the banquet welcoming their imperial majesties to Pakistan and Bhutto hosted a banquet two nights later in Lahore. "In the years that have elapsed since 1971 Pakistan has demonstrated to the world that it is infinitely bigger than its problems," Zulfi declaimed. "Both our countries seek an environment of peace, both are keenly sensitive to the prerequisites of stability."[14]

"If only he had not been such an obnoxious tool of Western interests," Zulfi wrote, after learning in his own death cell of the shah's ouster.

> If only he had been a little more human, he might have survived . . . as an enlightened Oriental Monarch of a great Nation. I knew his doom was at hand. I had a stubborn premonition that he was about to depart. . . . I tried to warn him of it . . . when he and Queen Farah came to Larkana in . . . 1976. I raised the subject dexterously, not once but twice, in Al Murtaza and afterwards at Mohen-jo-daro . . . if he followed what I was driving at, he did not show it. If he did understand, he must have said to himself "Look who's talking. The man is about to fall himself in the near future!" I say this because when I went to Teheran in the end of June 1977, the Shah definitely knew that a coup d'état was about to take place in Pakistan within days and with his approval.[15]

The shah and Bhutto would meet the next month at Izmir with their Regional Cooperation for Development (RCD) Conference host, President Fahri Koruturk. On the eve of that conference, Zulfi wrote,

> Iran, Pakistan and Turkey constitute a single civilization . . . permeated by a common faith. . . . Unlike the nations of West Europe, no two of us have gone to war against each other, in the relevant past. . . . Nor have our developing economies created the rivalries, or generated the antagonisms, prevalent between industrialized societies. . . . How do we stand today? . . . [T]he two superpowers have moved . . . to détente . . . a complex phenomenon. . . . But in the larger field of international affairs, détente cannot be meaningful for the bulk of the world's nations if it only means that competition, in the military field, is controlled and, in the political, restrained. . . . [W]hat we see today is a turbulent . . . multipolar world. . . . [M]ulti-polarity can potentially stir the hegemonic ambitions of even regional powers which are not subject to the restraints that nuclear parity imposes on the two superpowers. The result is . . . a tenuous line . . . between stability and chaos. The line can be crossed . . . by the military adventure of any assertive regional power. . . . This is the dominant characteristic of the current era. Turmoil and tensions seethe beneath a thin layer of tranquility. . . . If we miss the opportunity to mobilize and integrate our resources in order to face contemporary challenges, the world will take no note of either our heritage or our aspirations.[16]

Zulfi's fast-moving mind viewed Izmir as a golden international opportunity, and he was keen to carry both Koruturk and the Shah along with his grand global vision and innovative realpolitik strategy. Their tripartite union would, indeed, have been a formidable one, with Iran's wealth and Turkey's military strength "mobilized and integrated" under Pakistan's leadership. Zulfi's imaginings would have turned into a nuclear-powered political "reality" unlike anything the world had witnessed before. His problem was that neither the Shah nor Koruturk trusted him. Nor was either of those almost equally megalomaniacal monarchs willing to lower his own ego a peg in order to follow the lead of his more brilliant neighbor. Zulfi was, of course, the most mentally keen of that otherwise unimaginative trio. He knew it, naturally, and what doomed his dream from its first articulation was that they also knew it, trusting him no more than they ever trusted each other. Nonetheless, romantic rhetorician that he was, Zulfi tried his very best to move the unbudgeable, hoping to lift those wingless minds by his poetic flights of political fancy:

> Our three countries have a complementarity in resources and skills and a commonly held *weltanschauung* which would be the envy of many another region. If we, therefore, add a new dimension to the Charter of the RCD in the realization that we cannot separate our destinies and that, in the last analysis, economic collaboration without political and security arrangements is chimerical, we need not fear the reaction. . . . The systematic consolidation and formalization of our joint will to defend our civilization against all challenges—economic, political, ideological or military—is something different from adventitious arrangements which are apt to create suspicions in others. . . . It would reflect the vitality of our societies and be nourished by their energies and enthusiasms. . . . My perception of this association . . . is not oriented to military terms. It is focused on the psyche of the contemporary age. . . . [I]n the quest for ways to translate platonic levels of relationship into Aristotelian norms, I am swayed by the belief that military preponderence by itself, without the psychological and political pre-requisities, is incapable of attaining an equilibrium that will endure. . . . In the modern age, no nation can be sufficient unto itself. The Muslim nations need one another even more. . . . It is the vision of the larger unity that remains the anchor of my thoughts.[17]

Subconsciously, Zulfi may have realized as he penned those last words that his dream was foredoomed to sink, like an anchor, rather than take wing at Izmir. The tripartite summit that April merely recognized the need "to continue and further expand" all existing development programs designed to stimulate trade and economic growth among Turkey, Iran, and Pakistan.

A few days after Zulfi returned from Izmir, U.S. Ambassador Byroade met with him to discuss growing U.S. concerns over Pakistan's nuclear plans and military dreams. Before the end of April a Pakistani National Defence College deputation flew to Beijing and a high-level Chinese deputation toured Pakistan, helping further to strengthen Sino-Pakistan relations on the eve of Prime Minister Bhutto's May visit to China. On 26 May 1976 Bhutto was greeted in Beijing by Premier Hua Kua-feng, with whom he met in several "top-secret" sessions; Chairman Mao was too sick to see him for more than a few minutes in his office. "We know that China will continue to play a role to stabilise world peace and to combat hegemony and all forms of foreign domination," Bhutto told Hua in reporting on the recent meeting in Izmir, stressing how Pakistan, Iran, and Turkey all "face the Soviet threat." He quoted from the "will" of Peter the Great (a copy of which he brought and gave to Hua) and explained how eager Russians had long been to reach the "warm waters" around Karachi. He also reminded Hua that in 1971 he had called the USSR's UN ambassador "Czar Malik" before walking out of the Security Council. He informed the Chinese premier of his recent meeting with Kissinger in New York as well, and that Kissinger "got upset when I told him how the United States and the cause of the world peace had suffered due to détente. . . . I told Dr. Kissinger, and I repeated this in Izmir, that when a modus vivendi is reached between a stronger power and a weaker power, the initiative goes to the weaker power . . . the Soviet Union."[18]

While thus reaffirming how strongly anti-Soviet he was, Zulfi won China's support for constructing eight military munitions and five civilian factories in Pakistan. He also hoped to undermine Chinese faith in U.S. reliability: "Americans have elections every two years and four years, and they feel that they must achieve . . . things like SALT I and the Helsinki Accord. . . . [T]hat is why they are bound to make . . . errors of judgment in their haste and in their impatience." And he confided that "the situation in Bangladesh is unstable at the moment," but assured Hua that Pakistan would do everything in "our power . . . to help Bangladesh, because they were part of our country." As for relations between India and Pakistan, he frankly confessed in this minisummit that "nothing has changed. . . . The more India talks of peace the more her actions are dangerous. . . . You know that she has recently usurped Sikkim and her policy in Bhutan is well known to you. The King of Nepal is coming here after I leave. . . . [F]rom what we hear, Nepal is also not feeling very happy."[19]

By seeking more orthodox Islamic support both internally and externally during the early months of 1976, and bolstering the long-standing central pillar of his foreign policy, Sino-Pakistan friendship, while pushing his nuclear weapons program, Zulfi felt more confident about holding elections in early 1977. He wanted to be sure, however, of winning at least the two-thirds majority required for changing the Constitution to return Pakistan to a presidential system, rather than having to appear before the National Assembly every month or so to justify virtually every step he took. He had hired an academic expert on constitutional government, Professor Leslie Wolf-Phillips of the London School of Economics and Political Science, who was busy preparing the new presidential constitution that year in London, working for the Pakistan embassy there, devising his own "secret codes" and deceptive use of "appendices" to draft a document that no one would be able to read or understand until Prime Minister Bhutto himself was ready to spring it on his unsuspecting nation. Wolf-Phillips flew to Pindi that July to brief Bhutto on his top-secret labors. Pir Pagaro, Talpur, and Asghar Khan all continued to attack Bhutto as vigorously as they could, and though Pakistan's press kept most of their attacks out of news headlines, Zulfi was informed by his spies of every barb, every harsh word, every "lie and falsehood" aimed at himself personally or at his party.

A week after returning from China, Bhutto flew to Kabul for an official visit with President Daud. At the banquet he gave in Daud's honor, Zulfi rose to say, "I cannot follow the nightingales and surpass them," quoting one of Ghalib's poems, which he translated, "So many problems beset me that they become easy." Indeed, on this trip Zulfi waxed more poetic than he had been in recent years of diplomatic dinners. "I have come here to see that the future generation, the flower of tomorrow, is not going to be drowned by . . . dewdrops. . . . I have come and believe that our discussions have been fruitful. . . ." He invited Daud to visit Pakistan. "We will give him a great and spontaneous welcome, the welcome that the destinies of the two countries demand."[20] No better way, Bhutto knew, to win votes all along the Frontier than to reestablish cordial relations with Afghanistan. His political timing, as usual, was flawless, but his five early-June days in Kabul proved frustrating. "The Afghan side sought the release of the NAP leaders being tried by the Special Tribunal at Hyderabad before it would reciprocate by recognizing the Durand Line as the international border," Zulfi later recalled. "On my part, I insisted that *both* actions be taken simultaneously . . . the talks were inconclusive."[21]

With Wali Khan in prison and his son soon to be found guilty of conspiring to "murder" Sherpao, Bhutto could hardly have asked for much more help from Kabul in his forthcoming struggle to win votes in the NWFP and Baluchistan. Like Bhutto, of course, Daud was trying to help keep his regime alive by winning as much international support as possible, buying whatever time he could for consolidating his power against attacks from both wings. He managed to avoid a third coup in 1976, and flew in and out of Pindi that August for a four-day state visit "in an atmosphere of frankness and understanding" that resolved nothing. Little more than a year and a half later, Daud and most of his family would be brutally murdered within his own palace walls.

Two days after returning from Kabul, Bhutto addressed the closing session of the Minorities Conference held in Rawalpindi under the Ministry of Minority Affairs. "At the end of 1971, we were a broken nation," Bhutto began. "We have traversed a long, rugged road. The doubting Thomases have disappeared. The Cassandras and other prophets of doom are crestfallen because their dark forebodings have been belied. . . . We have had reforms in each and every field. . . ."

"We in Pakistan have withstood the onslaughts of little-mindedness, prejudice, and bigotry by drawing upon the strength of our traditional values," Zulfi declaimed with an eloquence unmatched by any of his rivals. "My friends, I offer you my felicitations that, in spite of the machinations of those who spout intolerant preachings you have manfully resisted alien influences. . . . Their propaganda, whether brazen or subtle, that you are excluded from taking a full share in the life of our society, has had no effect."[22] His ability to identify himself wholeheartedly with any group he addressed may have been his most remarkable political characteristic, for here he was all but openly denouncing Muslim fundamentalists, the maulanas and mullahs he had a few weeks earlier welcomed with devout, orthodox ardor as he opened the International Seerat Congress.

> My friends, Pakistan has a special interest in ensuring that the most beneficent treatment is accorded to its minorities. . . . The establishment of Pakistan itself had as its background the chauvinistic intolerance towards

> Muslims: then a minority community betrayed by a power hungry non-Muslim majority in undivided India. . . . [But] some interested elements who try desperately to make capital out of every issue . . . would very much like the minorities to rally under the banner of dissidence . . . will try to rouse your emotions and endeavour to lure you into a course of action prejudicial to your interests. . . . Let not these gentlemen dupe you with their sweet talk. Let them not dangle before you their apparently tempting offers of political support and seductive images. . . . Their interest lies in creating divisive conditions in the country. . . . Anyone attempting to disrupt the oneness of the majority community in Pakistan through such a manoeuvre will be dealt with very severely and invoke the penalty fit for an enemy of the nation.

Here spoke Zulfikar Ali Bhutto, not his kinder, gentler youthful persona.

Bhutto must have heard the harsh change in his tone and language, so he reverted to sweetness and light: "My government happens to be attuned to modern humanism. . . . Pakistan belongs to everyone of us, irrespective of caste, colour or religious creed," he added, echoing Jinnah's first speech to the Constituent Assembly in Karachi on 11 August 1947, quoting that famous passage: " 'You may belong to any religion or caste or creed—that has nothing to do with the business of the state. We are starting with this fundamental principle that we are all citizens and equal citizens of one state.' "

Then he rightly noted that previous governments had "expurgated" that entire passage from official publications of Jinnah's speeches.

> They claimed to be the political heirs of the Quaid-i-Azam, they pretended to speak on behalf of a party [the Muslim League] which they associated with his name. Yet such was their loyalty to the Founder of the State, such their allegiance to his message, that they dared to censor his words and had no hesitation in garbling a speech of his which is one of the texts of our nationhood. I am grateful to Providence that, in this vital respect as in many others, it fell to the lot of the Pakistan People's Party Government to salvage the Quaid's message from the debris piled upon it by the narrow-mindedness and prejudices of a class of people whose leadership has now been finally discredited by the nation. As I speak to you today, I feel that I am articulating the Quaid's vision, and I am obliterating from my mind, as well as yours, the intolerance, the hypocrisy, the chauvinism, the bigotry, the morbid insecurity which thanks to certain elements still trying to stir themselves held sway in Pakistan contrary to the Quaid's hopes and aspirations.[23]

With Wali Khan behind bars, the opposition rallied round a revived Muslim League banner, led for the moment by the Pir of Pagaro, but with Khar, Ramay, and Asghar Khan all waiting in the wings—potentially more popular, if not more powerful, leaders—as Zulfi knew from his numerous "intelligence" reports. By identifying his PPP and himself most closely with Quaid-i-Azam Jinnah and his dreams for Pakistan, Zulfi Bhutto raced, as usual, well ahead of his political opposition.

In July Zulfi invited Jinnah's only daughter, Dina Wadia, to Pakistan for the centenary celebrations being planned for December, sending her several "souvenirs" of her great father for her two children, both of whom chose, as did their mother, to reside outside Pakistan. "I am indeed very touched by your gracious invitation," Mrs. Wadia wrote Bhutto, saying she would have been "very happy to accept" but she had another distant "commitment at that time."[24] Social recluse that she was, Mrs. Wadia, long divorced from her Parsi husband Neville, who lived in Switzerland, never returned to Pakistan after having attended her father's funeral in Karachi.

"As elections draw near it is time that we start making preparations," Prime Minister Bhutto wrote to all of his chief ministers on 9 September 1976, with a copy to Rafi Raza, his minister for production:

> To start with, you should give immediate thought to the setting up of a Provincial Election hierarchy . . . with a clear line of command. At the Federal level, there will be a Supreme Federal Secretariat for controlling and co-ordinating the country-wide election work. This body will take final decisions on all vital matters. . . . You should chalk out mass-contact programmes . . . prepare Party propaganda literature . . . buttons, slogans, gadgets. . . . You should have lists prepared of outstanding speakers and poets. . . . Polling Officers and Polling Agents should be selected. . . . Lists of Party workers . . . should be prepared. . . . Efforts should be made to mobilize the potent and latent forces for the election campaign and special attention paid to women, students, labour, ex-servicemen, ex-POWs. . . . Please pay special attention to the youth. . . . The entire machinery needed for the election campaign has now to be put into first gear. . . . This is not the first time I am writing to you on this subject nor is it going to be the last. . . . I have promised fair elections to the people. We must prepare for them like a battle.[25]

Zulfi had not as yet publicly announced when the elections would be called but wanted to be sure that his own side's "battle plans" were well and carefully drawn before alerting the "enemy" to a date. He had all of 1977 to choose from, constitutionally, and hoped to pick the most auspicious, prospectively promising day. In any event he was certainly in no rush to notify his opponents of his choice more than the two months before he was required by law to do so.

As part of his party's preelection propaganda battle plan, thousands of little red-cover books called *Bhutto Says: A Pocket-Book of Thoughtful Quotations from Selected Speeches and Writings of Chairman Zulfikar Ali Bhutto,* modeled on the little red books of Chairman Mao's "wisdom," were distributed widely in Karachi, Lahore, and Rawalpindi. Chief Minister Jatoi of Sind himself penned the "message" in front of the book's introduction, noting that "His great achievements have brought new life to a half-dead nation."[29] Each small page carried its alphabetical quotation from the chairman's works, starting with "aggression" and running through "youth." Some letters were more amply represented than others; the fifteen "P" quotations opened with "Pakistan," "Pakistan People's Party," and "party members" and ended with "propaganda" and "prosperity." By year's end the Bhutto books had been translated into Urdu, Sindhi, and Pushto versions, published by the Motherland Press Bureau.

Zulfi realized that no amount of propaganda alone, indeed not even nominally enrolling more than half of the electorate in the bulging serried ranks of his party would ensure a two-third's majority if the economy remained as bad as it was because of "stagflation," smuggling, inefficiency, incompetence, and natural disasters. Something more dramatic, he thought, would have to be done, economically. By mid-July, therefore, Zulfi decided to spring his latest nationalization scheme to his people over television and radio. "I have to communicate to you tonight a . . . decision . . . meant to transform the hub of our national economy. It flows from that commitment which is supreme and everlasting in my heart and mind. That is the welfare, the happiness and progress of the people of Pakistan. . . . Our national economy, as you know, is primarily agrarian. One of the first campaigns which your government, therefore, launched was that of land reforms . . . We exempted small landowners from the payment of land revenue. We assured the farmer of minimum support prices for wheat. We established corporations for the export of rice and cotton in an effort to secure remunerative prices for these commodities in a volatile international economy. We arranged the supply of tractors. We reduced the price of fertilizers. . . . We are doing our utmost to improve the quality of seed . . . But, after all this sweat and expense . . . we found that the end product of agriculture was still beyond social control. Why? . . . Who manipulates the agricultural market? Who steals from the urban consumer the advantage of substantial government subsidies? . . . [T]he middleman, be he a cotton ginner or a paddy husker. For generations, the middleman in agriculture has sucked the farmer's blood and kept the consumer . . . at his mercy . . . He hoards stocks . . . He underweights the commodity . . . mixes one variety with another, forms a league with the smuggler and establishes a black market."[27] What a scoundrel—that "middleman"! And how fortunate for all poor Pakistani peasants and workers that their prime minister and party chairman had just discovered what the middleman's treachery was doing to the entire economy. First it was the British, then Pakistan's only field marshal, then the industrialists and capitalists, then the civil servants and large landlords who had "sucked the blood" of the poor people Zulfi worked night and day to serve. Now it was the fault of the middleman.

Pakistani middlemen, Zulfi argued, were worse than the British "colonial exploiters" of Lancashire and Manchester had been. "Unobtrusively, with stealth and graft, they have successfully managed to convert a service, ginning or husking, into a potent instrument of economic hegemony." Perhaps quite true, yet how strange that he, who carefully studied and checked his Larkana crop accounts every season, had never noticed such "stealth and graft" before election eve.

The case of wheat was slightly different from cotton and rice because wheat was subsidized by both federal and provincial governments, the former alone at the rate of more than $100 million annually, "not only to assure a price for the consumer within his means, but also to provide him wholesome . . . staple food. . . . I cannot contemplate with equanimity the spectacle of our men and women falling prey to undernourishment and disease because of flour milling outside our control."

For all the reasons Zulfi had enumerated virtually every privately owned cotton-ginning, paddy-husking, flour-milling unit or plant in the nation would "from today" come directly under government "ownership, control and management."[28] The new "laws" had already been "passed" by administrative fiat—flour milling, rice milling, and cotton ginning control and development ordinances—all officially signed. The new bureaucrats had been given their marching orders before their leader went on the air, for nothing could be left to chance; more than two thousand units owned by tens of thousands of Pakistanis, and worth approximately $1 billion were being seized as he spoke. "Compensation" was promised, "in accordance with our established . . . policy . . . in cash for the smaller undertakings and in bonds for the rest," to be "paid as soon as possible."

The explanation offered by Bhutto for the move was disingenuous:

> Let no one try to misinterpret this decision as the nationalization of an industry. . . . [W]e promised that we would not nationalize any more industries. . . . Governments discard assurances given today by devoting themselves to what they consider a worthier objective tomorrow. Aggrandisement may be evil. But aggrandisement by the poor, the underprivileged and the disadvantaged is pure justice, in accordance with God's decree and the lofty ideals of man. However, an adherence to pledges and fidelity to our assurances are woven into our creed. The decision taken today breaks no promise. The units which have come today under the state control do not belong to the industrial sector . . . they are an integral part of our agricultural economy.[29]

Zulfi had always been good at winning points in USC debates, yet even he seemed to grow a bit anxious as he delivered this economic thunderbolt to a startled nation, realizing that it might have a freezing effect upon all future investment: "This should leave no room for the insecurity of those with timid hearts and sagging souls who have capital to invest in Pakistan but are plagued by the morbid fear that [it] might be expropriated. I ask them: Which is a better haven for your capital than your own country?"[30]

A new stamp with Zulfi's smiling face appeared at this time; its printed message, "Economic well-being of the people is the real strength of the country." Hanif Ramay, one of the leading new spokesmen for the Pakistan Muslim League, openly attacked Bhutto's "irrational" economic policies and lack of "fiscal responsibility," as the prime cause of Pakistan's current "bankruptcy." Ramay's outspokenness, led to his summary trial and conviction by a "Special Tribunal" for alleged "misuse of official position and public funds."[31] His sentence was four and a half years in prison and a fine of Rs.50,000.

Prime Minister Bhutto never forgot that the army posed the greatest threat to his continued rule. Indeed, the most important reason that he allocated as much money as he did to build his personal Federal Security Force was to be sure that he never needed to call out the army to control internal civil strife. Such a move usually led to longer-range military rule, for Pindi's GHQ always considered itself far more efficient and competent to run Pakistan than any politician or party. To help insure the military's loyalty to him and his party on the eve of calling elections, Zulfi informed General Zia of his decision to allow junior officers in the army to "purchase" valuable lots of land in the D. G. Khan and Muzaffargarh districts of Punjab at virtually gift prices. "Since the JCOs are financially not well off and their continued service to the country should not go unrecognized," Zulfi wrote to Zia, "you will appreciate that this substantial concession has been granted out of our keen desire to promote the welfare of the junior ranks of the Army. . . . I trust that you will let it be known to all ranks of the Pakistan Army."[32]

Kissinger returned to Pakistan in early August to caution Bhutto against nuclear weapons proliferation. Zulfi refused to back down, however, and Kissinger diplomatically promised another few hundred thousand tons of U.S. wheat for flood-ravaged Punjab, after which Bhutto told the press he was "satisfied" with his meetings with the secretary of state. Zulfi would later insist that Kissinger "threatened" to make "a horrible example out of you," if he did not stop trying to amass a nuclear arsenal for Pakistan, but that November, after the Jimmy Carter victory, Bhutto wrote to his old friend: "The termination of your present high office saddens me. . . . I shall always cherish my association with you as a friend with esteem and affection."[33] Seven months later a shaken Prime Minister Bhutto told the Canadian ambassador to Pakistan that in their August 1976 meeting, Kissinger had been "all brimstone and fire," warning that if Pakistan went ahead with plutonium reprocessing "the Prime Minister would have to pay a very heavy price."[34]

In September of 1976 Zulfi expected to win his forthcoming elections with a comfortable two-thirds majority and to host the first Third World Summit, which would give him valid claim to the sobriquet some of his devotees already used in addressing him not only as Quaid-i-Awam but also as "Leader of the Third World." Pakistan was not invited to the fifth nonaligned summit held in Colombo that August, but Bhutto wrote to Prime Minister S. Bandaranaike to convey "our warm and sincere felicitations" on the eve of that meeting. "It is the small and medium states which are most vulnerable to foreign aggression, interference and domination," he wrote his small neighbor to the south of India, adding that "the call for a new international economic order is not simply a slogan for the developing countries: It is an imperative need for survival."[35]

On his way home from Colombo, Colonel Qaddafi stopped off in Karachi for a brief secret summit with his dear "Brother," Zulfi Bhutto. Libya's aid to Pakistan at this time was substantial. A week later, Bhutto wrote Mrs. Bandaranaike again to congratulate her on the "successful conclusion" of the Colombo meeting, stressing "your timely emphasis on collective action . . . by all the developing countries. . . . For the Third World . . . [the] essence of the present struggle lies in the confrontation between the rich and the poor, the so-called North-South polarization. . . . I believe that it is now imperative for the developing countries to give a definitive expression to their collective personality. . . . [A]t the highest level of the leaders of all the underprivileged nations of Asia, Africa and Latin America . . . I . . . hope you will endorse the call for a Summit Conference of all the developing countries. . . . We shall remain in close touch with you . . . to promote this Conference.[36]

Margaret Thatcher, then still leader of the Tory opposition to Prime Minister James Callaghan, visited Zulfi Bhutto in early September, and told the press in Pindi that his call for a Third World summit was "a very significant initiative." Libya's Qaddafi also announced his support for Bhutto's proposal, which was formally transmitted to Secretary General Kurt Waldheim by Pakistan's UN ambassador, Iqbal Akhund. Pakistan's Foreign Minister of State Aziz Ahmad presented the summit proposal to the "Group of 77" at their Conference on Economic Cooperation among Developing Countries in Mexico City that September. In transmitting the proposal, Aziz Ahmad explained that a ministerial meeting was not enough. "What is needed, as Prime Minister Bhutto has suggested . . . is the political will to exert our combined strength for changing a system that patently discriminates against the developing countries."[37] But the "Group of 77," like the UN, refused to be rushed into accepting Bhutto's proposed "summit." It would not be held in Zulfi's lifetime.

On 9 September 1976, when news of Mao's death reached him, Zulfi knew he had lost his strongest supporter in the Third World. "Men like Mao Tse-tung come once in a century," Bhutto emoted, "perhaps once in a millennium. They capture the stage and write the pages of history with divine inspiration. . . . Chairman Mao . . . was a giant among giants. . . . [T]he son of revolution, its very essence, indeed, its rhythm and romance, the supreme architect of a brilliant new order shaking the world." It was a eulogy the likes of which he hoped would be written about him. "He remains immortal, his thoughts will continue to guide the destinies of people and nations until the sun sets forever."[38]

With Mao gone, Zulfi focused greater attention on Saudia Arabia, inviting King Khalid to visit Pakistan that October, courting him as he had courted Qaddafi. Cheering Pakistanis lined the entire route from Pindi's airport to Islamabad, which was draped with Saudi and Pakistani flags and pictures of the king and Quaid-i-Awam Bhutto. No fewer than twenty thousand students in bright green and white costumes spelled out warm welcoming greetings to the Saudi monarch inside the new National Sports Stadium in Islamabad. The king and his equally regal host were entertained by Pakistan's best military band, playing both nations' anthems, and by agile gymnasts on flying trapezes. All of Islamabad was decorated with bunting and pictures of the king, with flowers and flags. The day after he arrived, King Khalid touched a silver trowel to the marble foundation stone of the King Faisal Mosque soon to be erected at the foot of Marghalla Hills in Islamabad. In Lahore, he and his host strolled through the beautiful Shalimar gardens hand in hand. From Lahore they flew to Karachi, and Zulfi hoped Khalid would feel up to visiting Larkana. The king, however, was a bit weary by then and October was hardly the month for hunting in Sind. Nonetheless, it was a successful state visit, and Khalid promised an additional $30 million on the eve of his departure from Karachi, presenting "a personal gift" of a Rolls-Royce to his host, which Bhutto accepted as a "gift to Pakistan."

"Ingratitude is sickness," Bhutto told his nation on the centenary of Quaid-i-Azam Jinnah's self-selected Christmas birthday.

> A nation that grudges honour to its leaders is a nation that thinks little of itself. . . . The Pakistan People's Party derives its inspiration from the Quaid's ideals and will never allow them to be submerged. The Quaid believed in democracy. We brought it back. . . . The Quaid urged us to cultivate faith, unity and discipline. We act on faith, we combat disunity and we do not countenance indiscipline. The Quaid despised chauvinism. He condemned obscurantism. He warned us against the dangers of provincial prejudice and schisms. . . . This describes exactly our outlook as well.[39]

By now, Rao A. Rashid Khan, Bhutto's "special secretary," whose "intelligence" duties focused primarily on preparing for the elections, agreed with Reuters correspondent Graham Stewart's assessment that the opposition was in a "sorry state" of disarray and Bhutto should have "no trouble winning the election" whenever he decided to hold it. The Pir of Pagaro was known to be reluctant to "discuss" election strategy with Asghar Khan, who was, moreover, repeatedly frustrated by the "unreliable" attitudes of Professor Ghafoor Ahmad, soon to become secretary-general of the reunited opposition, and Maulana Mufti Mahmood, former chief minister of the NWFP and head of Jamaitul-Ulemi-i-Islam (JUI), who would soon become president of the combined opposition. The reported "helplessness" of the opposition leaders even so much as to attend a meeting on time encouraged Zulfi to believe that the moment had finally come to call new elections. He would soon be forty-nine, and chose his birthday to announce a new gift package of economic reforms.

"My fellow citizens," Prime Minister Bhutto addressed the nation on 5 January 1977, "You elected a Government which had given you a pledge. . . . to stimulate social change that would transform . . . our society. It was, and is, our conviction that . . . unless our national wealth is more evenly distributed, we cannot achieve the social justice which is the vital tissue of the ideology of Pakistan. . . . I do not consider myself a scientist and my nation a laboratory in which I should perform experiments . . . I do not fancy disruption," he insisted, as if suddenly joined in argument with some other self. "I wish to enhance the nation's capability to produce wealth and to provide employment. Whether it can be done by private or by state enterprise is a matter of realistic assessment, not of any preconceived prejudice. . . . I am for the people . . . But the assessment of the people's interest is not an easy matter. . . . Your elected administration has to pass through strenuous exercise of balancing different considerations . . . trying to estimate how an important decision will look from Peshawar or Quetta or Lahore or Karachi . . . or Dadu . . . If our agriculture is unsound, our economy cannot prosper."[40] The new reforms included further limiting land ownership to no more than one hundred irrigated or two hundred unirrigated acres, with specified compensation for all land "resumed by Government"; that land was to be "granted" to then-cultivating peasants, "other landless tenants," or those owning fewer than twelve acres; and "failure to surrender land" could be punished with "rigorous imprisonment" of up to seven years or "forfeiture" of all property, or both. An "ordinance" spelling out the changes was to be "laid before Parliament" in two days, just before that body was to be dissolved for elections.

"If I were a politician like others," Bhutto declared on the virtual eve of his decision to announce elections, "I would make this reform an attractive item in the manifesto of our party. . . . But I do not act that way. . . . [O]thers . . . may be renowned professionals in dangling promises, raising bogus slogans, faking a concern for the poor only in order to achieve their selfish political ambitions." The timing of the announcement of the reforms, he explained, was simply to help "make our land green and more productive." And there were more exciting reforms as well, none of which had anything to do with "electioneering," for that was furthest from his mind. "From today we are abolishing land revenue and making agricultural income liable basically to the same taxation to which non-agricultural income is subject. At the same time . . . we are protecting the small farmer," totally exempting all income from twenty-five acres or less of irrigated and fifty acres or less of unirrigated land from any income tax. There were many other gifts.

"This measure is a final proof that . . . we have no favourites. We see no reason why a rich *Zamindar* should not contribute to public exchequer in the same way as a rich trader or an industrialist. We believe in the integration of our society. This is not a matter of sentiment for us. We consider it a compelling imperative. . . . The measures I have communicated to you tonight are . . . designed to change the ethos of our country." To mollify the very wealthy he also cut the "maximum taxable rate for the highest slab of income" from 60 percent to 30 percent, and promised to continue company rebates to stimulate corporate investment. He further promised subsidies for tube wells, fertilizers, tractors, and pesticides. It was his most munificent, and final, birthday gifting to his people, to his nation, for this was the last birthday on which Zulfi Bhutto would be free to address any one outside a courtroom or prison cell.

"My fellow citizens: I bow my head in gratitude to *Allah* for enabling me to fulfil my pledge to you to do away, as quickly as possible, with the vestiges of a feudal order in our society," Prime Minister Bhutto finished his message that night, almost intuiting, it seemed, that he would never speak to so large an audience again. "These five years have been years of the oil crisis, inflation, and recession abroad and of floods and drought at home. Is it not a case for legitimate pride that we did not let these calamities arrest the nation's forward thrust? Today's reforms are socially just and historically inevitable. They are the culmination of an irreversible process we initiated five years ago. With this culmination, my heart is filled with peace and with the satisfaction that, whatever the future may hold, I have not played false by my people."[41]

Had he any idea then what dread fate the future held in store for him?

15

New Elections and Their Tragic Aftermath

(early 1977)

On 7 January 1977, Prime Minister Bhutto announced to the National Assembly that general elections would be held in two months. Three days later President Fazal Elahi dissolved the assembly at the prime minister's recommendation, reporting that a new National Assembly would be elected on 7 March, and all four provincial assemblies would be elected on 10 March. A day after the announcement, all major parties opposed to Bhutto and his PPP joined forces and agreed to field single opposition candidates for all seats they contested, under the symbol of the plough, as the Pakistan National Alliance (PNA). The PNA included all parties of the older United Democratic Front, now bolstered by Asghar Khan's Tehrik-i-Istiqlal. Maulana Mufti Mahmood was unanimously elected president of the alliance.

The Election Commission set 19 January 1977 as the final date for filing nomination papers for National Assembly candidates, and because Zulfikar Ali Bhutto alone had by that date filed papers for the Larkana district, he was announced "elected unopposed." Nine other PPP nominees from Sind were also "elected unopposed," including Chief Minister Jatoi, Mumtaz Bhutto, and Education Minister Hafeez Pirzada. Jatoi rejected as "unfounded" charges by the opposition that its candidates had either been "kidnapped" or otherwise "prevented" from filing nomination papers. But Asghar Khan recalled, "When the last date of filing nomination papers passed, we learnt that no one had been allowed to file nomination papers for the seats being contested by the Prime Minister and the Chief Ministers of three Provinces. . . . Those of the Pakistan National Alliance who tried to file . . . for these seats were detained by the police. In one or two cases, candidates were abducted and were missing for weeks. The government probably hoped that as a result the opposition would be demoralized."[1]

The combined opposition kicked off its campaign with a huge public meeting in Karachi's Nishtar Park on 23 January. Karachi with its predominantly Muhajir population and its alienated business and financial leadership proved to be a major bastion of PNA support, giving nine of its eleven National Assembly seats to the combined opposition. Asghar Khan soon emerged as the most popular potential opposition "prime minister." His military stature and handsome face appealed to many urban and sophisticated Pakistanis, who by now found Zulfi Bhutto untrustworthy and far less attractive than he had seemed a decade ago, when he and his party both symbolized the "Sword" (*Zulfikar*) that would one day "liberate" Pakistan from Field Marshal Ayub Khan and other "corrupt" martial and feudal leaders. But the former air marshal really had no political base of his own to speak of, coming from the Frontier, and he was without inherited wealth or tribal title, unlike Wali Khan, who would have been the leader of the PNA were he not behind bars. Begum Wali Khan remained free, however, and won her husband's seat with the help of Sher Baz Mazari, leader of the recently formed National Democratic Party, which was a reincarnation of Wali Khan's outlawed National Awami Party. Even the Muhajir Karachi-wallahs and Frontier Pathans who responded most affirmatively to the PNA's passionate cry to vote Bhutto and his one-man party out of power knew as they voted that they were a minority among the some 31 million Pakistanis eligible to vote. For most of rural Sind and much of Punjab, Pakistan had but one Leader, Quaid-i-Awam Bhutto. That at least seemed quite clear up to the very day of elections in early March.

Zulfi Bhutto, much like Richard Nixon, could not rest content with, or perhaps even believe, his own popularity. At least one part of him always mistrusted, doubted, feared, suspected everyone, and thus always expected the worst. Was that because deep inside himself he knew, or feared, that he deserved defeat, mistrust, opprobrium, and much worse? Or was it that he wanted, needed, reached for so much power that nothing he could have easily won by free and fair elections, such as he so often promised to preside over, would ever suffice to satisfy his insatiable need?

"Although three weeks have passed since the announcement of the elections," Zulfi wrote Chief Minister Qureshi of Punjab on 2 February, "the Party's electioneering campaign has not picked up the desired tempo in your Province. . . . workers have not been mobilised. The Party apparatus is not functioning properly. Publicity lacks in vigour and punch. And to top it all, a bureaucratic stance is in evidence towards the elections. You know very well that elections are a political process which have to be conducted and won politically. Since there is no time to be lost, you should gear up the Party apparatus and mobilise all tiers of the Party from the Provincial to the village level. . . . In the end, I must stress that you should not wait any longer to come into the top gear."[2]

Chief Minister Qureshi and his cabinet colleagues proved much too successful in following their leader's firm order to shift into "top gear." Instead of losing Punjab, or merely winning it by a narrow margin, as most Pakistanis expected to happen, the PPP swept the polls, taking no less than 105 out of that majority province's 115 seats in the National Assembly. Ambassador Henry Byroade had been invited by the prime minister to his house to track election returns the night of 7 March, when television and radio would be reporting all night on the second "free" national election. "The results were coming in about

70 percent," Byroade recalled. "He was losing in Karachi. He was losing in Peshawar. Then the Punjab numbers started coming in and guys who were absolute thugs won by 99 percent. . . . Then [Bhutto] became absolutely quiet and started drinking heavily, calling Lahore, and he said, 'What are you guys doing?' . . . I saw Bhutto at 8 the next morning, and he wasn't himself. He hadn't had any sleep, obviously drinking. He was just sad."[3]

Unlike Zulfi himself, most of his current subordinates now were hardly politically astute. They were unsophisticated in the time-honored Western ways of winning elections by luring various constituencies with promises neatly tailored to fit their tastes, needs, hopes, appealing to ideals and popular dreams as well as crass material desires. They were simply skilled at herding stubborn beasts up the road or people into a polling place, at making sure their imprint was next to the sword and not the plough. Hanif Ramay understood the fine "art" as well as the "science" of politics, only Zulfi had not only thrown him out of Punjab's chief ministership but vindictively locked him away in a common jail.

Director M. Akram Sheik and Joint Director Muhammad Isa of Pakistan's Intelligence Bureau devoted virtually full time now to compiling dossiers, analyses, and detailed reports on every National Assembly candidate and his or her election prospects for the prime minister, most of which were forwarded to him through his secretariat, either by Special Secretary Rao A. Rashid Khan, or his regular secretary, Saeed Ahmed Khan. On 19 February 1977, Sheik and ISI (Inter-Services Intelligence) Director General Ghulam Jilani Khan jointly compiled a "*very conservative* assessment" of PPP prospects in each National Assembly race, "*as of today.*"[4] The Punjab estimates at that time were for PPP candidates to have a "clear win" in 33 contests, to "lose" 32, with the rest "doubtful," though 24 of those were "plus to PPP" and 12 "minus to PPP," and 15 "undecided." For Sind, there were 25 "clear wins," only 4 clear losers, with 14 undecided. The opposition refused to contest any seats in Baluchistan because the army remained in force there, which gave all 7 of its seats to the PPP, and the NWFP vote was estimated to be almost evenly divided, which meant that with "concerted effort" the PPP could win 13 or possibly even a few more seats. Despite so optimistic a "conservative" assessment, however, Rao Rashid Khan sent the prime minister a few days later "the names of those opposition candidates, who, if elected, would obstruct the democratic process in the country and cause problems. . . . Of course the guiding principle is that the People's Party should strive to defeat as many opposition candidates as possible but special effort needs to be made to keep the listed candidates out of the Assemblies."[5] The former inspector general of police for Punjab, Rao Rashid Khan, was taking his new political assignment most seriously.

Zulfi had to win a clear two-thirds majority to change his much-vaunted democratic constitution back to the presidential system he personally preferred for what he optimistically believed would be his last term in office, as well as his final half decade of life. Rao Rashid must have known of Bhutto's secret constitutional plans. After reminding his prime minister of the "special effort" needed to keep all the obstructionist opposition candidates he listed out of the assemblies, Rashid cautioned: "As regards conceding certain seats to the opposition candidates who have been helpful or are going to be helpful, the position is such that no such generosity can be shown as every seat is important."[6]

Zulfi's speeches, of course, were most brilliantly effective in mimicking his opposition, whose-nine party alliance he labeled the "Gang of Nine." The unwieldy combination, moreover, of Pakistan's most orthodox Islamic parties trying to keep in step with some of its most secular and modernist leaders offered a relatively easy target. He dismissed most of that "gang" of "gnats" arrayed against him as either "obscurantist" mullahs or "agents of capitalism" or "foreign reactionaries." He had never liked any of them, nor did he respect or fear them. Quite the contrary. He considered all of them, including Asghar Khan, "light-weights," indeed, "non-entities." Compared to him, perhaps some of them were, but Zulfi's hubris would soon to be followed by his tragic fall.

Asghar Khan was far less eloquent than Zulfi Bhutto, but he pointed out in most of his public speeches that central government expenditures on "luxury goods" had recently skyrocketed, while the cost of its police force had quadrupled, as had most crimes, in the past five years, and some 80 percent of all "nationalized industries" were currently producing nothing. Maulana Mufti Mahmood, in his campaign speeches, noted that many PNA candidates had been "kidnapped," including Maulana Jan Mohammad Abbasi of the Jamaat-i-Islami, who would have filed papers against Bhutto in Larkana had he not been locked up for two days by local "police." The fearless full-bearded PNA president also repeatedly pointed out that most PPP candidates had more Jeeps, new cars, and motor scooters than they knew what to do with, while his PNA people were usually obliged to walk or travel in old trucks or buses.

Maulana Maudoodi's fundamentalist Jamaat-i-Islami posed in some respects the most dangerous challenge to Bhutto's PPP, for the JI had a "large following" in all ranks of the army, though no one really knew exactly how large it was or how high in the topmost command its "sympathizers" went. The army was always supposed to remain apolitical, yet soldiers could vote. Director General Jilani of ISI had recently sent a "*SECRET*" message to General Zia, with a copy to the Prime Minister's House, warning of a hand-written Urdu poster found in Multan's barracks, calling in red ink for "*Army Revolution,*" and listing a number of grievances and demands. What Jilani pointed out, however, was that nowadays "the personnel are more susceptable to religious appeal. The Mullah is omni-present in our units. The Mullah has his own brand of religious affiliations. . . . Above all, there is the influence of the JI and JUI. . . . Multan itself is an ancient city and a seat of 'Pirs'. . . . [T]he JI and JUI have been making ingresses into the armed forces. The JI has the advantage and the benefit of the teachings of Maudoodi. The JUI has the advantage of Maulana Mufti Mahmud's [Mahmood] influence in . . . Multan and surrounding areas."[7] Zulfi read this important letter carefully, underlining parts of it, covering the margins with comments, but from the latter it is clear that he failed to grasp the significance of Jilani's message and that he appears to have entirely forgotten that Zia's own command prior to his escalation to top brass as chief of army staff had been in Multan.

"The background . . . of Multan is not the main cause," Zulfi scrawled along the letter's margin, as he always did in responding to such files, usually during his sleepless nights.

> The main reason for its circulation . . . in Multan is not because Multan is a city of Pirs . . . but because of the heavy concentration of the Army in Multan. Let us not exaggerate the *inherent* importance of Maudoodi's teachings. In the old days people were more backward and fanatical yet we joined the British Army in large numbers and . . . fought the British Wars in spite of inspiring Fatewas [religious decrees] to the contrary. Such teachings become dangerous only when the Chief of the Army gives them *official blessings* and *respect*. This will boomerang. . . . That is why I told him [Zia] in my first letter that I do not want a "Mullah Army."[8]

Zulfi still thought that General Zia could be trusted to do exactly as his prime minister directed.

Zia, of course, continued to behave as deferentially as ever toward the prime minister, smiling, bowing, quietly accepting whatever Bhutto told him with the seeming humility for which he was to become famous the world over. Another secret report had warned Bhutto the past November of how "well organized" the Jamaat-i-Islami was, and how much "black money" it received both in contributions collected by volunteer mullahs "door to door" and from foreign countries, primarily Saudi Arabia. Many thousands of copies of Maulana Maudoodi's *Tafhimul Quran* (Translations from the Quran) had recently been sold and could generally be seen in the private "libraries" of "most of the Army officers." On 1 July, 1976, Zia gave copies of that *Tafhimul Quran* as "prizes" to soldiers, who won a debate arranged by the Army Education School. The *Tafhimul Quran* was then proposed to be included in the examination "for promotion of Captains to the rank of Major," and Bhutto was sufficiently concerned about the "highly injurious" impact of such emphasis on religious study for all officers that he met with Zia on 24 November 1976, to discuss the matter. Zia wrote "My dear Prime Minister" that same day, returning the report and diplomatically noting, "as has been expressed by the Prime Minister as well, Islam is not the private property of an individual."[9] Not long after Zia circulated a "Restricted" memo on discipline to all units of the army:

> [O]fficers and men, like all Pakistanis, should have the knowledge of and practise Islam, and be inspired by it in the performance of their professional duties. We, in the Army, are not Mullahs and we do not need anyone's certificate for being the followers of Islam. . . . [T]he Army has re-established its image of patriotism and loyalty to the lawfully constituted government. I would like all ranks, and in particular all commanders to bear in mind, that we are professional soldiers who have been sworn not to get involved in any political activity whatsoever. We must,

> therefore, exercise the utmost vigilance and not allow any one to exploit the Army for ulterior political motives.[10]

In mid-February when Zia sent a copy of the memo to Bhutto, Bhutto believed that he meant it.

The PNA issued its manifesto in February, affirming that "the Almighty is supreme and powerful and all things belong to Him" and promising to "fully enforce" the "Holy Quran and . . . to enable . . . every Muslim to lead life in accordance with the tenets of Islam." At the same time it would "guarantee equal rights to all citizens of Pakistan," "inculcate feelings of love, brotherhood and sympathy among the people of the country," establish "a just system where justice was available to all," "destroy all forms and modes of suppression, . . . bring the prices of the commodities of daily use within the reach of the common man . . . [and] guarantee provision of food, dress, accommodation, education and medical care to each and every citizen." It sounded much like the initial manifesto and platform of the PPP.

The major campaign issue was soon reduced to which side would come closest to implementing its noble promises. Most Pakistanis remained, moreover, confused by the many faces of the opposition alliance. It was not a single party, and it seemed hard to believe that Asghar Khan, the Pir of Pagaro, and Mufti Mahmood, not to mention all the others, could really ever pull the PNA, not to speak of Pakistan, in the same direction. Zulfi Bhutto, on the other hand, was without question or doubt *the* leader of his party as well as his government, nation, and people. The "dictatorial" powers he now held, which the opposition used as a major complaint against him, proved to be his greatest asset in winning mass support for his still most "popular" party.

The PNA's oft-repeated promise to "enforce" Islamic law throughout Pakistan, if elected, and hence to ban the sale of wine and liquor, gambling of every kind, the payment of interest, and the use of "obscenity" was seen as a serious frontal challenge to Bhutto's leadership and the popularity of his party. To counter such attacks, Bhutto kept Minister Maulana Niazi busy, launching Pakistan's first National Seerat (Life of the Prophet) Conference in Rawalpindi on 19 February. "The People's Government . . . does not believe in sanctimonious pretensions to gain mundane popularity, nor in self-righteous heroics," insisted Bhutto, "but steadfastly pursues . . . the cherished goal of promoting Islamic ideals and values."[11]

By the seventh of March, when some 31 million Pakistanis were permitted to vote, if they wished, the clearest choice before them was whether they wanted Zulfi Bhutto to continue to run their country for another five years, or whether it was time to give some one else a try. Had he only been less greedy, less suspicious, less mistrustful—or insecure, he would most likely have won a majority, even if not two-thirds—of the seats in the National Assembly, and might still be alive today at the helm of Pakistan's People's Party government. No more popular leader has yet emerged from the soil of Sind or Punjab or the harsh and rugged Frontier, none more admired, even "worshipped," by the impoverished peasants and simple laborers than Quaid-i-Awam Bhutto. But no one was more feared or hated either, though many were more fearsome and violent, far more brutal and unkind to friends and foes alike than Zulfikar Ali Bhutto had ever been. Yet none of those feudal barons or criminal ganglords ever reached the top of that slippery pole of power.

Zulfi's wretched fate was the product of his singular success as well as his faults and failures. In the excess of his "victory" lay the time bomb of defeat. Small wonder that the news of his party's victory tally on that long, long night of ballot-stuffing and restuffing, of counting and recounting and discounting throughout Punjab, left him feeling so "sad," a deep foreboding that no amount of drink could assuage. Some 17 million of 31 million eligible voters cast ballots on 7 March 1977. The PPP received less than 60 percent of the popular vote but won more than 75 percent of the two hundred elective seats in the National Assembly. The PNA, with more than 35 percent of the popular vote to its credit, was allowed to translate that into less than 17 percent of the seats. Ten women and six members of minority communities were named by the majority party.

Hardly surprising that the PNA cried "foul" and charged "fraud," and immediately announced that it had no intention of contesting any provincial assembly seats on 10 March, for those too would be "rigged" and "stolen" by PPP "election thieves." So from March eighth onward the campaign battle moved from election platforms and polling booths to the streets of Karachi and Lahore, and to the gutters of smaller towns and the dusty roads leading into and out of every center of PNA or PPP power.

Zulfi insisted at first on holding provincial elections as scheduled. He denounced the boycott and appealed "to the mass of the electorate for the Provincial Assemblies to . . . turn out for the vote with full vigour and enthusiasm, no less than they showed at the National Assembly polls."[12] But less than half of those who had voted two days earlier bothered to return to their polling stations that Thursday. Now PNA leaders issued a call for all Pakistanis to "strike" on Friday to "demonstrate" the nation's "loss of confidence" in its present leadership.

Virtually all shops were closed that Friday, and from Karachi to Peshawar violence flared each time the forces of one party met those of the other. Most police stood aside as they watched the angry, shouting, slogan-chanting, fist-waving, brickbat-and bottle-throwing Pakistanis who were to become familiar in every city of Pakistan during the next four months of fierce postelection fighting, euphemistically called by some "political debate." Zulfi knew by now that he could no longer pretend that nothing had gone wrong at any polling stations, for too many had seen the ballot-stuffing and had suffered in the violence that had already claimed a dozen lives and hundreds of other casualties, as well as arrests. On 12 March he declared over nationwide radio and television that he was prepared to meet with leaders of the PNA to have a "dialogue" on "some questions" raised during the recent elections. The next day he wrote to Maulana Mufti Mahmood "in pursuance of the invitation" he had extended, "awaiting your response and I hope that it will be positive."[13] None of the opposition leaders trusted Zulfi Bhutto, however, because he continued to talk of the elections on 7 March as a "settled matter." Mufti Mahmood replied on 14

March that after meeting with coleaders of the alliance to consider the invitation, "it was decided not to enter in a dialogue, the terms of which are not clear."[14]

Zulfi knew that he was running out of time, but he was also running out of money. All of his recent economic ploys had backfired, and Pakistan's most recent cotton crop had failed. Inflation was rampant, investment was at a standstill, and capital was fleeing the country at so alarming a rate that he feared he would have no money with which to pay his own elite twenty-thousand-man Federal Security Force and the current base salaries of army men, not to speak of the 50 percent raise in pay he had promised virtually all ranks. On 13 March, therefore, he wrote with diplomatic restraint but obvious desperation asking his "good friend" the Shah of Iran to "guarantee" a $300 million loan that Pakistan had requested of Citibank of New York; its "immediate availability . . . is imperative for us." The shah replied on 22 March:

> I thank Your Excellency for your letter, in which you have referred to the question of a loan. . . . As you have correctly recalled, the decision to allow my Government to act as Guarantor for this loan was reached on my part last year. Our Ministry of Economic Affairs and Finance has, therefore, been under orders to fully cooperate in the arrangement. . . . I understand, however, that the Citibank has made the . . . final draft of the loan agreement subject to the receipt from Pakistan of information on the use of the funds, data on Pakistan's external debts and their payments, as well as agreement on other pertinent details.
>
> It is also the understanding of our Ministry of Economic Affairs and Finance that the Borrower is now His Excellency . . . not the Government of Pakistan, and that Citibank representatives have indicated this is being referred to their legal department for study. . . . [T]he above reasons, plus the fact that this is apparently the first case of its kind where a government acts as guarantor for another government, call for a careful examination. . . . [T]hey expect 3 to 4 weeks would be needed.[15]

Zulfi never got the loan, nor did he bother to answer the Shah's last letter to him. He probably never even finished reading it, for by late March he had no time to waste on Persian politesse. He had troubles enough simply seeking to launch some sort of "dialogue" with his PNA opponents. "I regret to say that . . . you have again avoided to clarify your stand regarding countrywide preplanned rigging of general elections reducing them to a complete farce," Mufti Mahmood wrote him on 17 March: "On the 7th March, 1977, the country was subjected to a farce in the name of general elections. . . . [T]he administration made every endeavour to subvert the national will and to ensure a new lease of life for a leader and a government which had been overwhelmingly rejected."[16]

The PNA demanded that Bhutto resign at once and that new elections be called, to be administered by "agencies enjoying the confidence of the people and the PNA," by which was meant the judiciary and the army. The police used tear gas to disperse a PNA march in Karachi and arrested hundreds of demonstrators. The Karachi High Court Bar Association demanded a "judicial enquiry" into charges of election "malpractice." On 18 March six leaders of the PNA—including Asghar Khan, Mian Mahmud Ali Kasuri, and Mir Ali Ahmad Talpur—were arrested for creating "lawlessness" and for "unconstitutional activities." The riots in Karachi the next day were so violent that the army had to be called in to "restore law and order" and a curfew was enforced. Soon there were demonstrations in Lahore, Multan, Gujranwala, and other cities of Punjab as the PNA's attack against Bhutto and his "rigged elections" spread across most of the country.

On 19 March, Zulfi wrote Mufti Mahmood, reiterating that "my Party has secured an overwhelming vote of confidence from the electorate, which no false charges of rigging, no matter how strident and sweeping, can throw into dispute." He charged that "every day your colleagues and supporters are inciting violence. . . . Only suffering for the poor people results from this negative approach . . . your present activities have caused prices to rise higher. . . . Your Alliance is releasing hate and anger when all responsible political elements should direct their attention to the ways of building a harmonious society."[17] Nonetheless, Bhutto was conciliatory, open to the discussion of any or all issues connected with the electoral process as long as it remained within "constitutional" limits. "I cannot allow license to any forces of chaos and disorder in the country. . . . Acts of destruction will not attain what amicable discussion can achieve."[18]

Zulfi hoped to negotiate a political settlement, certain that his opponents were merely bargaining for a larger share of the pie, more seats in the National Assembly. He was ready to scrap his dream of using a two-thirds majority to change the Constitution. He would be content to live within the parliamentary boundaries of the 1973 Constitution as long as he was left at the helm. But he was highly suspect among the most powerful PNA leaders and his words and promises fell on deaf ears. Some of his close advisers, including his secretary, Yusuf Buch, had urged him to call fresh elections immediately, and it may be that had he done so, and had there been no ballot stuffing or tampering of any kind, confidence could have been restored. Yet his simply conceding the need for a second electoral round might have made many Pakistanis lose all faith in their prime minister's judgment, if not his strength or probity. He was, as he would soon learn, in a no-win situation: whatever he did and said, or failed to do and say could be used and construed against him. And with each passing hectic day and sleepless night, he looked and sounded more haggard, more exhausted, less sure of himself, sadder. The harder he tried to clutch it, the more the sands of power seemed to slip through his fingers.

Bhutto urged the chief election commissioner, Justice Sajjad Ahmad Jan, to review as swiftly as possible all "charges" of electoral "malpractice" of any kind. Special tribunals were to be established all over Punjab for that very urgent reason alone, and the law secretary drafted a special amending ordinance to legalize such procedures, for, as Bhutto put it in his letter of 21 March to Justice Sajjad, "If some individuals have violated the law, it is my earnest desire that they be proceeded against and brought to book as quickly as possible."[19] And he meant it because each day of street protests and military rule in cities like Karachi was costing Pakistan millions of rupees it could barely afford while the world was wondering what had gone wrong with this once most popular "leader of the people," who less than one month ago had aspired to lead the entire Third World. Indira Gandhi had sent him a personal "congratulatory message," which had never appeared in the Indian press "for reasons best known to the Indian Government," as Ambassador S. Fida Hassan reported to his prime minister on 10 March. The pat from New Delhi provided cold comfort to Zulfi, for on 24 March Mrs. Gandhi herself was swept aside by India's electorate; she bowed out gracefully, surrendering her dictatorial emergency powers to Morarji Desai and his Janata opposition. By then over ten thousand PNA protesters had been arrested throughout Pakistan, according to PNA leaders. On 25 March, PNA President Mufti Mahmood and General-Secretary Ghafoor Ahmad were arrested.

Prime Minister Bhutto addressed his 168 newly elected members in the otherwise empty National Assembly in Islamabad on 28 March. He reissued his offer of "dialogue" with the PNA, calling upon his opponents to give up the "politics of vandalism," to stop allowing themselves to be the "tools" of "great power tricksters." He also promised that there would be no further nationalization of "anything," and that "justice" would be done to "all."

That evening he told visiting Turkish Minister Halil Basol how deeply "disappointed" he felt that a year after the Izmir conference there had been "no follow-up" to his proposed "firm military alliance" of Pakistan, Iran, and Turkey, an alliance that could "become the pivot for all Muslim countries." He hoped that some dramatic breakthrough internationally, something as exciting as the Islamic summit in Lahore, might help restore his much deflated prestige, especially among the more devout members of his maulana-led opposition. In the "present world," Zulfi told his Turkish visitor, it was "difficult to find even one sincere friend." Ah, how true, his worthy guest agreed, as true in Istanbul as it was in Islamabad. Not to speak of Teheran, which Zulfi barely mentioned, smarting from the Shah's very formalistic handling of his urgent request for help in procuring Citibank funds. Zulfi tried his best to win greater Turkish support by mentioning how "insulting" it was for a "civilisation" as old and great as Turkey's to be "given lectures on human rights,"[20] such as those recently received from President Carter.

But Turkey's own precarious president was no more eager than the Shah of Iran to pull Zulfi Bhutto out of the fire into which he had fallen. Marches in Lahore to protest the "rigging" of elections attracted larger and larger crowds, and though police still fired only tear gas, they would soon feel obliged to use live ammunition. On April first, Prime Minister Bhutto appealed again to his nation for "calm" and called upon his opponents to "negotiate a solution," promising then to repeal the emergency he had declared, under whose Section 144 he had arrested them all. He would now release them, and retract all his draconion Defence of Pakistan Rules (DPR). Only no one believed him, for it was, after all, April Fools' Day. On 3 April he wrote to his neighbor Prime Minister Morarji Desai, congratulating him on his election and suggesting "early talks" to "develop trust, harmony and good-neighbourly co-operation" between India and Pakistan, and informing Morarji of his call for a "Third World Summit." Five days later came the diplomatic reply: "I have noted your reference to your proposal for a Third World Summit . . . but we have yet to appreciate its full implications."[21]

On 9 April police opened fire on a crowd in Lahore, and official sources reported that "8 persons were killed and 77 others injured." The next day a PNA press release gave the number of deaths as thirty-seven. Five more deaths were reported that day from police bullets in Karachi, and twenty-eight prominent leaders of the opposition were arrested in Pindi marches. On 11 April the General Council of the PNA called upon Pakistanis to stop paying taxes or duties to the "present unconstitutional government." A strike called on that day brought Pakistan to a virtual halt, except for protest marches in most major cities. The prestigious Lahore High Court Bar Association announced its decision to "terminate the membership" of any lawyers in its association who had been elected on PPP tickets. Finance Minister Hafeez Pirzada now "disclosed" that the government had just appointed a committee to reconsider the "position" of recently nationalized rice-husking, cotton-ginning, and flour-milling operations. The next day Karachi's High Court Bar Association in a resolution deplored the police firing within the precincts of Lahore's High Court on 9 April, "killing many innocent people and injuring several dozen others, including advocates and retired judges." On 13 April Zulfi told the BBC that he "would not hesitate" to call out the army, if he deemed that "necessary" to "restore order."

As if in response to Bhutto's threat to use the army, General Gul Hasan and Air Marshal Rahim Khan, who had both been instrumental in bringing Bhutto to power in December 1971, wired him their resignations as ambassadors to Greece and Spain that same 13 April 1977. Rahim Khan charged Bhutto with having "made a mockery of democracy in Pakistan. . . . You have not honoured your pledges. . . . The previous regime held fair and free elections but you have imperiously ignored that fine precedent and allowed them to be rigged instead. I cannot sit idly by and see the country being dragged into another civil war by power-hungry men. I am, therefore, resigning in protest against your oppressive and dictatorial regime."[22] Gul Hasan wrote substantially the same thing. Both officers were in close touch at this time, coordinating their resignations as well as the speeches they were soon to make in Britain against Bhutto's regime. The resignations, Zulfi must have known, were the gravest of portents, not because of the political or diplomatic weight of either man but because they had never ceased to be exemplars of Pakistan's military mind.

In mid-April Zulfi met with old Maulana Maudoodi, hoping to win his backing in convincing the PNA to call off its escalating "strike" and daily repeated demands that he resign before "free and fair" fresh elections were scheduled. Attorney General Yahya Bakhtiar met the same day with PNA leader Nasrullah Khan to suggest holding immediate "fresh elections" for all provincial assemblies, and if the PNA won a "majority" in all of those, then "dissolving" the National Assembly. Nasrullah Khan dismissed the proposal as "entirely irrelevant." Begum Wali Khan now issued "an ultimatum to the ruling party Chairman" to accept all PNA demands by 20 April or face the consequences. But Zulfi still thought he could wheel and deal his way out of the predicament, for did he not have General Zia still at his beck and call?

Two days after meeting Maulana Maudoodi, Zulfi announced his decision to introduce "complete prohibition" throughout Pakistan, and to ban all gambling, nightclubs, bars, movie houses, and other anti-Muslim activities, thus bringing Pakistan's laws into complete conformity with the Quran and Sunnah "within six months." He understood by now that the most powerful, implacable opposition confronting him was the mullah- and maulana-led force of tens of millions of devout Pakistanis, both inside the army and out, who believed that the laws of Islam were higher and far mightier than the laws of any land. And if it was necessary for him to stop drinking within Pakistan to remain in power, he was even ready to try that, just as he had abandoned construction of a huge gambling casino on Karachi's seashore—where its concrete whale-like skeleton remains a forlorn symbol of Zulfi's unrealized dream. He would give up all the wine and the waltzing, for power was far sweeter and more important. He thought that a ban on alcohol would convince the mullahs who hated him that he should stay on the throne. Even as he believed that by donating a solid gold door for the shrine of Lal Shahbaz in Sehwan all his sins would be forgiven.

Zulfi went on the air again on April twentieth to call upon Pakistanis to "banish the politics of irrationalism, eschew the fomenting of passions and open their minds to the realization that there is no problem which cannot be solved through a reasoned approach."[23] That same day the Saudi ambassador, Sheikh Riyadh Al-Khatib, was a visitor. Zulfi said that he was "not concerned" about the "fate of his government," and would be "hesitant" to accept any foreign "help or assistance" in resolving Pakistan's current "domestic problem," but Saudi Arabia had "always occupied a very special position in Pakistan" and King Khalid had told him that "he would never like to see Pakistan or its people suffer"—hence this meeting. Bhutto told the ambassador that he "had made every effort" to resolve the "political crisis," taking several "initiatives," but that "because of guidance from abroad," the opposition had "adopted an utterly negative attitude." Whether Zulfi now believed that U.S. "guidance" through the CIA was responsible for what was happening to him or whether he merely wanted the ambassador to think so is unclear.

The ambassador assured the prime minister that he had been "praying for a reasonable solution" to Pakistan's political problem, and would do whatever he possibly could to "mediate." He had sent a cable to his foreign minister and was awaiting a "signal" to fly home to give his "assessment" of the situation to his king. Now Zulfi told of Kissinger's having urged him the previous August to give up the idea of acquiring the reprocessing plant from France or "pay a very heavy price," and how he in reply had said he would never "succumb to brow-beating or pressures from the United States." Foreign Minister Aziz Ahmad also attended this meeting and diplomatically suggested that "King Khalid might consider sending a message to PNA leaders asking them to enter into constructive dialogue with the Prime Minister so that the stability and security of Pakistan did not suffer." All three agreed that the ambassador should fly to Riyadh, carrying a letter to that effect for King Khalid, with a "personal letter" from the prime minister as well, to alert the king to the "gravity of the situation."[24]

The next day, 21 April, the Saudi ambassador returned to the Prime Minister's House, having received his signal to return to Riyadh from his foreign minister, but because of the PIA strike, there was no available plane. He requested a C-130 military plane to fly him from Chaklala to Riyadh on 23 April. Bhutto would send him, however, on a smaller PAF Falcon. He now informed the ambassador that he had decided to "impose martial law" in Karachi, Lahore, and Hyderabad, where the "law and order situation had gravely deteriorated"—even "essential services," including flights, had been "disrupted," as the ambassador knew. Zulfi told the ambassador that Asghar Khan and Mufti Mahmood were both being "used" as unwitting pawns by "plotters" within the PNA whose "inner plan" was "to eliminate Mr. Asghar Khan after bringing him into power, replacing him with Mr. Wali Khan, who would then proceed to bring about the dismemberment of the country. Baluchistan, NWFP and Sind would secede leaving Punjab alone." The prime minister said that he had "authentic information" that the Soviet Union had placed "unlimited funds" in the hands of one opposition leader, whose "strategy" was to "create troubles in parts of Punjab and Sind—particularly in Karachi." That was why he had imposed martial law: "to restore normalcy and to deal with the external threat." He went on to explain that he had consulted "top-ranking officers" in the army and informed them of the

> correct position. . . . [I]t was not a matter of his own future. He was a politician, with a real belief in the democratic process . . . not a military ruler engaged solely in perpetuating his own regime. . . . It was a matter of the security and the integrity of the country. . . . [H]e had dedicated his life to Pakistan. He had neglected his personal life for the sake of the country. He had saved it from grave danger in 1971. . . . If the people of Pakistan disregarded his services, he would be happy to leave and live a respectable, quiet life. . . . But the matter was not as simple as that . . . the present chaos was part of a major conspiracy and a repetition of what had been done in 1971. Pakistan was threatened and its subversion would mean the end of Islam in South Asia.[25]

"They" were after him again. Zulfi wanted Saudi Arabia to understand that the very existence of Islam in South Asia was now being threatened. He had asked his military commanders only to "do their Constitutional duty" in taking over Pakistan's biggest cities. The commanders of the armed forces had "agreed entirely" with him, he said, and assured him of their "loyalty and support" in "getting rid of the 'cancer.'" A month later Gul Hasan was reported to have told a Pakistani meeting in Birmingham that General Zia was "a sycophant and a coward and I am sure he must be seeking Mr. Bhutto's permission even for taking his meals."[26] Others would say that Zia was simply giving the prime minister enough rope to "hang himself."

The Saudi ambassador noted that at their meeting a day earlier the prime minister had told him that the "United States had supported the Opposition with funds," and now he was talking of "massive Soviet involvement." The puzzled ambassador wondered what might be the "aims" of the two superpowers? Zulfi was quick to explain that the "object of the United States was to weaken Pakistan"; that of the Soviet Union was "to dismember it." He also added at this point that India played "some role" in this "ballgame," and that now Pakistan was "again" facing a "three-pronged attack."[27]

Special Secretary Rao Rashid Khan had written his prime minister earlier on that fateful April twenty-first to report that "the PNA has now come on the streets in full strength to force the Prime Minister's resignation. It is unfortunate that the PIA, Railway and other labour unions have now joined the agitation . . . paralysing the whole life in the country . . . The law and order situation in the main cities . . . has deteriorated to an extent that nobody's life and property seems safe. The Police . . . have become incapable of handling the situation. . . . The PPP position in these cities is weak and, therefore, it has not been able to mount an effective retaliation. . . . Only the Army can now control the situation."[28] Zulfi accepted that piece of Rao Rashid's advice immediately. He did not, however, implement other suggestions in the same urgent letter, which helps to explain the sudden breakdown of virtually all of Pakistan's central government in less than a week.

"The Prime Minister has so far carried the terrible burden of decision making all by himself," Rao Rashid continued, though he knew how dangerous it was ever to criticize Zulfi Bhutto to his face, so he immediately added, "This is a burden which is beyond human capacity. It is time that the Prime Minister creates a permanent forum of his trusted colleagues who should meet every day for briefing and help him in reaching . . . momentous decisions. . . . So far the Government has functioned in compartments. Various individuals have been consulted at various times and the circle has been constantly changing. The decision making, therefore, has suffered from a partial or complete ignorance." Rao Rashid also noted that "alienation" of "Government servants" had resulted from taking away of their security of service by the PPP Government. It may not perhaps be too late to restore the constitutional guarantee." But it was too late, of course, much too late for Zulfi Bhutto to start running his government by honoring all those traditional institutions of Pakistan he had long since alienated or abolished, or by introducing and using modern management methods of ascertaining what in fact was needed, wanted, or merely "happening" in this large and disparate nation. He ruled more by personal passion or pique, more by instinct or intuition, or with the help of strange "secret" advice from his countless spies than by any rational insight or consensual appreciation of long-festering problems. All Zulfi could do now was to call out the army, and to call Mustafa Khar back from his London holiday to become his "new" special adviser.

On 22 April five Pakistanis who violated the curfew in Karachi were shot dead. Asghar Khan reported that another twenty people had been "killed in disturbances" in other parts of the country that day. Nasrullah Khan, one of the few PNA leaders still out of jail, called the imposition of martial law in Karachi, Lahore, and Hyderabad "high treason," and was arrested two days later as he led a protest march. In Lahore a boy was shot dead that day. Martial curfew was extended to Sialkot and Bahawalpur. The Lahore High Court declared martial law in Punjab "illegal." Zulfi still hoped to have a dialogue with jailed PNA leaders, including Wali Khan, to whom several secret overtures had been made in past months. He ordered Mufti Mahmood, Professor Ghafoor Ahmad, Begum Wali Khan, and several others brought from more distant prisons to the Sihala police "rest house" in the hills of Murree outside Islamabad-Pindi, where he planned to meet with them. On 27 April, Asghar Khan was brought to Sihala as well, and the Pir of Pagaro, who had not been arrested but was in Islamabad, "consulted" with them, at Bhutto's behest.

"The police rest house at Sihala has spacious grounds," Asghar Khan recalled. "The accommodation, though not luxurious, was reasonably comfortable . . . a pleasant change from the conditions in which we had been kept in various prisons. . . . Our first visitor was . . . the Saudi Ambassador who came to see us on the day following my arrival with a message from King Khalid asking us to reach a settlement with Bhutto. He was a frequent visitor to Sihala and did his best to persuade us to give up our stand."[29]

Bhutto addressed a joint session of his party in Parliament on 28 April and blamed the "current violent agitation of the PNA" on a "massive international conspiracy." U.S. "political bloodhounds" opposed to Pakistan's gaining nuclear power as an independent Islamic state "want my head." Abu Dhabi's Sheikh Zayed, president of the United Arab Emirates, sent Foreign Minister Al-Suweidi to Pakistan that day with a warm letter from the sheikh, who called Zulfi his "Brother" and remained among his closest "friends." U.S. Secretary of State Cyrus Vance wrote Bhutto the next day to urge him "silently and dispassionately" to discuss any "grievances" he had with the United States rather than shouting them aloud as he had done in Parliament. Vance himself would soon meet with Aziz Ahmad in Paris. After meeting Ambassador Byroade again, Zulfi agreed to calm down and stop taking public "shots" at the United States. On the last day of April, when the Pir of Pagaro had promised to lead a "long march" of "millions" of Pakistanis opposed to PPP rule to the Prime Minister's House in Pindi, Zulfi personally showed up at the Pir's suite in Pindi's Intercontinental Hotel, even as his FSF and the army saw to it that very few of the long march's scattered "millions" ever got through police barricades or roadblocks.

Zulfi explained to the Saudi ambassador, with whom he met again the night of 30 April, that he had "known Pir Pagaro from childhood, and as fellow Sindhis and friends" they had "talked of personal matters." Bhutto told the Saudi ambassador that the "situation" was "fast returning to normal," and that "with the blessing of Allah" he would be "able to overcome the present crisis." He planned to visit King Khalid in the near future, and was told that he "would be most welcome."[30]

Early in May, Asghar Khan addressed a message to the chiefs of staff and the officers of the defence services:

> It is your duty to . . . differentiate between a "lawful" and an "unlawful" command. . . . Bhutto has *violated* the Constitution and is guilty of a grave crime against the people. It is not your duty to support his illegal regime nor can you be called upon to kill your own people so that he can continue a little longer in office. Let it not be said that the Pakistan armed forces are a degenerate Police Force fit only for killing unarmed civilians. . . . There comes a time in the lives of nations when each man has to ask himself whether he is doing the right thing. For you that time has come. Answer this call honestly and save Pakistan. God be with you.[31]

On 12 May, Bhutto informed the National Assembly that Asghar Khan's letter to the armed forces was an act of "high treason." The following day he announced his plan to hold "a referendum" on his "continuance in his present office." He insisted that holding new National Assembly elections would prove "disastrous" for Pakistan. The National Assembly proclaimed Friday, no longer Sunday, Pakistan's weekly holiday, and banned all forms of "gambling." Two persons were killed in Multan, and ten injured, when police opened fire on a procession led by the opposition.

"The Prime Minister's courageous decision of holding a referendum has taken the opposition alliance by a complete surprise," reported Rao Rashid Khan. "Even the opposition leaders feel that the Prime Minister's own personality is, by and large, undisputed and that the people are most likely to give their vote of confidence."[32] As "expected," the PNA rejected the idea and threatened to boycott any referendum. Rao Rashid felt, however, that if the popular turnout for a referendum was anywhere close to what the vote had been for the National Assembly, the opposition's "bluff" would be called and it would be totally "discredited." All information "media," therefore, had to "carry out a scientific and credible campaign to convince the public of their duty to turn-out in large numbers and vote for the Prime Minister." Rao Rashid Khan had done such a good job as special secretary that he was promoted to director of the Intelligence Bureau, where he had access to more information to feed to his prime minister, including a warning in early June that the "Pir of Pagaro" had been "consulting" an "astrologer," who "told him that the Government would fall on the 7th of June." Another "local astrologer" told him, however, that "nothing" would happen on the seventh but that the "22nd or 23rd of June are unlucky for the Prime Minister."[33]

Zulfi received many letters from mystics, one from a maulana who now urgently offered to "RECITE PARTICULAR VERSES OF HOLY QURAN on your face" as he had done "with successful results" in "the early period of your rule."[34] A self-proclaimed "SPIRITUALIST, elder statesman" had a month earlier "guaranteed" within "ONE DAY" to "save and redeem" Zulfi, otherwise warning "You are *DOOMED* (repeat *DOOMED*)" and "History SHALL NEVER pardon you."[35] Zulfi read and kept such letters but had no time during this crisis to see as many mystics as he used to see in years past. He barely had time to annotate all the files and secret reports that kept flooding in, filling his bedroom floor at night and half his bed as well. One "*Secret*" letter from Karachi transmitted a message "telephoned" from a London astrologer who wanted Bhutto to know that "everything would be alright. He said that you should be firm with the people who are causing trouble and . . . after 17th July you would emerge much stronger and would be in a position to eliminate elements who had been responsible for disturbances in the country."[36] Another admirer addressed him "Great Colossus!" and also urged him to "Be firm, brave and bold! . . . You cannot placate these wrathful gods, set atop the opposition. Every time you concede, they take it as a symptom of failing strength."[37]

Soviet Ambassador S. A. Azimov encouraged Bhutto not to resign his office, for in the "Soviet view" the "Prime Minister's resignation was not in the interest of Pakistan nor was it in the interest of peace, stability and security in South Asia." Zulfi replied that the ambassador was "very intelligent," and Azimov informed him that "the Mullahs had no genuine support and . . . people were being paid and bribed to agitate." He also told Bhutto that "Soviet leaders believed that destiny was blessing the Prime Minister and that the time had come for him to take the historic decision to quit CENTO." Zulfi stated that he would do so only if Pakistan received "definite" promises from the Soviet Union of "continued and uninterrupted supply of military hardware. . . . The needs and requirements of the Armed Forces had to be fulfilled."[38] Not that he was ready to "quit CENTO" immediately, for the "matter needed careful and cool consideration." Azimov understood, and was patient, but did his best to encourage the prime minister by feeding his already hot resentment against U.S. "interference." He informed Bhutto that his "friend" Byroade before the elections had "sent to the State Department" his own "assessment" that PNA would win 60 percent and PPP only 40 percent of the vote. According to the ambassador, the United States was "afraid of Pakistan's independent foreign policy, and its support for the Arabs in their just struggle against Israel." Azimov characterized the U.S. "approach towards developing countries" as that of a "cow-boy," quick to use "force."

Zulfi announced the next day, 10 May, that he had decided not to send his foreign minister to the forthcoming meeting of CENTO in Teheran, nor would he attend it himself, leaving that job to Pakistan's envoy to Iran. Secretary of State Vance very much regretted that decision, commenting a few days later that he had hoped Pakistan might "seize the opportunity" for face-to-face discussions in Teheran. But Bhutto by now considered the United States his foremost international opponent. His tolerance for criticism had never been high, of course, but he also believed that U.S. "elephants" were "after him," as he told General Faiz Ali Chishti.[39]

On 11 May, Zulfi was flown to Sihala for a private meeting with ailing Mufti Mahmood. Was he encouraged by sudden Soviet promises of "support" or by continued Libyan, Chinese, Saudi, and UAE support, or by what he thought to be a "turning" of the popular tide of protest against him? He brought no new concessions to Sihala, and next day the Mufti wrote to report "unanimous" agreement among his colleagues that further dialogue would be "fruitless," explaining, "I agreed to meet you in the hope that an honourable and speedy end may be found of the present impasse. I did this in good faith and disregarded any false sense of vanity or pride . . . in the hope that better sense would prevail and the country would be saved further blood-shed and misery. Unfortunately . . . you have tried to create the impression of a political dialogue in the hope that in the meantime the people's will to resist . . . would be weakened."[40] The PNA was still insisting that Bhutto resign first and then call fresh elections.

Zulfi feared that if he resigned, he would not live long enough to return to his position either as premier or president. And he may, indeed, have been right in that pessimistic judgment, not of his own popularity but of his opponents' collective resolve to keep him out of power. Bhutto used his former tough home minister, Sardar Abdul Qaiyum Khan, who had recently "abandoned" him to "join" the opposition, as his conduit for launching a new secret initiative with the PNA leaders. Qaiyum was kept in a Pindi rest house, and was called in by Zulfi a week after his futile meeting with Mufti Mahmood. Zulfi urged him to bring all the others "round the table" for a "package deal" dialogue that was to be "brief," and could include "new elections" to the National Assembly but no resignation by the prime minister. On 19 May, Qaiyum flew off, and "sold" his colleagues on the "deal" by shrewdly insisting that they negotiate with Bhutto as "de facto" prime minister, nothing more, and then they would win fresh elections and "throw him out." By the last week in May both sides were tired of the deadlock that had developed and heartsick at the carnage it had wrought: more than three hundred innocents dead, more than thirty thousand opposition supporters behind bars, and most major cities controlled by armed troops. So the opposition agreed, and on 26 May the Saudi ambassador returned to the Prime Minister's House to report the good news.

All nine PNA leaders had agreed that Mufti Mahmood, Nasrullah Khan, and Ghafoor Ahmad would serve as their dialogue team, the last to be brought to Sihala from his cell in Dadu. The one leader Zulfi preferred not to have to face was Asghar Khan, and Qaiyum had made it clear to the others that it might be "best" to keep the former air marshal off the team. The Pir of Pagaro, of course, remained free, and in their warm reunion on 30 April at the Intercontinental Hotel, he had told Zulfi that the two of them were the only "true leaders in Pakistan," the rest of the country's politicos being only "paper leaders."[41] The Pir had no better opinion of Asghar Khan than Zulfi did, and Bhutto did his best to play on such lack of solidarity within the PNA leadership every time he met with any one of them. "They were all masters of devious tactics and were not trustworthy," Zulfi told the Saudi ambassador. Now that Qaiyum had brought his opponents around, Bhutto stiffened his position, making several new demands, smelling "victory," and not wanting to surrender more of his power than was absolutely necessary. First of all he demanded "written," not just "verbal acceptance" by the PNA to enter into "talks with the Government," of which he must be "recognized" as the constitutional prime minister, not simply a "de facto" one. Bhutto suddenly insisted, moreover, that he "needed the permission of his party" to "enter into talks," and also wanted to "consult ranking members" of the "Armed Forces." The ambassador promised to go to Sihala at 10:00 A.M. on the next day; he reported back on 27 May, bringing a "draft" of Mufti Mahmood's letter to Bhutto. Zulfi was not, however, satisfied with the draft, and insisted that unless the PNA was more "reasonable," there would be "no agreement." He then wasted an hour of his own and his Saudi mediator's time in free-associating about the Pir of Pagaro and Mustafa Khar, whom "he had brought up like his own son and had treated . . . like one. . . . But since Mustafa Khar had lied," Zulfi explained, he had "had to throw him out." And after he had made Khar into the "Sher-e-Punjab," Mustafa had "intrigued" against him and even became vice-president of the Pir of Pagaro's Muslim League. Now he had come back and "fallen on his feet" and "begged to be forgiven for his past mistakes." Zulfi, of course, forgave him, but rhetorically inquired of the tired Saudi ambassador, who was much too diplomatic to interrupt the disjointed monologue of this prime minister, "Why had the 'Sher-e-Punjab' . . . suddenly decided to leave the Opposition?" The answer, quite clear to Zulfi's racing mind, was that "Mustafa Khar was an extremely selfish man and had felt that PNA, despite all the recent happenings, had limited support."[42] Yes, that was it. And that was why they had agreed to come to the "table" to talk. Then why give them anything to talk about? When Zulfi finally stopped, the ambassador softly suggested that "the dialogue should be approached with an open mind and should be handled with great care." But it was too late for Bhutto to follow such wise advice.

With the "Lion of Punjab" "at his feet," Zulfi hoped soon to rout his opponents, whether or not he sat around a "table" with any or all of them. He had never liked that "Gang of Nine." And who were they, after all, to "negotiate" with Zulfikar Ali Bhutto? Those few who were born to tribal power had no Oxford or California degrees, and those with degrees had no family traditions or money. He alone combined the two, hence he alone was fit to rule Pakistan. Didn't they know that? Of course they knew. That was why they were now so keen, so eager, so ready to sit around the table with him. He understood their "game" far better than any of those "paper leaders" ever did, or ever would.

The Saudi ambassador "requested" an interview with Bhutto on 30 May, and Zulfi invited Ministers Hafeez Pirzada and Kausar Niazi, his negotiating teammates, to join them. The next day Bhutto met in Pindi with the armed forces chiefs of staff and the corps commanders of the army. Then Zulfi told the ambassador he was now ready for "talks" with the PNA team after midday prayers on Friday, 3 June. By that date Bhutto had released all the incarcerated leaders, except for Asghar Khan, who was let out of his prison at Sahiwal only after he learned that he had been "released" by listening to the broadcast over BBC world news inside his cell at 10:00 P.M. on the fourth of June.

The PNA suspended its agitation as the talks started on 3 June, and a full bench of the Lahore High Court declared that the government's imposition of martial law in three cities was "unconstitutional." That decision was appealed to the Supreme Court. Yahya Bakhtiar and Hafeez Pirzada both advised Bhutto to "lift" the martial law decree, and after a second round of talks on 7 June an executive order to do so was issued, with "immediate effect." Minister Pirzada and Ghafoor Ahmad were now assigned the task of working out "details of the formula" for an agreement between the government and the PNA. They met again on 9 June, but now "some basic differences" about when to hold fresh elections surfaced. Zulfi had no intention of holding national elections again in 1977, and the PNA wanted "free and fair" elections last month or the month before. After three rounds of talks the differences were almost as irreconcilable as ever. By the end of the fifth round on 10 June, the parties were deadlocked.

"The whole country is expecting . . . a settlement," Intelligence Director Rao Rashid Khan warned his prime minister that afternoon.

> Unfortunately the credibility of the Government is so low that the blame of the deadlock would be placed squarely at the door of the Government. The PNA supporters will at once take to the streets. In the Punjab and in the big cities of Sind the agitation in the short term would be extremely violent. . . . [T]he army will have to be very quickly called in aid of civil power. This would raise complications. After the decision of the Lahore High Court the army would be extremely reluctant to side with the Government and expose themselves again to the public ridicule and humiliation. Even if the senior Generals behave responsibly, some kind of revolt in the rank and file, especially if they have to resort to large number of killings, can be expected.[43]

Rao Rashid Khan urged the "desirability" of some sort of "political settlement." He suggested either of two options: the first, to call fresh elections in August, without stepping down or bringing the PNA into government; the second, to postpone elections till November—but it would be impossible to keep the PNA out of government that long, hence he deemed the first alternative "preferable." The clever director of intelligence believed that the PPP would win in August, provided its "tickets" to candidates were "awarded more equitably and wisely" and there was "politically minded administration in the Provinces, especially in the Punjab." Zulfi was inclined to agree with this shrewd advice, especially now that he had the "Lion of Punjab" back at his feet, ready to spring into action in Lahore.

Internationally, Bhutto was faced with equally troublesome options, for President Carter had just turned down the Pentagon's recommendation to sell 110 A-7 fighter-bombers to Pakistan. Though the French continued to "promise" that they would "adhere" to their nuclear reprocessing plant agreement, they had as yet shipped no part of it to Pakistan, and Zulfi now feared they would capitulate to Washington's "pressure." The evening of 10 June he addressed his party in the National Assembly again and warned that Pakistan might "quit" CENTO because CENTO was "discriminating" against Pakistan and the United States was trying to "intimidate" him. Zulfi had met with Secretary General Umit Bayulken of CENTO in early April, and had been briefed by him on his recent trips to London and Washington. The secretary general felt it would be "understandable" if Pakistan were to leave CENTO, but had the "impression" after meetings in Washington with the new secretary of defense, Harold Brown, and with people at the State Department, that the

United States "valued Pakistan as an ally," and "hoped to find a satisfactory solution" to the "reprocessing plant question."[44] But that was two months ago.

Then on 31 May, Foreign Minister Aziz Ahmad had met with Secretary of State Vance and Carter's newly appointed ambassador to Pakistan, Arthur W. Hummel, and briefed them both in great detail about his prime minister's "concern" and "evidence of American intervention" in the recent elections. He also told Vance about the Soviet ambassador's recent meetings with Bhutto in an effort to get Pakistan to leave CENTO and join the Soviet Asian Collective Security System, thus giving the Soviet Union "access" to the Indian Ocean. That was, of course, also why the Soviets were "involved in subversion" in Baluchistan. Hence, if Pakistan "became weak," Aziz Ahmad explained to Vance and Hummel, there would be "a Soviet avalanche" in "this region." Zulfi had hoped that so frank a report on Soviet attempts to win his hand might revive Washington's waning passion for Pakistan, and Vance had, indeed, assured the foreign minister of his "deep conviction" that cordial relations between the United States and Pakistan were "important" to both countries and to the "stability" of the entire South Asian region. The liberal new ambassador also informed Aziz Ahmad that he would "follow" the "path of friendship and co-operation" and do his best to "promote better co-operation" between the United States and Pakistan during his tenure in Islamabad. That was less than two weeks ago, and now the president had turned down the A-7s deal, and every other day either the *New York Times* or the *International Herald Tribune* carried a report of the French "decision" to "slow down" or "scrap" the reprocessing plant. And every day, the Voice of America repeated in its broadcasts to Pakistan that the "Prime Minister must resign" and "new elections" must be called. Zulfi appealed to his Chinese friends for help, but with Mao and Chou both dead, he knew there was little likelihood of much aid from Beijing.

On 11 June the PNA demanded "fair elections" within thirty days. All negotiators returned to the table four more times in the next four days, and after the ninth round of talks ended on 15 June, both sides announced that they had reached "agreement on the basic issues." Subcommittees met the next day to hammer out "details" of the final agreement, and Bhutto that evening told his National Assembly audience that fresh national elections would soon be held in the "larger interests" of Pakistan. That same day he swore Mustafa Khar back into the government as his special assistant, and met once again with Soviet Ambassador Azimov.

Zulfi told Azimov about Aziz Ahmad's Paris meeting with Vance, who had said he would like to "open a new chapter" in Pakistan-U.S. relations, but now it seemed as if the "new page" was no different from the "old page." If the PNA leaders were "serious about quitting CENTO," as their election manifesto advocated, Bhutto was now ready to take a "joint national decision" with them on that point, and had therefore instructed Hafeez Pirzada to "take up the question of CENTO" in his meetings with Ghafoor Ahmad. Now he asked Ambassador Azimov whether the Soviet Union would "react as a 'great power' or as a 'human power.'" The Soviet ambassador "warned" Bhutto that the United States would be "starting a new campaign" against him "between August–

November 1977," and he should remain "very vigilant." The Soviet Union was "watching the situation very closely," and "deeply regretted" the "diabolical game" that the United States was playing against the "Prime Minister's Government." Soviet leaders, both in Moscow and Islamabad, Ambassador Azimov declared, "hoped and prayed that the Prime Minister would find a way out of the present difficult situation and reach a solution in the supreme national interest of Pakistan. . . . [H]e had no doubt that the Prime Minister would use his ingenuity, skill, experience and political acumen to overcome the present difficulties."[45] The ambassador then told Bhutto that "Soviet representatives" in Washington had been informed by the U.S. State Department that "Mr. Bhutto was not willing to leave CENTO," but was "sending signals to Moscow only to pressurize the United States, and that there was no substance in the Prime Minister's statements." Azimov also urged Zulfi to take the "more fruitful" initiative of quitting CENTO on his own, and not wait for the PNA's cooperation because that would "strengthen the Prime Minister's position." Such a step would "open up more opportunities for . . . Soviet-Pakistani cooperation," Azimov added.

Zulfi was angered to learn that the State Department had informed Moscow of the "top-secret" information he'd asked Aziz Ahmad to convey to Vance, after he had received it in equally "secret" confidence from Azimov. And he was frustrated with Azimov's vague rejoinder to his "substantive proposal"; the ambassador "had not said anything new, and . . . it did not constitute a concrete reply." Zulfi now insisted that the United States was "misleading Soviet representatives," by informing them that his statements about planning to "leave CENTO" were made only for "tactical reasons." Washington knew that he was the only "Pakistani leader" with "the courage" to do it, which was why it was so worried about him. If the Soviet Union was ready to extend "solid and concrete assurances for military and economic assistance to Pakistan," then he would "be able to take" the decision to quite CENTO. That would be the "biggest possible slap to the United States," Zulfi explained, and could "also be regarded as a success" for Soviet foreign policy. The ambassador was delighted and promised to inform Moscow immediately, but also wanted to be able to inform his superiors of the date on which new elections would be held.

Zulfi told Azimov that 7 October 1977 was the date that had finally been agreed upon in the last round of negotiations. He confessed that the PPP had been "over-confident" on the eve of the last elections and had "not expected such large-scale foreign interference," but promised by October to be "fully prepared to meet any eventuality." And by October he expected "many contradictions" to "surface within the PNA . . . he would not even rule out disunity." He sounded like the old Zulfi: politically astute, self-confident, optimistic, certain that "even at present the masses" were loyal to him and his PPP. He had "categorically rejected" the PNA's demand to "share power" in an interim government, moreover, and still believed that he would remain at the helm of Pakistan as well as his party until at least the end of the first week in October.

Turning to India, Zulfi said to Azimov that he had known Prime Minister

Desai since "childhood," when the new head of India's Janata coalition had come to see his father in Bombay. He pointedly called Azimov's attention to a recent statement by Morarji Desai about how in 1971 the United States had "driven India into the arms of the Soviet Union." The clever ambassador "praised" the "Prime Minister's judgment" and "superior knowledge" of India, to which Zulfi replied that "one thousand years of Hindu-Muslim struggle furnished ample evidence of the Hindu character. The Muslims knew Indians well. . . . [I]f the Soviet Union wanted 'to control' India then the Soviet Union should 'bless and befriend Pakistan.'" The ambassador was quick to insist that his government had "no desire to control India . . . the Soviet Union only wanted friendship with both India and Pakistan." Zulfi laughed, and said he had used the "wrong word." He then repeated how much he should like the assurance of a "steady supply of military assistance" to "help Pakistan . . . shed CENTO to teach the US a lesson." Ambassador Azimov cautioned Bhutto that the United States and its "allies" had "a new plan against Pakistan" for "August and November," and might "resort to other actions and other means." He urged the prime minister to be "very careful" about his "personal security. . . . Even an attempt of assassination could not be ruled out."[46] But Zulfi felt confident that his elite Federal Security Force now kept him well covered from every angle.

Two days later Bhutto flew to Riyadh to meet with King Khalid and express his "gratitude" for the positive and "meaningful role" Saudi Arabia had played in resolving the "differences" between his government and the opposition. The king said he considered it his "duty" to bring about "reconciliation among brothers." He hoped that the "crisis in Pakistan" would be resolved as a result of the "agreement" reached this week. Zulfi said he had not come to Saudi Arabia to "criticize the PNA leaders" or to apportion "blame or fault" to anyone, but went on to say that "both the Soviet Union and the United States had interfered in internal affairs of Pakistan." He blamed the United States much more, however, because it had "great influence" in Pakistan, unlike the Soviet Union. He said that Kissinger had told the Pakistani ambassador in Washington after his August 1976 visit to Pakistan that "if the Democrats won the election in the United States, they would make 'a horrible example of Pakistan' if Pakistan acquired the nuclear reprocessing plant. Dr. Kissinger had explained to the Ambassador that President Carter was determined not to allow any further proliferation of nuclear weapons." Bhutto added that his government had ample evidence of U.S. intervention in Pakistan's internal affairs; it had not only financed the PNA's election campaign but also advised the opposition on the strategy and tactics of agitational politics. Zulfi concluded his audience by telling King Khalid that the "main purpose" of his brief visit to Riyadh was to "enquire after the King's health." The king "thanked" him and wished him "great success in all his endeavours."[47]

From Saudi Arabia Zulfi flew to Libya, Iran, Abu Dhabi, and Kuwait. His warmest meeting was with "Brother" Qaddafi, who promised him "full support." The Shah was very "cool," however, and by now Zulfi believed that he was working "hand in glove" with the United States and several PNA leaders to "destabilize" his government. The Sheikh of Abu Dhabi helped restore his confidence and optimism, at least enough to encourage Zulfi to propose "another Islamic Summit Conference" in Pakistan, and a "Treaty of Non-Aggression among all Muslim countries," by which he hoped to derail the worrisome mobilization of Iranian troops along the Baluchistan border. Success of a second Lahore conference, an international "Third World" forum, by the end of this year or early next year—perhaps to celebrate his fiftieth birthday—would finally put to rest internal mullah opposition against him by firmly establishing his position as the premier host and champion of the Islamic world. Then petty maulanas like Mufti Mahmood and Islamic professors like Ghafoor Ahmad would treat him with proper deference and the respect he deserved. In Kuwait, Zulfi met with Yasser Arafat as well as the amir, and before returning home he stopped in Kabul for another brief summit with President Daud.

While Bhutto was abroad the PPP-PNA subcommittee that was to have ironed out all "details" of the "agreement" deadlocked over fundamental differences that had never been resolved. PNA leaders now trusted him no more than they had on March eighth, fearing that he had simply used the time he had won by his willingness to "negotiate" with them to consolidate his power, both internally and internationally. He not only had brought Mustafa Khar on board to recapture Punjab in October but had launched Mumtaz Bhutto on a most "successful tour" of Sind, preparing the political ground in smaller towns and villages for what would surely prove to be another "Bhutto victory" among his people. From press reports of him in Libya, Abu Dhabi, and Kuwait, and his optimistic statement to the National Assembly on 23 June after his return about the "good exchange" with Sardar Daud, it was increasingly clear that he would try to conclude a settlement with Afghanistan on the eve of elections, perhaps even releasing Wali Khan and certainly Bizenjo, which would give him a fair shake at possibly "winning" both Frontier Provinces. It was Zulfi dancing the political waltz again, which he carried off better than all of the PNA leaders put together. He was as nimble footed at slipping around opposition protests and internal problems as he was brilliant at mesmerizing the masses, who would soon be listening to him in Lahore and Karachi and Peshawar as well as Pindi as he promised "*Roti, Kapra aur Makan*" to every Pakistani who voted for the PPP.

The opposition resolved to wait no longer, presenting its final draft "accord" to Bhutto the day he returned from his Islamic summit tour. Ghafoor Ahmad insisted that the leaders wanted his response within twenty-four hours. They demanded immediate introduction of a central power-sharing "Supreme Implementation Council" that would "supervise" all functions of government and all official agencies. It would have "equal representation" from among the leaders of the PNA and PPP, chaired alternately by the president of the one and the chairman of the other. Zulfi called the demand "impossible," and the PNA's draft accord "most humiliating and insulting." He had, after all, just met with Islam's leading kings, presidents, shahs, sheikhs, and amirs, all of whom treated him with honor and respect, if not always as a friend, yet here was this "Gang of Nine" rushing up with an "ultimatum" that "demanded" he step aside tomorrow, to take "orders" from them. Pirzada met again with Ghafoor on 25 June and presented an entirely different proposal, offering to establish a "Government-PNA Committee" to supervise the elections. The next day, Minister Mustafa Khar, special adviser to the prime minister for political affairs, warned that the "PPP is in a better position to launch a movement against the PNA" than vice versa. Its "Lion" had started "mobilizing" Punjab.

By now Zulfi claimed to believe that the "United States had actually formed the PNA," as he told Turkish Ambassador Nihat Dinc on 27 June. He said it was "simple" and "inexperienced" Sardar Sherbaz Mazari of the National Democratic Party, who had written to all the other leaders initially "suggesting" the alliance. Zulfi told the ambassador, "in confidence," that in the forthcoming elections he would "attack the US . . . expose the US to the Pakistani nation and to the world. . . . [P]olitics generated its own dynamics. There were some leaders who were intimidated whereas there were others who believed in waiting and hitting back at the opportune moment."[48] The ambassador "advised against it," but Zulfi rarely heeded words of caution. A decade earlier he had used his slingshot with deadly accuracy against his "field marshal dictator"; now he would take aim at a "superpower." He would teach Jimmy Carter and Cyrus Vance a few "lessons."

First, Bhutto would have to finish dealing with the PNA, which wanted "dissolution" of the National Assembly as well as power sharing in a new "super cabinet" council. He gave in on the former and refused the latter, thinking he knew Mufti Mahmood and Ghafoor Ahmad well enough to feel they might accept it. But none of the PNA leaders were willing at this point to agree to any formula other than the power-sharing formula they had devised. Both teams met the evening of July first to try to thrash out their differences, and did not stop until 6:30 A.M. on July second, reaching a "final accord" only after Zulfi did a last-minute switch-step, agreeing to chair a PPP-PNA "super cabinet" with a casting vote in case of an evenly divided "disagreement." That accord was to be signed after the PNA Central Council approved it. The council narrowly rejected it instead, raising eleven other "points" for consideration by the government, all of which were so minor that Zulfi was ready to agree but to save face insisted he would first have to consult his own party and cabinet.

On the evening of 4 July he called his most trusted cabinet members and General Zia together and told them he would "break the deadlock" next day. Mumtaz and Hafeez Pirzada joined him for dinner at the Prime Minister's House in Pindi that night, drinking and dining till after 1:00 A.M. Pirzada turned to his host before going home and said, "Congratulations, Sir, the crisis is over." "I asked him what made him say that," Zulfi recalled. "He said that the steam had gone out of the Opposition. I laughed and told . . . Mumtaz . . . to wash off some of Pirzada's perennial optimism. He replied by saying that, in order to do that, he would have to take Pirzada to Sukkur Barrage during high flood. All three of us laughed. Within thirty minutes, we heard the other laughter."[49]

" 'Wake up! Get dressed! Hurry!' my mother called out sharply, rushing through my room to wake my sister," Benazir remembered. Begum Nusrat's words were etched traumatically into the 24-year-old Oxford graduate's memory. " 'The Army's taken over! The Army's taken over!' . . . A coup? How could there be a coup? . . . General Zia and the Corp Commanders had come personally to see my father two days before to pledge the Army's loyalty to him."[50] Her father was "on the telephone," speaking to General Zia, whom Benazir quotes as telling Zulfi, " 'I'm sorry, Sir, I had to do it. . . . We have to hold you in protective custody for a while. But in ninety days I'll hold new elections. You'll be elected Prime Minister again, of course, Sir, and I'll be saluting you.' " Zulfi's elder son, Mir Murtaza, soon to become founder-leader of his own Al-Zulfikar band of PPP activists, was ready to "resist" but his father cautioned, " 'Never resist a military coup. . . . We must give them no pretext to justify our murders.' " It made Benazir "shudder" to think of how Mujib and most of his family had been assassinated in their Dacca home two years before.

"I have been blamed by some for not killing Mr. Bhutto on the night the Army took over," noted Lieutenant General Faiz Ali Chishti, commander of the Rawalpindi corps that carried out the coup, code named "Operation Fairplay."[51] It was but one of several GHQ "contingency plans" and was executed smoothly, without any need to use the tanks that had been "deployed," or, indeed, to fire any of the recoilless rifles that had been issued by Chishti to his "most trusted officers." Chishti was a good soldier—methodical, careful, well-trained in all respects. Zia called him "Murshid" (mentor), and would soon put him in charge of three federal ministries. "Murshid, do not get me killed!"[52] Zia cautioned him that night, knowing full well, of course, that under Pakistan's Constitution of 1973 any attempt at a military coup was deemed treason, punishable by death. Chishti had "no regrets," but from that fateful dark night when almost all the leaders of the PNA as well as the prime minister and his closest PPP advisers and cabinet ministers were rudely awakened and driven off into military custody, Zia would have reason to fear that someday he would, indeed, be killed for the postmidnight operation he had launched in the name of "Fairplay."

16

Zulfi's Fall—From Martial Coup to Martyrdom

(5 July 1977–4 April 1979)

Zulfi was driven in his black Mercedes with an escort convoy of armed soldiers fore and aft up to Murree's best hilltop military "rest house" on the morning of 5 July 1977. His prime minister's staff awaited him there with breakfast. He was greeted by all with appropriate respect and told that he was simply being held under "protective custody." He had no illusions about the precariousness of his position. Nonetheless, he still believed he could intimidate Zia by reminding him of his oath to "defend the Constitution," and of the treasonous action he had taken in ordering the coup. He continued to underestimate the ambition of the man who had usurped his power, the intensity of Zia's hatred for him personally, and the contempt Zia felt for Zulfi's hedonistic lifestyle, for Zia always behaved with extreme "humility" whenever he spoke to or met with the man he had deposed.

Although Zia would have found this transition much easier had one of Chishti's "trusted officers" opened fire on Zulfi Bhutto before they reached Murree, Zia was a most patient man. Certainly patient enough to wait for the death sentence that would now have to be carried out through the slower, more serpentine process of "legal" execution, rather than the swift, clean military method. Once having resolved to take the forceful action he did this fifth day of July, Zia well knew that unless he lived to see Zulfikar Ali Bhutto dead he himself would not live very long. Even a "simple soldier" or a "simple *Momin*" (disciple of God) as he so often called himself, understood that much of Pakistan's iron rules of *Siasat*. General Zia ul-Haq was no simple soldier but would prove to be Pakistan's most effective dictator-politician. He would survive at the top even longer than had Ayub Khan, though his flaming fall would be both faster and fiercer.

"*Assalam-o-Alakum*" (Peace be unto you), General Zia began his address to his "great nation" that afternoon in Urdu. "You must have learnt by now that the Government of Mr. Zulfikar Ali Bhutto has ceased to exist and an Interim Government has been established in its place. . . . I am grateful to God

Almighty that the process . . . has been accomplished smoothly and peacefully. All this action was executed on my orders."[1]

Chief Martial Law Administrator (CMLA) General Zia explained that the "former Prime Minister" and some of his colleagues, including Mumtaz, Pirzada, Kausar Niazi, Rao Rashid, and Masood Mahmood, and most leaders of the PNA, including Asghar Khan, Mufti Mahmood, and Ghafoor Ahmad, were all in "protective custody." In earlier issuing his proclamation of martial law for the "whole of Pakistan," Zia had suspended the Constitution, dissolved the National Assembly and fired all federal ministers, ministers of state, advisers to the prime minister, and provincial governors and chief ministers. Yet now he told the nation that he wanted Pakistan to "remain in the hands of the representatives of the people who are its real masters," because "I genuinely feel that the survival of this country lies in democracy and democracy alone." Zia called it an "inexcusable sin" for the armed forces to "sit as silent spectators" while "political leaders failed" to reach agreement, and when "I saw no prospects of a compromise . . . feared that the failure of the PNA and PPP . . . would throw the country into chaos." Then he wanted "to make it absolutely clear" that "neither have I any political ambitions nor does the Army want to be distracted from its profession of soldiering. . . . My sole aim is to organize free and fair elections which would be held in October this year. . . . [A]fter the polls, power will be transferred to the elected representatives of the people. I give a solemn assurance that I will not deviate from this schedule."[2]

At 2:00 A.M., when Asghar Khan had been taken into custody by the troops who surrounded his house, the former air marshal had not been sure whether it was under "orders" from the "Prime Minister or whether the army had staged a coup," but soon found himself inside Pindi's corps headquarters mess with Rao Rashid and Ghafoor Ahmad, Mufti Mahmood, Nasrullah Khan, and Kausar Niazi. It was then that Asghar Khan rightly realized that "the possibility of an accord being reached between the Government and the PNA was not to his liking and Zia-ul-Haq decided to act without delay to obviate that risk."[3]

"My dear countrymen," concluded Zia in the first of his many addresses to the nation, "I have expressed my real feelings and intentions, without the slightest ambiguity. I have also taken you into confidence about my future plans. I seek guidance from God Almighty and help and cooperation from my countrymen to achieve this noble mission. I also hope that the Judiciary . . . will extend wholehearted cooperation to me." He would, of course, soon very much need the help of Pakistan's judiciary, and was "pleased to announce" that the chief justices of all provincial High Courts had at "my request consented to become the Acting Governors" of their provinces. He "appealed," moreover, to all officers and men in the armed forces "in the true Islamic tradition" to "forgive those who have ridiculed or harassed them in the past." Was it strange of him to appeal to his soldiers in such manner, or did that act, like his assurances regarding elections reflect perhaps some darkly hidden deception, some fear and hatred he felt most deeply for the proud and powerful man he had taken prisoner—the man who had so often ridiculed and harassed *him* in the past year and a half?

Asked by Edward Behr of *Newsweek* shortly after the coup how he felt about Bhutto, Zia replied, "He has been very kind to me but I have no regrets . . . I have no personal venom against him."[4] Was it not odd for one man to use the word *venom* about his feelings toward another? Questioned whether he thought Bhutto would "go ahead and campaign" in the autumn after what had "happened" to him, Zia responded, "He should. He's a very determined man with a great sense of history. He's also a very tenacious fighter and a great politician." How would Zia feel about returning to "normal soldiering" after running the country? "I am a soldier, and I will go back to soldiering." Then Behr reminded him, "That's what the late President Ayub Khan said and he then held onto power for eleven years." Zia invited Behr to "call me after the elections" to "confirm that I mean it."[5] Zia promised that "political parties will be given all the facilities to shout their heads off," and "we will only impose minor restrictions, like a possible ban on processions." It would be a "clean" campaign with the armed forces present only to ensure "impartiality."

Zia flew up to Murree with Chishti on 12 July, and both generals met with Bhutto alone in his rest house. "General Zia . . . said that he was only a temporary custodian," Chishti recalled, and that "it was going to be all Mr. Bhutto's again and he should manage it as deemed fit. . . . Mr. Bhutto asked how long he would be in custody, to which General Zia said for some days only, during which he should rest and recoup"[6]

Mustafa Khar had initially been "difficult to find," Chishti noted. Soon after martial law had been declared, however, Khar wired Zia, requesting an "audience." Zia was too busy but passed the message along to his corps commander, who met with Khar at the officers' mess in Pindi's Chaklala military airport. "He told me he was a worried man and completely upset because his wife was expecting a baby," wrote Chishti, who refused to permit Khar to leave the mess, placing him under custody there but permitting him to phone his wife and "stay the night in the Mess" before going up to Murree the next morning. Intelligence reported that as soon as Khar arrived at the hill station rest house, Zulfi accused him of being "disloyal." He asked Mustafa to look into "his eyes," upon which Zulfi snapped that he could "see disloyalty and mischief" in the dark eyes of the "Lion of Punjab." "Khar . . . has been the most treacherous and the most ungrateful of them all," Zulfi later wrote from his death cell.[8]

Mumtaz, Khar, and Pirzada would all be driven over daily in an army Suzuki van from their barracks quarters to visit Zulfi in his "private" rest house at Murree as the sun started to set. Khar's pregnant wife, Tehmina, who stayed close to him in Murree, reported that they would

> sit and discuss past foibles, current affairs and future policies. Mr. Bhutto was furious at what Zia had done, his arrogance never left him. He was convinced that the Generals were on a sticky wicket. There was no other reason for the VIP treatment he was receiving at their hands. . . . Mr. Bhutto held court . . . still displaying Prime Ministerial airs. He was indiscreet. He would openly abuse the Generals and accuse them of treason. He swore revenge and vowed to take them to task when he returned to

power. His men felt that Mr. Bhutto was being reckless. The house was obviously bugged.[9]

Zia held his first press conference on 14 July, and again vowed that "my intention has all along been . . . to lay a tradition that the Army will not meddle in politics; let the politicians decide things by themselves. . . . We were determined to do our constitutional duty and perform our constitutional obligation, namely, the support of the Government. . . . Unfortunately, as we have seen, the law and order situation deteriorated. . . . It is the job of the Army, in fact its national duty to act if somebody is looting or destroying public property or is trying to disrupt the pattern of normal life. . . . All of us stressed upon the Government, in the name of God, to solve the problem politically."[10] Zia insisted that

> We have treated the former Prime Minister and his colleagues as well as Opposition members with grace. . . . We are providing them a comfortable stay. They need to do a lot of thinking for the future because the destiny of the country is in their hands. They have got to look after the future of this country, not I. I am there only for 90 days. . . . I am a very humble man. . . . To say that there is collusion between me and the former Prime Minister is ridiculous. Can a man of his calibre share power with me? . . . Other people say: No he is not a Bhutto protégé but a PNA man because he went to say his prayers; because he is talking about Islam. You can draw your own conclusions. . . . Incidentally, maybe I am wrong, but none of the political parties had Islam as Number One item on their manifestos. . . . *Nizam-e-Mustafa* and *Nizam-e-Islam* came up during their political activity. It was, in my opinion, forced by the people of Pakistan on the political leaders. . . . I say Islam has to be the cornerstone of this Islamic Republic of Pakistan. . . . I am a Muslim by faith, by birth and by actions.[11]

In some respects, Zia proved to be a much shrewder politician for Pakistan than Zulfi Bhutto had ever been. His "humble" political image and style was at the opposite end of the political spectrum from Zulfi's flamboyance. His top priorities, the army and Islam, were in many ways more important to most Pakistanis than "*Roti, Kapra aur Makan,*" for all those who joined the army were assured of bread, clothing, and housing, and those who did not could always pray to God for them. Well-to-do Pakistanis in large cities, moreover, by now considered law and order, if not their highest priority, certainly essential to daily existence. They were sick of violence, and most of them were tired of political promises, usually forgotten the day after polling ended. Here at last was a "simple" soldier, a devout and modest man, who wanted only to keep the country calm and secure, to restore Islamic values of pride, dignity, and decency to its troubled and all-too-turbulent populace. Why then for only "90 days"? Why turn the country back after so brief an interval to the greedy, arrogant, squabbling, inept politicians?

"We have no intention of any witch-hunting," Zia promised in the mid-July press conference. "It is up to the next lot of elected representatives of the people, or the people themselves, to take action against the politicians. The Courts are still functioning and we have not stopped anyone from going to the Courts to take the politicians to task. Then why do they want me or the Military or the Armed Forces to hang a few politicians? Why should I? Isn't it as much of a concern of the public as it is mine? It should be done by them, if it is to be done."[12]

Foremost among those "few politicians" that "*they*" wanted him to "hang," of course, was Zulfikar Ali Bhutto. But part of Zia's political appeal was that he not only maintained an aura of humble simplicity, which almost made the world forget what dictatorial powers he had usurped in direct violation of Article 6 of the Constitution he had sworn to protect and defend, but also sought to project himself as a nonvindictive, if not a nonviolent man. Let the hanging be done by the "Courts" and the "public," "if it is to be done," he now softly suggested, even more gently adding, "But I would dissuade them from this. What we want at the moment is a calm, peaceful and congenial atmosphere for holding elections. After the new legislature has come and whosoever forms the Government, let them do it." That was fair, was it not, to let "them" do the hanging?

Hanging was much on the mind of General Zia ul-Haq because his "Murshid" had failed to make a clean and swift job of it. Now it would have to be done slowly by that most cumbersome of processes, taking the "bastard" (as Zia by now thought of and later called Zulfi Bhutto) to court, unless he could be more expeditiously hanged by a military court under martial law. That might raise shouts of "injustice" the world over, however, and could potentially backfire. A prudent general had to cover both flanks and his rear as well as his front. "After the new legislature has come and whosoever forms the Government, let them do it," Zia offered, showing many teeth when he smiled. "And as representatives of the people it is up to you to see that they do take some action. But so far as the pre-election period is concerned, why should we create difficulties for one person or one party and not the other?"[13] Of course, *they* must "take some action"; why should "*we*" worry now about "one person"? Or "one party"? Zia's singular political power was derived not only from the tanks and guns he commanded, formidable though those were, but from his pretence of always remaining above the fray of partisan political haggling and scuffling. He "represented" the national ethos, even as the army defended and protected it.

"Our acts and deeds will demonstrate whether we were usurpers or traitors and whether we violated the Constitution or not," General Zia concluded. "Once, by the Grace of God, the elections are held and the new legislature comes into being, they may consider all that . . . our aim . . . is straight-forward. We want to hold free, fair and just elections in October. The date I have intentionally not given because there may be some adjustments to be made."[14] Free and fair elections would, in fact, be held again in November, not *October*—eleven years later—some two months after General Zia ul-Haq's C-130 went down over Bahawalpur, with him and all those inside swiftly consumed by the flames of its exploding gas tanks.

That day, as Zia spoke to the press in Pindi, Zulfi penned two "Top Secret and most Confidential" letters in Murree, one to the Soviet ambassador and one to the Chinese ambassador, both of which were hand delivered by his servant to the embassies in Islamabad. "I have come to know that the coup leaders told the Chinese Ambassador that one of the reasons for the coup was because I was going to move towards the Soviet Union," Zulfi wrote Azimov. "Soviet leaders should also know that I had informed the relevant authorities of my decision to leave CENTO when the National Assembly was to re-convene on the 9th of July. The coup took place four days earlier! As for the Accord, it was going to be signed within two days at the most."[15] To the Chinese ambassador he wrote much the same letter but also wanted to make it "abundantly clear that in my foreign policy, the People's Republic of China occupied a central position . . . no power on earth can shake or injure our relationship. It is of mutual interest and vital to the whole region. As far as CENTO is concerned, I told your leaders last year and I have said this repeatedly, that my decision to leave CENTO and to join the mainstream of non-alignment is based solely on national considerations, uninfluenced by the policy of any other country. . . . Pak-China friendship is taller than the Karakoram and deeper than the oceans."[16]

Zia went up to Murree again on 15 July, and met with Mufti Mahmood, Nasrullah Khan, and Ghafoor Ahmad for more than an hour before seeing Asghar Khan. He informed those PNA leaders that the army had intelligence of a PPP "armed procession" planned to be held in Lahore on 5 July, which was why the army took the action it did that morning: to "save the country from civil war."[17] He then went off to meet Bhutto alone, but did not, of course, tell him that story. Zulfi, however, gave Zia a piece of his mind, "chastizing him," reminding him that the Constitution called for the "death penalty" for any officer who "abrogated" it or tried to "topple his Government by force." It was hardly the sort of chat designed to endear him to Zia's heart, though it shows how little Zulfi suspected as yet of what his fate would be, and how fearless—some would say foolhardy—he was. His "personal staff" was removed the next day, which may have chastened him a bit. But Zulfi never quite imagined throughout most of that stormy July he spent in Murree's hilltop rest house that his days of prime power and glory were over, and even those of a far more wretched, ignominious existence were by now numbered.

Former FSF Director Masood Mahmood was also kept in relatively comfortable isolated detention from the fifth of July, given plenty of paper and told to "write" everything he remembered about any "illegal things" the former prime minister had "ordered" him to do, including arranging the ambush "murder" of Raza Ahmad Kasuri's father. For Masood the alternative to preparing those briefs that would help "hang" his former chief, the "approver" testimony to his fatal orders, was his own hanging. Zia's men had made that very "clear" from the first dark night they brought him in. And Masood knew Zia much better, after all, than Zulfi Bhutto did, for Masood and Zia were both men well trained in every martial art. Like Zia, Masood was a "devout Muslim," a "humble" man who also had reason to hate and contemn Zulfi's arrogant, insulting ways in dealing with him, and his "crude vulgarity" in dealing with most women, including PIA "stewardesses" and the "young wives" of many of his colleagues. Masood Mahmood knew, moreover, where all the "bodies were buried," when and why prisoners had been sent to Dhulai "camp" in Azad Kashmir, or had been incarcerated in the "black hole" of Lahore's old Fort, or "otherwise disposed of." So Masood had a very busy month, digging deeper each day into the mountain of dirt that would soon bury his old boss, "the worst lecher" he had ever met in his life, he asserted. When Zulfi had lost faith in old friends, he had ordered Masood to "take care" of them, which to Masood's mind was a clear "message" to "remove" them. "Jam Sadiq Ali is a mouse," Zulfi had once told him, Masood recalled. "Can't you find a cat to take care of him?"[18] Masood immediately informed his friend Jam Sadiq Ali, who flew from Karachi to London the next day.

Afzal Saeed Khan, public secretary to the prime minister had also been arrested. He too was given paper enough to help hang the man he had long intimately served in the most delicate money matters, disbursing large sums kept in "secret" funds for the prime minister's "discretionary" use, most of which were used for "election work." Other larger amounts donated by Shaikh Zayed and placed in trusts for the prime minister and Begum Nusrat to manage, were also used for the "party." Sixty-year-old Afzal Saeed, who described himself as "honest and straightforward" as well as deeply "religious," explained how "hard" a "task master" the former prime minister had been, tolerating no "delay" in his secretary's work, never sparing him "even for a minor lapse."[19] It was not surprising that he too was quite ready and willing to serve as an approver in the criminal cases of peculation as well as murder that were being hammered together all the rest of that rain-filled summer to erect at least one scaffold high enough for Zulfi Bhutto.

Even Mustafa Khar was flown, on 24 July, from his rest house in Murree to Lahore's stately old neo-Gothic Court House, there to give his deposition; much was known of the many "differences" he had had with his old "friend" ever since Zulfi first appointed him governor of Punjab. Then their fights had "increased" after Khar was made chief minister, until those "differences became intolerable," as Khar explained it in the High Court. "It was true that when somebody showed political differences with Mr. Bhutto he used to be jailed and based on his personal likes and dislikes, revengeful action used to be taken against political opponents but sometimes he also showed tolerance."[20] Small wonder that Zulfi later thought of Khar as "my Brutus!" Yet Khar preferred not to remain in Pakistan and stand up to say what he knew—for he "knew many things" about Zulfi that were known, indeed, by no other "brother" of that soon-to-be-much-accused man—but chose instead to accept Chishti and Zia's offer a few months later to fly to London with his sixth wife and their new daughter. "We could not tell a soul about our flight into exile," his now-former wife recalled.[21] Was it shame or fear that kept Mustafa so darkly cloaked and silent after having said so much in his Lahore deposition? At least he would never have to look again into the eyes of his soon-to-be-executed political mentor.

"I declare once again that elections shall, *Insha'Allah,* be held in October," General Zia assured the nation on 27 July 1977.

> The world shall witness this great nation electing its representatives in free and fair elections which will lead to the establishment of a national government to which the Armed Forces of Pakistan shall hand over power that they now hold in trust. . . . [T]he representatives that emerge from these elections should be such as are imbued with the spirit of service to Pakistan—men who are not governed by greed or avarice, who do not fight the elections to satisfy their lust for power. . . . In so far as my personal view about the restoration of democracy is concerned, I believe that democracy has not been given a chance to flourish in this country during the last 30 years. True, there has been much talk of democracy but no positive measures were taken to help it take root. . . . Despite this, I firmly believe that the tender plant of democracy shall not only take root in our country, but would eventually become a full tree.[22]

He then urged voters to cast their ballots only for "one who is a true Pakistani and a *Momin* [devout Muslim]."

Zia's earlier apolitical posture had clearly by now been abandoned, and in warning his countrymen not to reelect any one motivated by "greed or avarice" or "lust for power," he obviously meant Zulfi Bhutto or his party. Not that anyone would have another chance to do so, despite Zia's "belief" in democracy—at least not while Zia was alive. That plant would remain much too "tender" to take root in Pakistan's tough militarily governed soil for the next decade and a bit longer. But now that "peace and tranquillity" had been restored to Pakistan by the armed forces, with the help of "God Almighty," Zia was pleased to report that "within a day or two" he would release all political leaders from protective custody. But from August first till the end of Ramadan only "limited political activity" would be permitted: "no political meetings, processions or political activities in the streets." He wanted everyone to keep calm, and his earlier martial-law ban on the publication of "political statements" or newspapers would remain in force.

The army would prove no more effective in curing Pakistan's many economic ills than the People's Government had been, but Zia now promised, following the policy of the "Holy Prophet (peace be upon him)" to pay a worker's "wages" before the "sweat" had dried from his brow, that his "Interim Government" would

> protect the workers from every exploitation. Some factory owners took undue advantage of the ban on trade union activity imposed by Martial Law, but . . . the purpose of stopping trade union activity was to keep the factories and mills free from political wranglings. It was certainly not our intention to prevent them from obtaining redress of their grievances. . . . But all the same, it is just as equally true that protection to the rights of workers does not mean that we ride roughshod over the rights of industrialists. . . . The nation expects its workers to . . . work hard and to

> increase national production . . . not become political agitators and destroy factories.[23]

No, it would not be easy for this simple soldier to run the nation any more successfully than did the complex politician he would soon hang. Still he tried, sounding rather like Zulfi once had, as he appealed to those who "feed the nation": "I urge my brother farmers to increase production by harder work," for the "economic situation" was "bad." Nor would all of his military power and devout pleas do much to improve it until the Soviet Union was inadvertently to "save" Zia by invading Afghanistan two and a half years later—bringing billions in U.S. arms and aid to his rescue—which a decade thence helped to topple the Soviet Union itself.

The day after his national broadcast, 28 July, Zia flew with Chishti to Murree, where they both met Bhutto and told him he was free. On the morning of 29 July, Zulfi was flown to Pindi by helicopter, and had tea with Chishti there before boarding the "Prime Minister's Falcon," which Zia now "placed at his disposal" to fly him back to Larkana. When he stepped down from that plane onto his home turf, he was greeted by cheering crowds of admirers, most of whom had waited up all night for one mere glimpse of him, for a touch of their beloved Quaid-i-Awam. Much to Zia's chagrin and puzzlement, nothing he had said or done since July fifth nor all the countless times he had repeated God's name in all the countless prayers he had recited in Arabic, appeared to diminish Bhutto's popularity with Pakistan's masses, the passionate love affair he had with these millions who cast their votes for him and would to this day have given their lives for him, even while martial law remained in full effect. How was it possible?

It continued to shock and amaze many more sophisticated Pakistanis as well. Zia, after all, was a "simple" soul, but many of the others who hated Zulfi Bhutto as much were highly educated, cultured men of letters and philosophy, modern minds trained in science as well as the arts. They too found Bhutto's mesmeric link to peasant *haris*, his mystic Sufi "saintlike" appeal to Pakistan's poorest and hardest-working folk in city and countryside alike—to withered old women in rags as well as nubile beauties in silken saris that hid nothing of their voluptuous charms—"puzzling" at best, "impossible to explain" for most. "Sir, he was a *Prince*," one sophisticated journalist opined, or "You see, he was like—like a 'Savior'! I don't know how else to put it." A bright-eyed "young" man (about forty now) in Lahore who had lost faith in Bhutto as well as his party after the "rigging," put it rather nicely; after saying how disenchanted he had become because of the PPP's gross abuse of its power, he shook his head and added, "But it was Mr. Bhutto who taught us to stand up, and protest!"

And now Zulfi was free again, free at last, back among those who loved and adored him. No military force could have kept the people from flocking to Al-Murtaza that day, or could have controlled the crush of more than a million who waited at Karachi's main station to receive him there, when he returned to his other home, at 70, Clifton, a few days later.

"What was normally a half-hour trip from the train station to our home took my father ten hours," Benazir, who awaited him there, recalled. "His car was dented and scratched by the time it arrived at 70 Clifton. My brothers, sister and I didn't dare go outside the gates to greet him for fear of being crushed."[24] His children naively thought that the multitudes and other clear signs that "Papa" Bhutto's popularity had been dented by the coup far less than his car had been dented by the throngs augured well for their future. But every time Zia heard of the "surging" crowds and "disorderly mobs" that were now flocking to see and hear their Quaid-i-Awam, in Lahore as well as Karachi and Larkana, he only grew more resolved to hang the man who would, he feared, otherwise hang him as soon as he was reelected prime minister. Even to Zia's "simple" mind it was obvious that Bhutto and his party would reclaim a majority of Pakistan's votes in free and fair elections, though they might win twenty or thirty fewer seats in the National Assembly than they had in March.

Zia picked his good friend Maulvi Mushtaq Hussain, whom he had appointed chief justice of the Lahore High Court on 13 July, to chair the five-judge Election Commission he appointed on the first of August. Justice Hussain announced then that national elections would be held on 18 October, and the nomination filing process would start on 7 August. Zulfi had twice before chosen junior chief justices over the more senior Maulvi Mushtaq on the Lahore bench, hence he knew immediately that Zia's chairman of the Election Commission would hardly favor him or his party in any future filing or polling disputes with the PNA. What he did not as yet realize, however, was that Maulvi Mushtaq himself would preside over the panel of High Court judges that would soon hand down his death sentence.

On 14 August 1977, the thirtieth anniversary of Pakistan's independence, General Zia opened his nationwide address with a most solemn warning from the Holy Quran, which he translated: " 'When we decide to destroy a population, we first send a definite order to those among them who are given the good things of this life and yet transgress: so that the word is proved true against them: then we destroy them utterly.' " That sacred script was "uppermost in my mind today," the general explained. "The purpose of recitation of this *ayat* is not to over-awe you with the wrath of God and to create despondency but it is only to underline the need for a hard struggle to save and strengthen this country. . . . We have been under an illusion that this country is a gift from God Almighty and He alone is responsible for saving it. I want to remind you again that God helps only those who help themselves."[25] The "simple" soldier was also a devout Muslim, and he pondered deeply his daily readings in the Quran. He feared very much the "wrath of God" himself, and now was seeing and hearing again how charismatic and supremely popular the "Quaid-i-Awam" remained, the man whom he had treacherously toppled in the night more than a month ago. He, Zia ul-Haq, had been "under an illusion that this country is a gift from God Almighty," and needed most of all, therefore, to "remind" himself that "God helps only those who help themselves." Among those who had been given so many "good things of this life," surely that proud and haughty "raja of Larkana" was the worst who "yet transgress;" therefore, Zia knew that he must "destroy" Zulfikar Ali Bhutto "utterly." No easier way would be vouchsafed to him now. Zia was not a man to flinch from hardship, or to hesitate about destroying anyone "utterly," once he knew the "word"—*guilty*—was "proved" against him. For that he would need some help from Maulvi Mushtaq, however, as well as from himself and God.

"Justice demands that for an assessment of the performance of any government . . . both its good points and shortcomings should be kept in view," Zia continued. "The point to ponder is as to where does the poor common citizen of the country stand in all these political wranglings. . . . Who will worry about his livelihood? Who will take him out of his world of fear and apprehension?"[26] "Fear and apprehension" were very much in the troubled mind of the self-appointed chief martial law administrator—who also retained his titles of general and chief of army staff, to both of which he had been promoted by the prime minister he had arrested—who knew that "justice" would demand an "assessment."

On 15 August the Federal Investigation Agency (FIA) that had been kept very busy for the past month and a half handed down its first "indictment" of Zulfikar Ali Bhutto for "direct responsibility" in the 1972 "murder" of Dr. Nazir Ahmed, a *Jamaat-i-Islami* member of the National Assembly, whose violent death had not been solved. Now both Bhutto and Mustafa Khar, then governor of Punjab, were charged with having conspired to issue "orders" to "murder" Ahmed. But that case was so old that even the FIA knew its charges would be very "hard" to prove. Besides which, Khar was ready to fly off to London, promising never to return to contest any elections, and Chishti strongly advised Zia to let him go. Zia had never been especially concerned with Khar, and though he doubtless despised him as much as he did Zulfi Bhutto, Khar had not ordered him about like a "monkey" or made fun of his teeth. And Khar had not once "threatened" to "hang" him for "treason." Only Bhutto had had the audacity to say he would do that, once he was "back" in power. The Nazir Ahmed "murder" was kept, therefore, as a second string in Zia's bow of "justice"; Nawab Kasuri's "murder" seemed simpler to "prove," even though that meant giving Masood Mahmood his freedom. Zia had never disliked Masood. On 21 August the evidence against Bhutto in the ambush "murder" of Nawab Mohammad Ahmad Khan was heard by a Division Bench of the Lahore High Court.

The crowd that hailed Zulfi Bhutto in Lahore that August was larger and more enthusiastic than any crowd ever to have greeted a Pakistani politician. He was more than ever admired and loved—it soon became all too perfectly clear to General Zia—even in the heart of Punjab, as he had always been throughout Sind. The coup, despite all of Zia's Quranic moralizing and pious platitudes that had followed, seemed only to have backfired insofar as Zulfi Bhutto's popularity was concerned. Shouts of "Jiye Bhutto!" and "Bhutto Zindabad!" filled the air from Lahore to Islamabad and Rawalpindi, penetrating even the tightest top-secret war-rooms at GHQ and the prime minister's secretariat, where countless "confidential" files were being diligently combed. Five heavy volumes of *White Papers* would be published in 1978 to "prove" how bad a man Zulfi Bhutto was—how much money he had "stolen" from government coffers; how many rules and regulations, laws and orders he had ignored, broken, or grossly abused in his "lust for power" that compounded the moral felony of his "greed," not to speak of his "lechery" and insatiable craving for the finest Scotch and the best French wines.[27] None of his behavior, even when documented, seemed to diminish his political appeal for most Pakistanis on the streets of Lahore, Pindi and Karachi. To men, women, children of all ages, and to villagers throughout the country he was still their beloved Quaid-i-Awam, Pakistan's very own Mao and Qaddafi combined, its most popular living leader.

Zia announced on 24 August that "reports of inquiries" currently being conducted by his regime into certain "irregularities" committed by the former prime minister and his government would be made public before martial law could be lifted and elections held. He still promised, however, that "free and fair" elections would be held on schedule—in October.

Zia met with Bhutto again on 28 August at 2:30 P.M. General Chishti was also there, as chairman of the Election Cell. "We stood outside waiting for Mr. Bhutto," Chishti recalled. "I found him off colour. . . . We went inside and General Zia asked him if he would like to have a cup of tea. He said no, he was fasting. . . . Mr. Bhutto asked General Zia how the progress was towards the October elections. General Zia replied that he would be discharging his duties for the restoration of democracy. . . . Then he asked me how the Election Cell was doing. . . . We had met quite a few political leaders since July. . . . When I named Ghulam Mustafa Khar. . . . Mr. Bhutto remarked that so far he had not authorized anybody to meet the Election Cell."[28] Zulfi returned to Larkana after his two-hour meeting with Zia, at which he had finally learned of Khar's treachery. He also sensed that Zia was worried enough about his own treachery and the surge in PPP popularity to use any excuse to call off elections.

By 30 August, Zia had managed to persuade four retired judges to "appeal" to him, as chief martial law administrator, to "prosecute and punish" Bhutto for "crimes" he had committed while in high office, before holding general elections. At the same time the PNA adopted Bhutto's foreign policy as its major platform, promising to follow a "positive non-aligned policy," to strengthen ties with all "Muslim countries" and encourage greater "foreign investment" from those states, and to give "top priority" to the purchase of the nuclear reprocessing plant from France and the development of nuclear power for Pakistan.

On September first, Zia held another press conference, at which he too affirmed his intention to "abide" by the "deal" made by the "previous Government" with France for a reprocessing plant. He also spoke of the Chinese as "brothers," and told of his last pilgrimage to Saudi Arabia, when King Khalid "had graciously desired me to take a seat next to him." It almost sounded as if Zia were trying to follow in the footsteps of Bhutto, yet nothing he said or did quite sufficed ever to make him popular, not to speak of loved, by his own people. He was feared and obeyed, of course, and many soldiers and a few mullahs, and at least one maulvi "Justice," respected and admired him, but Pakistanis in their millions never took Zia ul-Haq to their hearts while he was alive, or shed a tear when he died, as they did for Zulfikar Ali Bhutto.

At 4:00 A.M. on 3 September 1977, commandos of the army surrounded the high-walled compound at 70, Clifton, broke through its solid-steel front gates, and then pounded at the front and rear doors of the handsome two-story home till sleepy servants opened them. Troops raced upstairs to round up the terrified children (except for Mir, who was in Larkana) and their parents at gunpoint. Bhutto calmly showered and dressed before following the director of Zia's Federal Investigation Agency down to the black car that drove him away, under heavily armed escort to Murree on the morning of July fifth.

Zulfi spent the next ten days under arrest, denied bail while awaiting his September thirteenth scheduled hearing before the Lahore High Court on several charges of "murder" and other "high crimes" filed against him by the advocate-general of Punjab. Yahya Bakhtiar was of no help to him now, though Zulfi asked his old friend and former attorney general to "appear for him" in the Lahore High Court. "I said, 'No, Sir,'" Yahya recalled. "With Maulvi Mushtaq, who was the Chief Justice, he doesn't believe in fair trial. I think in all his career as a judge Maulvi Mushtaq has not decided one case on merit, not one case! So I told him I will not appear . . . I kept out of it."[29]

Initially, Hafeez Pirzada, who was still free, took over as interim leader of the PPP, stating the day after he had visited Zulfi, that "arrests" of party members would "not affect the Party's mission." Begum Nusrat felt too ill at this time to take on the active leadership she would assume two weeks later, after Pirzada and Mumtaz were also arrested. At twenty-four, Benazir was as yet too young, and her brother Mir Murtaza had never been interested in nonviolent political activity, but prepared now to take his battle underground, gathering recruits as well as the high-power weapons and ammunition Al-Zulfikar would need from his father's well-stocked "hunting" arsenal in their Larkana compound. Shah Nawaz, then nineteen, had long been in poor health, deeply depressed, "addicted" since early adolescence. Sensitive young Shah Nawaz turned away from politics and Pakistan, seeking escape, it appears, from the all too violent world of political vendetta in his own Shangri-la on the French Riviera, where he was found dead of an "overdose"[30] eight years later. "Sunny" Sanam, Zulfi's youngest child, was the only one successfully to escape the powerfully seductive or tragically repulsive legacy of her father's strange complex life and vengeance-filled political universe. She wisely opted for a quiet private life in London.

"The Armed Forces . . . are the only stable institution," General Zia reminded the nation on "Defence of Pakistan Day," 6 September 1977, the twelfth anniversary of the start of the 1965 War.

> They are imbued with the highest sense of patriotism . . . to defend the national frontiers. . . . [But] let me tell you, my dear countrymen, that a country does not survive by the defence of its geographical borders alone. Its ideological frontiers have also to be fully protected. Otherwise, a country is likely to slide into internal chaos and confusion. In the past, the

> Armed Forces have always undertaken unhesitatingly all tasks assigned to them to save the country from internal collapse. It was in pursuance of this spirit that Martial Law had to be imposed in the country to save it from a civil war and confusion.

Zia kept repeating that justification for his coup each time he had any reason to address large groups, for how else could he account for his own treachery? He still insisted, moreover, less than a month before he called the election game off, or rather announced its "postponement," that "I, as the Chief of the Army Staff and Chief Martial Law Administrator, assure my countrymen that by the grace of God Almighty . . . the Armed Forces will fulfill their promise of holding free and fair elections."[31]

"I hate anybody projecting as a leader," General Zia told a Saudi Press Agency correspondent, who interviewed him a few days later. "If you want to serve the Islamic *Ummah* and humanity, do it as a humble person. . . . Amongst Muslims, we are all humble brothers . . . not leaders." It was one of his rare public admissions of how much he "hated" Zulfi Bhutto. The Saudi then asked him, however, what "role" he visualized for Pakistan as "Chairman of the Islamic Summit Conference," if another Arab-Israeli war started. "We will go the whole hog in moral and material support for Arabs and Palestinians."[32] Going the "whole hog" seemed rather odd usage for a devout *Momin,* as Zia liked to call himself. Would it mean all-consuming commitment, or total Islamic abstention?

On 13 September, Justice K.M.A. Samadani of the Lahore High Court granted bail to Zulfikar Ali Bhutto, much to the chagrin of Zia, who now feared that he could not trust even that Punjab high court, though Maulvi Mushtaq was its chief justice. Justice Samadani, however, was a Sindhi. Zia knew that God helped only those who "help themselves," which now seemed to mean that in addition to having taken on the heavy burden of administering the country, he would also have to take on the task of meting out "justice" to his "hated" enemy by bringing him up for murder in his own more reliable martial "law" court. The same day that Zulfi flew home, first to 70, Clifton in Karachi, then on to Larkana, where Mir Murtaza and Mumtaz were waiting for him, Zia invited all other political leaders to Pindi, continuing to try his best to deceive them, as well as the outside world, about his true political intentions and "humble" ambitions.

"I thought it advisable to ask you for a joint meeting to exchange views about the conduct of elections," Zia began, still insisting that

> personally I have regard for all political leaders of all political parties. I have neither any personal likes or dislikes nor is there any reason for the Martial Law to be partisan. However, since I acquired the responsibility for the conduct of the coming elections, certain . . . rules for the game have to be prescribed. . . . You are all honourable people. . . . You may launch as vigorous an election campaign as possible but take care of not hitting your political opponents below the belt. . . . Let me assure you that the Government will cooperate with all alike except those who may

> attempt to take the law into their own hands. . . . The elections will . . . be held on 18 October 1977 as planned.[33]

Now that he finally announced a day as well as a year and month, he was ready to call off the game entirely.

Zulfi had been warned, upon his release from prison on 13 September that "an order for my detention under some preventive law or martial law was being prepared," as he wrote Begum Nusrat that night, appointing her "acting chairman" of the PPP in the event of his subsequent arrest. His "Emergency Directive" as he called that last letter to his wife written in freedom, gave her his full "authority" to run the party, if anyone "questioned" or challenged it. He feared that Zia had now "decided" to "perpetuate himself," and thought that if elections were "postponed," there would be "disastrous" consequences for the country, which "will face its worst crisis." Anticipating that Zia would do his best to "divide and disintegrate" the PPP, he warned that "a witch-hunt has already commenced," and that Zia would try to "consolidate his position" by "exploiting the name of Islam and introducing his version of Islamic laws." Zia had already introduced amputation of a hand for "stealing," whipping as well as death by stoning for adultery, and several other traditional Arabic punishments as part of his military "law" code.

Though Zulfi expressed "confidence" that "the General would be fighting a losing battle against his own people and democratic forces" that would eventually "defeat" him, he dared not venture a "prediction" as to just "how long" that would take. He did not ever suspect, not even in his most pessimistic moments, that Zia's tenure would last more than a decade. Zulfi knew his own popularity and thought that as long as he was free to move about Zia would lack the resolve to hang him, and that he could intimidate or somehow still strike a deal with the man he had brought from obscurity to the top—one man whose "humility" Zulfi Bhutto not only believed in but judged to be a thoroughly accurate presentation of Zia's intellect, talent, and personality. The PPP had an executive committee whose functioning Zulfi understood much better, than he understood Zia's as yet, hence he added to his directive that any "problems" caused by "intrigues and pressure" within that body were to be resolved, with "my authority and support," by Begum Nusrat as acting chairman. "God Almighty will bless our mission," he closed. "It is in the supreme interest of the masses of Pakistan and the unity of the State."[34]

"I personally could not believe they would hang him," Mumtaz recalled. "But he himself felt it was hopeless, right from the beginning. We were arrested the same time that night in Larkana, taken to Sukkur. . . ."[35] It was shortly after midnight on 16–17 September when a company of commandos climbed like black cats over the walls of Al-Murtaza, knocking out all the guards before they could raise a cry, hammering their rifle butts at the front door till it almost flew off its heavy hinges. "Tell them it's not necessary to break the door down," Zulfi told his elder son.[36] A few minutes later he and Mumtaz were driven away together in the dead of Sind's darkest night to Sukkur jail. Mumtaz went to sleep, even as they drove, but Zulfi found no rest that night, nor the next day, as the sun rose blood red in Sind's scorching sky. When he asked by what "right" he was being rearrested, having just been released on bail by Lahore's High Court, he was told that it was under "martial law," so he knew it was all over. His last waltz had been danced. And even as he smoked his fine Havana cigar he tasted only bitterly acrid fumes of impending doom, for with the dawn full awareness burst over him that Zia would never let him go free. Zia, like himself now, had long known that Pakistan was too small a country to leave both of them alive within it much longer.

"How can you sleep?," Zulfi asked his uncle-brother Mumtaz. "Don't you know that you and I are not going to see the outside world again? Don't you know it's curtains?"

"Look. What can I do?" Mumtaz replied, phlegmatically. "If I get hysterical and start crying, they're not going to let me go!" Ironically, Mumtaz would outlive them both, for Zia never hated Mumtaz Bhutto the way he hated Zulfi. What, after all, had Mumtaz done for him? He had never promoted him to be chief of army staff over the heads of six more-senior general officers!

"You've been sleeping all night," Zulfi said, shaking his weary head in disbelief at the sangfroid of his companion in military custody. "You're not even bothered by it!"[37] That truly amazed him. At 9:00 A.M. on 17 September, Zia's army commandos drove the former prime minister and the former governor-chief minister of Sind in separate cars toward Karachi, till the car bearing Mumtaz "broke down." They moved, "together again" for a few more hours on that long lonely road, into Pakistan's first capital, which both of them had so long considered "home." Now Karachi was just a brief stop en route to Punjab's harshest isolated prison fortress, Kot Lakhpat, several miles beyond the busy heart of Lahore.

"The aim of military intervention was to save the country from impending civil war," Zia repeated in his press statement that afternoon, "and to hold free, fair and impartial elections to further the cause of democracy in Pakistan. . . . In that hour of crisis I showed utmost courtesy and decency to all the political leaders irrespective of the fact whether they were ousted from power or they were aspiring to come into power. . . . I also made public my intentions not to take sides. . . . That only showed my good faith and sincere intentions." Ah, but now, imagine what "the independent judiciary" had suddenly "brought to light." There had been "serious irregularities" committed by those "who had presided over the destiny of seventy million people during the last five and a half years." Amazingly enough, the humble general explained, "these acts had remained hidden from public view because of wilful suppression of facts and one sided projection of Government policies." Remarkable! Yet now "a plethora of evidence which escaped public eye" all those years had, thanks to him, of course, been "unearthed." Unfortunately,

> all civil institutions [no military ones!] in the country were systematically destroyed. Civil services were politicized . . . public funds were used for personal luxury and party benefits. The life, property and honour of the law abiding citizens were made unsafe. Primitive, inhuman and barbaric

> methods were employed to crush all dissident elements and political opponents. Germs of class hatred were sown. . . . Elections held in March 1977 were rigged . . . and Government funds . . . and Government agencies were used in support of Pakistan People's Party candidates.[38]

It had taken Zia and his FIA two months and ten days more to learn exactly what the entire leadership of the PNA had been shouting ever since 8 March, four months prior to his coup. But now that it was finally "clear" to him, Zia could not possibly hold the free and fair elections he had promised till "the true face" of "all candidates" was shown to the entire electorate. That was the reason he had today "ordered the detention of Mr. Zulfikar Ali Bhutto and some other PPP leaders," who would "be treated well in the best traditions of our religion and given a fair trial in Military Courts."

But even Zia, devoutly and humbly persuasive though he tried his best to appear, could not quite convince anyone of why, if he wanted to give Bhutto a "fair trial," he found it necessary to take the many "challenges" (*challans*) already presented against him to the Lahore High Court out of that civil body's legally experienced hands and transfer them to some military court. Asghar Khan and Mufti Mahmood soon learned that Zia was doing his utmost to wean away several leaders of the PNA, including the Pir of Pagaro, who filed against Zulfi in Larkana but never once dared so much as to "visit" that constituency. Bhutto would have easily won all three seats (in Lahore and Thatta as well as Larkana) for which he had filed, had Zia held the fair elections he promised. But several of Zia's friends in the PNA "begged the General to stay in the saddle," Asghar Khan reported, thus assisting him to retain the power he had usurped and "rob the people of their rights."[39]

"The political situation in our country has reached such a point that I had to address you before I had intended," Zia told the robbed people of Pakistan that bleak October first.

> You will recall my speech in which I had unconditionally made a pledge that fresh elections would be held in October. . . . Even now I hold on to the position that except in a crisis the Army should not be holding power. . . . The right to rule belongs to the people alone, who should run the Government through their chosen representatives. . . . But will it be proper to apply the term "democracy" to whatever has happened during the last three months? . . . Everyone is indulging in the character assassination of his opponent. Even the actions and deeds of certain individuals which are already under investigation by courts of law have been made the subject matter of political debate. . . . Is this how the political parties visualize their roles towards the restoration of democracy? . . . [I]n order to prop up their leadership, certain elements have made inflammatory speeches. . . . Are elections synonymous with violence? . . . We have great respect for the institution of elections, but I cannot allow the country to face disaster for their sake. . . . Today the unfortunate position is that the common citizen is suffering from a sense of fear and anxiety and is a victim of uncertainty.[40]

Yes, the last sentence was all too true; the reason for it as obvious as the unsmiling dark-eyed face that filled every television screen in Pakistan, and the teeth through which those sad words were so solemnly spoken.

"My opponents will drum up the charge that the intoxication of power leaves nobody unaffected," Zia rightly anticipated as he announced his decision to "cleanse" the nation's politics by putting off elections until "the process of accountability" was completed. But accountability is an ongoing process, and Zia had at least read enough to expect that "the educated ones will resort to the lessons of history to prove the point that no military ruler had given up power voluntarily . . . but I am not bothered about their ways of thought. . . . I have consulted my senior colleagues. I prayed to God that He should enable me to take a decision . . . in the best interest of the country." Zia now also announced that all "accused persons" would be tried in civil courts, not the military court he had initially considered the only safe place in which to try the prime minister he had so treacherously arrested.

By October first, Acting Chief Justice Mushtaq Hussain had transferred Sindhi Justice Samadani out of the Lahore High Court bench, and Zia himself had forced the courageously independent chief justice of the Supreme Court, Yaqub Ali Khan, to retire on 22 September, appointing his friend Sheikh Anwarul Haq to the position on 23 September 1977. Anwarul Haq had no legal training but entered the "judicial service" by its back door as an administrator; he was first appointed a district magistrate and later was elevated to the Lahore High Court and then to the Supreme Court, thanks to his mentor and old friend, Chief Justice A. R. Cornelius.

Three days before he was forced to retire, Chief Justice Yaqub Ali had permitted Begum Nusrat to file a petition in Pakistan's Supreme Court challenging the constitutionality of her husband's detention. One of many "accusations" brought against Zulfi Bhutto by Zia's regime was that he had "tried to discredit and destroy" the judiciary. "An individual who puts into grave jeopardy the constitutional structure and contemptuously calls it 'the old legal order' is morally estopped from speaking on behalf of the Judiciary," Bhutto replied.

> An independent Judiciary is the antithesis of Martial Law. An independent Judiciary can only function under the umbrella of the Constitution and not under the shadow of the gun of a brown Duke of Wellington. An independent Judiciary exists side by side with an executive chosen by the people and a legislature elected by them. But the people's Executive is in jail. The assemblies . . . have become as silent as the graveyards. Can one flower flourish in a garden turned into a desert? I was the author of the Constitution of 1973. . . . [T]his is the classical case of the aggressor blaming the victim of aggression for aggression. It is a part of barrack logic. . . . [T]he Respondent [Zia] alone has destroyed . . . the judiciary by his illegal action of 5th July, 1977. The only way to restore legitimacy and save Pakistan is to roundly reject his action of 5th July. . . . Any attempt to justify that action will . . . take us back to . . . Doom.[40]

His best hope, Zulfi knew, of defeating Zia was to win the Supreme Court action because that would remove all constitutional ground out from under the general's every step taken since one minute past midnight on the fifth of July, when he launched his coup. Zia's "attempts to subvert the Constitution," Bhutto argued (in his 31 October modification of the 26 October Rejoinder to A. K. Brohi's earlier Reply on behalf of the Federation), made "him guilty of the offence of high treason."[42] Zulfi now charged that Zia's "personal venom and vendetta against me has caused irreparable damage to the image of Pakistan by deliberately making false and malicious accusations of a very petty nature and propagating these through our embassies and international Press throughout the World." He characterized Zia's "worst and meanest" charges against himself as "political pornography," arguing that "as a political leader of national stature it was my irksome duty to place the truth on the record of this Hon'ble Court." By canceling the elections, Zulfi wrote, Zia was "turning his personal vendetta against me into a vendetta against Pakistan," and was "suffering from 'Bhutto-phobia.'" Zulfi insisted that "God is on our side because we are on the side of the people and the right."

Zulfi's long and passionately argued Rejoinder noted that "it is not good to drink alcohol but it is much worse to drink the blood of freedom fighters. Why does the respondent [Zia] conceal his own dossier? . . . Will it praise him for his role in Jordan? [Brigadier Zia was on loan to the Royal Jordanian Army when its troops opened fire on Palestinians during "Black September" of 1970]. . . . [T]he dossiers of Mr. Mumtaz Ali Bhutto and Mr. Abdul Hafeez Pirzada, false as they are . . . will look like the dossiers of saints," compared to Zia's, Zulfi declared. One of Zia's methods of weaning Zulfi's followers away from him was to allow them to see secret dossiers Bhutto kept on his colleagues as well as his opponents. The compilation of such smut-filled "intelligence dossiers" was hardly begun in Zulfi's era, however, nor is it a sickness unique to Pakistan.

"Martial Law is a paper tiger," Bhutto concluded with premature optimism, more accurately adding that

> Martial Law is darkness at high noon. It is neither an order nor a system. . . . [T]his form of lawlessness takes us back to the law of the jungle, where only the strongest survive. Most of our people are weak and backward. They will perish in such an arrangement. Force, naked and brute, moody and mad cannot be made the sole criterion of our honour and our respect. . . . This Hon'ble Court cannot be silenced. Its destiny demands that it upholds the Constitution and the law. . . . Martial Law and a new order are terms which cannot be reconciled. . . . [T]hink of the damage being done to Pakistan. Did millions die to live in terror, did they die to be flogged and lashed? . . . In your hands, my Lords, lies the decision to make or mar. . . . The voice of the people has been silenced. Today only you hold that pen which is mightier than the sword.

Here again, as if by miraculous rebirth, was the voice of young Zulfi—liberal, ardent, freedom-minded Bhutto at his finest. What a challenge to toss up to the highest bench in his land, and what an opportunity for Pakistan's Supreme Court to assert its faith in the primacy of civil over military governance, of elective representative rule over martial dictatorship. Had Yaqub Ali still presided, he might have led his colleagues in handing down a historic verdict, releasing all political detainees, reversing the coup. But Yaqub had been forced to retire.

Chief Justice S. Anwarul Haq penned the judgment that was handed down on 10 November 1977. A. K. Brohi had appeared for the Federation and argued that as a result of Bhutto's "massive rigging" of the 7 March elections, his government had "lost whatever constitutional validity it had earlier possessed." Chief Justice Haq agreed with Brohi. The ensuing "widespread disturbances" amounted to a "repudiation of Mr. Bhutto's authority to rule the country," and the "spectre of civil war" was averted only thanks to the timely action by the army on 5 July. Zia's attorney general, Sharifuddin Pirzada, supported Brohi's submission, arguing that Zia's action that day was not a "coup" but was "valid," based on the Old Roman "doctrine of state necessity" as the only "proper" means of ousting "the usurper who had illegally assumed power as a result of massive rigging."[43]

In an earlier historic challenge to martial law's constitutionality under Yahya Khan, the *Asma Jillani* case, Brohi and Sharifuddin Pirzada had both taken the opposite side of that legal argument, helping to convince a prior Supreme Court of their arguments. Now Chief Justice Haq wryly noted that "Mr. Brohi has indeed faced an uphill task before us to question the correctness of this judgment." The venerable advocate had obviously changed many of his earlier noble civil libertarian ideas now that he was soon to join Zia's regime as its minister of law. Brohi argued that "the more unsettled the time and the greater the tendency towards the disintegration of established institutions, the more important it is that the Court should proceed with the vital, albeit unspectacular task of maintaining Law and order and by so doing act as a stabilising force within the community."[44] "Peace in our time," had been Neville Chamberlain's rationale for his capitulation to Hitler at Munich. Because General Zia had made it "absolutely clear" after his coup that he had no "political ambitions," moreover, his chosen new chief justice decided that the "objectives" of martial law were merely to "create conditions suitable for the holding of free and fair elections." Now, surely, that was no "abrogation" of the Constitution. Only "certain parts" of it were now "held in abeyance on account of state necessity."[45] In the unanimous view of Pakistan's nine-member Supreme Court, therefore, Begum Bhutto's petition to release her husband and his colleagues from prison was "dismissed" as not "maintainable."

Zulfi's trial for "conspiring to murder" Ahmad Raza Kasuri had by now been under way for a month in Lahore's High Court, where Acting Chief Justice Maulvi Mushtaq never so much as attempted to suppress or hide his personal animus against the man who as president and prime minister had demonstrated similar feelings of disdain and dislike for the Maulvi. It never, of course, occurred to Mushtaq Hussain that he should recuse himself from the trial. Instead he denied several pretrial motions that he do so. Zia had selected him to preside over this drumhead court, and Maulvi Mushtaq picked the other four judges, all of whom he felt confident could be trusted to find their most famous "principal accused" as Bhutto was termed, "guilty" as charged. Five members of the FSF, including Director General Masood, were also on trial for the "conspiracy to murder" Kasuri, who was called by Prosecutor Shafiq Anwar as the "State's" first witness and took five full days of court time to tell his side of the story. Maulvi Mushtaq treated Kasuri with unctuous courtesy, in sharp contrast to the harsh way in which he invariably addressed Bhutto. Masood Mahmood's testimony then took almost two weeks, but as the "prime Approver" Masood alone could link Zulfi Bhutto to the "murder"; only he had received "orders" from the prime minister, "transmitting" them to all the other FSF "conspirators" on his team. Zulfi's secretary and security officer, Saeed Ahmad Khan, also testified as an approver. He too would be granted his freedom, thanks to his evidence, which helped convict the prime minister whom he had served so faithfully for years, delivering all his "messages," including that most brutal one to J. A. Rahim and his son, Sikander. Masood testified that both Bhutto and Saeed Ahmad "reminded" him several times to "get on with the job" of seeing to it that his FSF assistant, Mian Muhammad Abbas, quickly produced "the dead body" of Kasuri or "his body bandaged all over." Mian Abbas himself was not the actual "hit man" but supposedly transmitted Masood's instructions to his FSF subordinates, promising that those fatal "orders" from above would be "duly executed."

"Although I am described as principal accused, archcriminal," Zulfi later argued in appealing his conviction before the Supreme Court,

> actually if there is any link or thread, it is a very thin or thick link or thread depending on my discussion with Masood Mahmood. Secondly, Saeed Ahmad reminding Masood Mahmood of something that I had said. . . . Thirdly, if I had a motive otherwise independent. If I can succeed in establishing before this court that there is no link. . . . [w]hatever happens afterwards is not my concern. I am not involved. . . . At one stage, although it was not suggested in the legal sense, it was said that Masood Mahmood might have had his own motive for committing the offence. On that remark, the State counsel said half his case had been proved! Half of your case had been proved: against whom? Maybe against Masood Mahmood, maybe against the confessing co-accused, but not against me.

Barrister Bhutto was shouting. He had not forgotten what he had learned at Lincoln's Inn. "I would say, subject to correction, even if we admit, as in Wollmington's case, that 'I took the gun, I shot my wife', still it falls on the law, on other aspects of the law. If the test is not met, the case falls. Hence, the question of half the case being proved on that remark is not based on law."[46]

But Zulfi himself knew by that cold December of 1978 that such fine points of English law had nothing to do with his fate at the hands of Zia ul-Haq, who on 16 September 1978 had proclaimed himself "president" of Pakistan. As his *White Papers* flooded Pakistan, Zia bluntly spoke of his former president-prime minister prisoner as a "murderer," insisting to shocked foreign officials and journalists who could barely believe that he would thus speak of the man he still held under trial, "I have seen the evidence with my own eyes!" Zia asserted there were "secret files," dossiers on "Bhutto's opponents," along whose margins Zulfi had scrawled, "Eliminate him!" Not one of those "files" has ever come to light, however, nor is there reason to believe that one ever will. That particular phrase sounds far more congenial to Zia's military mind than to Bhutto's more sophisticated intelligence. If Zulfi ever used that word about a person, moreover, instead of a public problem or noxious condition, he would never have done so in writing.

Benazir, who faithfully attended her father's High Court trial, was most relieved to learn that none of the many shells picked up by investigating police around the site of Kasuri's ambush fit any of the FSF weapons that were supposed to have been used in the attempted murder, and she rushed to tell him that great "news." "'Papa, we've won! We've won!' I said to him, telling him about the ballistics report. I'll never forget the look of kindness on his face while he listened to my excitement. 'You don't understand, do you, Pinkie,' he said gently. 'They are going to kill me. It doesn't matter what evidence you or anyone comes up with. They are going to murder me for a murder I didn't commit.' "[47]

Maulvi Mushtaq and his full panel, of course, found Zulfi Bhutto "guilty" of "murder" as charged, and sentenced him to "death" in March of 1978. A few days later, a British reporter asked Zia, "Will Mr. Bhutto be hanged?" The general refused, however, to "comment on this because the case is still *sub judice*. But all I will stress on is that justice must be done and nobody is above the law according to my conviction. Whether it is Mr A, Mr B or General Zia-ul-Haq himself, justice must be done."[48] Zia was then asked about his "commitment" to "free elections," and would that also "be done"? "We are committed, totally committed to review the democratic procedure in this country, and for that the first basic step is the election. . . . [W]e are all endeavouring to create an atmosphere in which we could hold free and fair elections."

"Politics is not the illegal seizure of the State machinery," Zulfi wrote to Zia from his stinking, dark death cell in Pindi's now-demolished prison.

> Politics is not the conversion of a flowering society into a wasteland. Politics is the soul of life. It is my eternal romance with the people. Only the people can break this eternal bond. To me politics and the people are synonymous. . . . You or your coterie have no right to take away my spiritual and imperishable links with the beloved people of my country. It is an inseparable part of my heritage. My blood is in the blood of Pakistan. I am a part of its dust, a part of its aroma. The tears of the people are my tears. A smile on their beautiful face is a part of my smile. . . . My destiny is in the hands of the people. Only the people have the right to severe or seal their affinities with me.
>
> Have you not done enough? Surely your thirst for senseless revenge is not completely unquenchable. Is your appetite for primitive vendetta so insatiable as to stop at nothing? . . . [W]hereas you take so much pride in

being "a soldier of Islam" (an expression which you stole from my speech at the Islamic Summit Conference) are you true to anyone and what have you learnt from the sacred principles of Islam? Does Islam teach you to break your oath, does Islam tell you . . . to fabricate false cases and hound your Mohsin [mentor], does Islam tell you to bow before Hindu India? . . . My Islam and the Islam of the poor and exploited masses of Pakistan teaches us magnanimity and fidelity. Islam teaches the virtues of justice. . . .

What justice can I expect from you? The conspirator who dislodged his own Prime Minister by the force of arms is estopped from sitting in judgment against him. This has been more than abundantly established by the macabre persecution, by the shameless victimization and malignant vilification you have done in the last six months. Each minute, each hour and every day of your despotic rule is a living testimony of your hatred and enmity towards me. . . . My persecution, which is without parallel in the contemporary history of the sub-continent, is a matter of public knowledge. The views you have expressed in your interviews. . . . What you have told Foreign leaders is also within my knowledge. The day of accounting awaits every individual. I cannot take accounts from "a mighty General" like you, but before the Final accountability on the Day of Judgment, I have no doubt that you . . . will be taken to the door of Shahbaz Kalander for the blood you spilt. . . . I would be a dishonest coward if I did not tell you how determined are my people to see the dawn of that day. Please do not misunderstand. I am in no position to threaten you. I am only attempting to inform you that you have deeply violated the finer sentiments [of] our people and aroused their eternal hostility.

General, please do not overstep the bounds under the intoxication of power. You have no right to disqualify me from the public life of this country. You have no authority or mandate to take such a fatal decision. . . . [Y]ou sublimely think that . . . [you] can with naked arbitrariness disqualify from Public life the former elected President and Prime Minister of Pakistan and the true leader of the people? It is preposterous. . . . From beginning to end the whole scheme is perverse. We will meet one day. You pursue me now. Wait till I pursue you.[49]

Zulfi penned that last letter to Zia in his cell, just weeks before his death. He did not expect an answer, nor did he hope for any final act of clemency from the man who feared him much more now that he was his prisoner than he ever had when he "served" Prime Minister Bhutto as his chief of army staff. As chief, Zia had owed Zulfi only his job, but as the "president" who usurped his power by abrogating the Constitution, he owed him his life. Which was the reason Zia would never pardon him, or show any mercy, though he solemnly "promised" several Arab heads of state, including the sheikh of Abu Dhabi, that he would do so.

There had been nine justices on the Supreme Court of Pakistan when the last trial of Zulfi Bhutto started in October of 1977. By the time the Court was ready to hear Bhutto's appeal from that verdict, Justice Qaiser Khan had retired and Justice Wahiduddin Ahmed was too "sick" to remain on the bench. Yahya Bakhtiar had initially hoped for a 5–4 decision in favor of "reversal" of the High Court's verdict, counting not only on the three justices who finally voted for Bhutto's acquittal, Justices Dorab Patel, G. Safdar Shah, and Muhammad Haleen, but also on Justices Wahiduddin Ahmed and Qaiser Khan. With the latter two gone, the verdict handed down by Chief Justice Anwarul Haq on 6 February 1979, was to dismiss the appeal, confirming the Lahore High Court's verdict and death sentence by the narrowest of margins, 4–3.

Chief Justice Anwarul Haq took more than eight hundred pages of foolscap to write the "casting vote" decision that sent Zulfi Bhutto to his death. He dismissed all the "errors" and "illegalities" in the Lahore High Court's trial as totally irrelevant to the verdict. More relevant, however, was that the four Justices who voted to hang Bhutto were from Punjab; the three who voted to free him were not. I went to "interview" the by-then-retired Anwarul Haq in his home in Lahore a decade later, but he refused to say "anything more" about his judgment or his "impressions of Mr. Bhutto" than he had in his written verdict. "But if you want to know about my training and qualifications as a justice of the Supreme Court," he told me, "you should go see our great Justice Cornelius, who was my mentor. He's very old, and has been too sick of late to leave his bed, but he lives in Faletti's Hotel, and he might be willing to speak to you."[50]

The venerable retired justice was, indeed, on his deathbed in that once grand and famous, now shabby and faded, hotel near the heart of Lahore. Yet he seemed cheered to have a "visitor" and reminisced for almost an hour, chain-smoking as he lay on his bed, indulging in "my only pleasure." He talked a bit about "law," but ventured no opinion of Mr. Bhutto or the infamous trial other than to say, "I'm sure they did the right thing." Before leaving, I asked if he had been born in Lahore. "Oh, no, I was born in India, Indian Christian, you see, but at the time of Partition I was in the judicial service in India and some Muslim friends told me to come to Pakistan. It would be much easier for me to get promoted here, and they were right, of course. I could never have become chief justice in India, without any barrister training, you see. Then when I got here, I decided to throw in my lot with the Punjabis and settle down in Lahore—because you can't trust a Sindhi."[51]

Ah, still another variant of *Siasat!* Provincial conflict and mistrust, ancient geo-historic fears, taboos and terrors; the sources long forgotten, but with legacies of pain ever green in hearts brimful of parochial prejudice. Punjabi distrusting Sindhi, Baluchi suspicious of both, and half the tribesmen along the North-West Frontier holding rifles loaded at the ready, cocked to fire at the other half. Where, then, was that Land of the Pure called *Pakistan,* for which one Quaid had worn out his pain-racked life and the other was now about to be hanged before dawn on 4 April 1979?

"The longer Martial Law remains the shorter will be the remaining life of Pakistan," Zulfi wrote from his death cell.

> Sind will say *Khuda Hafiz* [good-bye] before Baluchistan and NWFP. The Indians made two leaders select the Janata Prime Minister, one was Jai Prakash Narain and the other was a Sindhi, Acharya Kripalani. Also, [L. K.] Advani is very powerful. The dream of their life is to get back Sind, the Sind of Shah Latif and the Karachi Port, the exploited and "raped" Sind, the Sind of the Sufis. The Hindus of Sind always considered "Sufism" to be a bridge between Hindus and Muslims . . . all these forces are . . . more active than ever before. . . . Pakistan is decomposing very fast. In Europe there was "*Balkanization*." Here, there will be "*Bangladeshization*." The process is in motion. Thanks to Zia's follies it has been accelerated. . . . If I am not a part of Pakistan, in that case Sind is not a part of Pakistan. . . . [M]y roots in the soil of this land are very deep, much deeper than of those who came across the border due to disturbances or fear of disturbances. . . . My genesis to political fame is written in the stars. . . . The last days come for every actor on the stage. There is no exception to this immutable rule. There are no nightmares more dreadful than the last days of a usurper, of a man who stabs his own benefactor. Brutus did it to Caesar to prevent despotism. . . . "Brutus is an honourable man and so are all of them! Yes, all of them are honourable men . . . the engines of oppression and symbols of ingratitude." . . . For over a year and a half I have been in solitary confinement . . . I have not left the death cell to have a shaft of sunshine or the embrace of fresh air. . . . I am the Rana [lord] of Shah Latif. In that sense I am the Rana of not only Larkana but of the whole of Pakistan, and if no longer of the whole of Pakistan, certainly of Sind. . . . We are on a razor's edge.

The sun had gone down and the final appeal had been denied, and Zulfi knew there would be no clemency, no mercy, no "presidential" pardon, though most leaders of virtually every nation on earth [except India] had appealed for "mercy" to Zia on his behalf. "All our instruments and levers have been taken away, neutralized or liquidated. But we must fight back to the last drop of our blood. . . . We must be the last to throw the ace," Zulfi cried out from his dark cell. "Zia . . . is frustrated. . . . We are dealing with a very strange and mercurial person, a liar and a double crosser. He is . . . desperate."[52]

"Zia told me 'It's either his neck or mine,'" Minister Roedad Khan recalled. "He said, 'I have not convicted him, and if they hold him guilty, my God, I am not going to let him off!'"[53] Then secretary general of the ministry of the interior, Roedad was responsible for digesting the many mercy petitions and pleas that came to Pakistan from heads of state the world over. Those appeals were summarized so that they could be submitted to President Zia in only a few pages, permitting him to "read" them more easily, but that boiling-down process took a "lot of time," or so it seemed to Zia, who was "impatient" to get the hanging over and done. So the president's office kept calling Roedad to ask, "Why are you taking so long?" Because he could not be hanged until the last of those mercy petitions, all to be rejected, had first been "digested."

"I went with Hafeez Pirzada to the jail and we were told by the officer in charge, 'You can stay today as long as you like it's your last interview with him,'" Yahya Bakhtiar recollected. "The decision had been made but not yet announced, and the guards were alerted." The next day, despite the prison guard's orders, Yahya went again to visit his weak and weary friend. "We had a cup of tea . . . he was pessimistic in his heart of hearts."[54]

"I found him in great pain," Benazir wrote to prison authorities after one of her weekly visits to her father in his Pindi death cell. "An infection of the gums which were full of pus and bleeding. . . . [H]is mouth was swollen and he kept having to wipe the pus and blood with his handkerchief. So severe . . . it is visible to the eye in the poor light. . . . If immediate attention is not paid to clearing the gum infection, the blood and pus will pass . . . into his blood stream."[55] Because of Benazir's appeal, Dr. Zafar Niazi, one of Pakistan's most eminent dental surgeons, who had been Zulfi's dentist for more than a decade, was permitted periodically to visit his most famous patient. He was unsuccessful at relieving the acute periodontal pain, whose "source" may have been "pancreatic" or "liver disease" that had plagued Zulfi's last years. "We tried to get him released from that small and terrible prison cell," Niazi recalled, "but . . . Zia hoped he would die a natural death in there . . . knowing how little he could eat."[56]

Begum Nusrat and Benazir were both detained in Sihala's police "camp," several miles from Pindi's district jail death cell, where the gallows were being tested on 3 April 1979. That morning they were unexpectedly taken outside, and driven to Rawalpindi's black prison for their final meeting with the chairman of the Pakistan People's Party they now jointly led as his political heirs and legatees. And far to the south, in the heart of Sind's Larkana District, troops cordoned off the Bhutto ancestral home and burial village, Garhi Khuda Bakhsh, where the remains of at least six generations of Bhuttos had blended into the blood-colored soil and rusty dust of a civilization whose Sindhi roots reach back more than five thousand years to an urban "Mound of the Dead" (*Mohenjo-daro*).

At 2:00 A.M. on 4 April 1979, Zulfikar Ali Bhutto was taken from his death cell to the gallows in Pindi's dark prison fortress and hanged by the neck until pronounced dead.

"'No!' the scream burst through the knots in my throat," wrote Benazir of that darkest hour. "Papa! Papa! I felt cold, so cold, in spite of the heat, and couldn't stop shaking. There was nothing my mother and I could say to console each other. . . ."[57]

At 4:00 A.M. Zulfi's body was flown from Chaklala airport to Larkana, where rain had fallen, most unseasonably, that night. Zulfi's first wife, his cousin Sheerin, was brought to Garhi Khuda Bakhsh at dawn from the neighboring village of Naodero, to join Mumtaz's relatives and servants at a fresh grave dug next to Sir Shah Nawaz's by Bhutto clansmen. As rosy fingers of light blushed the eastern sky, two army helicopters landed near the Bhutto village. Soldiers swiftly transferred the coffin from one of the dark choppers to an ambulance waiting to drive it to the grave. Neither Begum Nusrat nor Benazir had been informed in time to fly down to witness the burial. Nor did any of his other children know as yet that "Papa" was dead; both boys were in London, and Sunny was back at Harvard.

Nazar Mohammed, a lifelong Larkana servant of the Bhuttos, was permitted to help bury his Quaid, and later recalled that "his face . . . shone like a pearl. He looked the way he had at sixteen. His skin was not of several colours, nor did his eyes or tongue bulge out like the pictures I'd seen of the men that Zia had hanged in public. . . . I turned Bhutto Sahib's face to the West, towards Mecca. . . . His neck was not broken. There were strange red and black dots on his throat, however, like an official stamp."[58]

Fearing possible riots, Zia had rushed to have it "over" and "done" with before dawn. Contrary to the Pakistani prison "code" for hangings, he had ordered Zulfi Bhutto's murder in the dead of night. Then Zia felt for the first time in twenty-one months that he could breathe easy. And soon he would announce gleefully to his henchmen, "The bastard's dead!"

"Zia ul-Haq. The General who would ruthlessly rule Pakistan for the next nine years," Benazir reflected in her *Autobiography*. "For almost two years, I had done nothing but fight the trumped-up charges brought against my father by Zia's military regime. . . . Yet not until yesterday [3 April 1979] had I allowed myself to believe that General Zia would actually assassinate my father."[59] At 26, Benazir took up the mantle of her father's leadership, first of his party, and later of his nation. The next "election," which Zia still promised he would "soon" announce, might be fought by the PPP under "Chairman" Begum Nusrat, but Benazir was her father's true political heir, and wherever she went or spoke, people cried aloud, "Jiye Bhutto!" Little less than a miracle it seemed, to most Sindhis at least, for though Zia and his minions reported to the world that "Bhutto was dead," here was Bhutto "reborn," and so much "more beautiful." "Bhutto lives!" was the political battle cry that swept over Pakistan, from Karachi to Peshawar, from Larkana to Lahore, long before the "promised" election date in November, which Zia finally did announce. Only to cancel, however. Again and again. Each time the chosen day drew near, it became clear to the "Momin" general-president, who would soon become the United States' "front-line defender of freedom" against the Soviets in Afghanistan, that he could never win a "free and fair" election against Zulfi Bhutto, who "lived" through his dauntless daughter Benazir.

On 17 August 1988, General Zia ul-Haq with most of his top command, and U.S. Ambassador Arnold Raphel and Brigadier General Herbert Wassom, went down to fiery deaths in a mysterious plane crash shortly after takeoff from Bahawalpur Air Base. A few months later, free and fair elections were held throughout Pakistan under the watchful eye of interim President Ghulam Ishaq Khan, and Benazir Bhutto soon stepped into her father's office, as prime minister. That reincarnation of Zulfi Bhutto's first People's Party Government lasted less than two years. In August of 1990, President Ghulam Ishaq Khan, with full support of the army, "dismissed" Prime Minister Benazir Bhutto, and announced new elections before year's end, which the PPP lost.

There remains a powerful, persistent, possibly growing, but certainly undying, mystic belief held by millions of Pakistanis, not only Sindhis but Punjabis, Baluchis, and Pathan Frontiersmen as well, that "*Shaheed*" (Martyr) Zulfikar Ali Bhutto was never hanged, that he never died. "Zulfi Bhutto lives on," they say, "and he always will!"

Notes

Chapter 1: Sindhi Roots

1. Z. A. Bhutto's Prison Cell Holograph (hereafter cited as Bhutto PCH) in Bhutto Family Library and Archives (hereafter cited as BFLA), 70, Clifton, Karachi.

2. For early Bhutto family history I have primarily relied upon Sir Shah Nawaz Bhutto's unpublished memoir (hereafter cited as Sir Shah Nawaz's Memoir), which begins, "This is not an Autobiography. . . ." It is held in the BFLA.

3. For background information about Mohenjo Daro and the early Indus Valley civilization, see Stanley Wolpert, *A New History of India,* 4th ed. (New York: Oxford University Press, 1993), chap. 2, and appended bibliography.

4. The quotation is from Fray Sebastian Manrique, a Portuguese who visited Sind in 1640–41, and is noted in Yasmeen Lari, *Traditional Architecture of Thatta* (Karachi: Heritage Foundation, 1989), p. 8.

5. Sir Shah Nawaz's Memoir.

6. Khushwant Singh, *Ranjit Singh, Lion of Punjab* (London, 1962).

7. See Patrick Macrory, *Signal Catastrophe: Story of the Disastrous Retreat from Kabul, 1842* (London: Hodder & Stoughton, 1966).

8. *The Life and Opinions of General Sir C. J. Napier,* ed. Sir W. Napier, 2:218 (London, M.

9. See Fawn Brodie, *The Devil Drives: A Life of Sir Richard Burton* (New York: Norton, 1967).

10. Sir Shah Nawaz's Memoir.

11. Benazir Bhutto, *Daughter of the East: An Autobiography* (London: Hamish Hamilton, 1988), pp. 26ff. Hereafter cited as Benazir's *Autobiography.*

12. Ibid., p. 28.

13. Mumtaz Ali Bhutto interview at his home in Karachi, 10 January 1990.

14. Begum Sahiba Nusrat Bhutto interview at the Prime Minister's House in Rawalpindi, April 1989.

15. Sir Shah Nawaz's Memoir.

16. Ibid.

17. Ibid.

18. Ibid.
19. Ibid.
20. Ibid.
21. Ibid.
22. Ibid.
23. See Stanley Wolpert, *Morley and India, 1906–1910* (Berkeley and Los Angeles: University of California Press, 1967).
24. See Stanley Wolpert, *Jinnah of Pakistan* (New York: Oxford University Press, 1984).
25. For the historic background to Bengal's first partition, see John H. Broomfield, *Elite Conflict in a Plural Society* (Berkeley and Los Angeles: University of California Press, 1968).
26. Sir Shah Nawaz's Memoir.
27. Ibid.
28. Ibid.
29. Ibid.
30. Ibid.
31. Interview with Mrs. Manna Islam, Z. A. Bhutto's sister, at her home in Karachi, April 1989.
32. See Stanley Wolpert, *Massacre at Jallianwala Bagh* (New Delhi: Penguin, 1988).
33. Wolpert, *Jinnah of Pakistan,* p. 121.
34. Sir Shah Nawaz's Memoir.
35. See P. N. Chopra, chief ed., assisted by Prabha Chopra, *The Collected Works of Sardar Vallabhbhai Patel,* 3 vols. (Delhi: Konark, 1990–93.
36. Sir Shah Nawaz's Memoir.

Chapter 2: From Larkana to Bombay

1. There were supposedly two horoscopes cast for Zulfi Bhutto, both reputedly in the possession of Husna Sheikh in London.
2. Solicitor Dingomal's report during an interview with him at Mumtaz Ali Bhutto's home in Karachi, January 1990.
3. Mrs. Manna Islam interview, Karachi, April 1989.
4. For Lal Shahbaz and Sindhi Sufism, see J. P. Gulraj, *Sindh and Its Sufis* (Lahore, 1979), pp. 87ff.
5. Piloo Mody, *Zulfi, My Friend* (Delhi: Thomson Press (India), 1973), p. 13.
6. Ibid., p. 14.
7. Benazir (in her *Autobiography,* p. 28) says "eight or nine years," but Zulfi's sister, Mrs. Islam, said "ten or twelve years," and Mumtaz put it at fifteen, recalling that "he was 13 and she was 28."
8. Mumtaz Ali Bhutto interview.
9. Governor of Sind, Sir Lancelot Graham, to Viceroy Lord Linlithgow, Karachi, 18 February 1937, in Chopra, *The Collected Works of Sardar Vallabhbhai Patel,* vol. 1, *Towards Freedom 1937–47.* (New Delhi: Indian Council of Historical Research, 1985), p. 262.

10. Hafeez Malik, ed., *Iqbal: Poet-Philosopher of Pakistan* (Washington, D.C.: Public Affairs Press, 1963).
11. Wolpert, *Jiinnah of Pakistan*, pp. 145ff.
12. Bhutto PCH.
13. Mody, *Zulfi*, p. 30.
14. Syed Shamsul Hasan Collection, courtesy of Khalid Shamsul Hasan, "The Shade," Karachi.
15. Maulana Abul Kalam Azad, *India Wins Freedom: The Complete Version* (New Delhi: Orient, Longman, 1989), pp. 162ff.
16. Bhutto PCH.
17. *White Paper: The Performance of the Bhutto Regime*, vol. 1, *Mr. Z. A. Bhutto, His Family and Associates* (Islamabad: Government of Pakistan, 1979), Annex 6, p. A-31.
18. Mody, *Zulfi*, p. 37.

Chapter 3: Brief California Interlude

1. Young Bhutto's typed "Reflections" (on the United States) is held in BFLA. (Hereafter cited as *Reflections*.)
2. Al Cechvala interview at his home in Glendora, California, 30 May 1989.
3. Mary Ellen interview at her home in Encino, California, June 1989.
4. Dr. Leili Bakhtiyar interview at her home in Brentwood, California, 2 May 1989.
5. "One World," in *Politics of the People* (a collection of articles, statements, and speeches by Z. A. Bhutto; hereafter cited as *PoP*), ed. Hamid Jalal and Khalid Hasan, vol. 1, *Reshaping Foreign Policy, 1948–1966* (Rawalpindi: Pakistan Publications, n.d.), p. 4.
6. Ibid., pp. 7–8.
7. Ibid., pp. 15–18.
8. Mody, *Zulfi*, p. 41.
9. Z. A. Bhutto to Ambassador M. A. H. Ispahani, Washington, D.C. This letter, written on 11 September 1948, from 3421 S. Flower Street, Los Angeles, held in BFLA.
10. Undated, written on "Embassy of Pakistan, Washington, D.C." stationery; held in BFLA.
11. Newspaper clipping (n.d.), held in BFLA.
12. Dr. T. Walter Wallbank, General Studies, University of Southern California, 6 October 1948, "To Whom It May Concern," held in BFLA.
13. "To Whom It May Concern," 8 December 1948, held in BFLA.
14. Z. A. Bhutto, "Response" to the Zia *White Papers*, typed, held in BFLA, p. 313.
15. Telephone interview with Joe Phillips in Grants Pass, Oregon, 29 June 1989.
16. Edward M. Mannon died at age 59 in early May of 1989, just one week before I called his office in San Francisco, hoping to interview him.
17. Telephone interview with Inder Chabra at his home in San Francisco, 29 May 1989.

18. Telephone interview with Dr. Leo Rose at his office in Berkeley, California, September 1989.

19. *PoP,* 1:32.

20. "The Indivisibility of the Human Race," University of California at Berkeley; the date given is 12 November 1948, but it should be 1949. In *PoP,* 1:22.

21. *Reflections.*

22. Begum Sahiba Nusrat Bhutto interview. The subsequent Begum Nusrat quotations are from the same interview.

23. Begum Hushmat (née Habibuddin) Hussain interview at her home in Rawalpindi, January 1990.

24. *Reflections.*

25. Carolyn to Zulfi, 13 October 1951, Berkeley, held in BFLA.

26. Mody, *Zulfi,* pp. 47–48.

27. *Reflections.*

28. Ibid.

Chapter 4: From Oxford to Karachi

1. Zulfikar Ali Bhutto, *My Execution* (London: Musawaat Weekly International, 1980), p. 7.

2. Mody, *Zulfi,* p. 51.

3. Mumtaz Ali Bhutto interview.

4. Begum Sahiba Nusrat Bhutto interview. Subsequent quotations are from the same interview.

5. Salmaan Taseer, *Bhutto: A Political Biography* (London: Ithaca Press, 1979), p. 31.

6. Muzaffar Husain interview at the home of mutual friends (Laris), in Karachi, 24 January 1990.

7. All these legal files are held in BFLA.

8. Eric Hurst's letter, 10 July 1989, from his home in Stratford-upon-Avon.

9. From barrister Hurst's statement about barrister Bhutto at the international seminar The Legacy of Prime Minister Bhutto, which he addressed in Karachi, in April of 1989.

10. S. N. Grant-Bailey to "Sir Sardar Shahnawaz Bhutto," August 28, 1953, Temple, London, held in BFLA.

11. *Biographical Encyclopedia of Pakistan* (Lahore: Moh. Saddiq Biographies Research Institute, 1969–70), p. 80.

12. Justice Dorab Patel interview at a mutual friend's home (Begum Ikramullah) in Karachi, 9 January 1990.

13. Solicitor Dingomal interview.

14. Governor Justice Fakruddin Ibrahim interview at the Governor's House in Karachi, 24 January 1990.

15. M. A. Jinnah, speech in Karachi Club, 9 August 1947, in *Speeches of Quaid-i-Azam, Mohammad Ali Jinnah as Governor General of Pakistan* (Karachi: Sind Observer Press, 1948), pp. 4–5.

16. Ibid., pp. 6–10.

17. See Jamna Das Akhtar, *Political Conspiracies in Pakistan: Liaquat Ali's Murder to Ayub Khan's Exit* (Delhi: Punjab Pushtak Bhandar, 1969).

18. "Unification of West Pakistan" by Z. A. Bhutto, Bar-at-Law, in a newspaper article held in BFLA.

19. Z. A. Bhutto, "On One Unit," press statement, Larkana, 24 November 1954, in *PoP,* 1:58.

20. Hamida Khuhro interview at the home of mutual friends (Laris) in Karachi, 24 January 1990.

21. Z. A. Bhutto, "The Distinction Between Political and Legal Disputes," in *PoP,* 1:45.

22. Z. A. Bhutto, "Pakistan: A Federal or Unitary State," Ibid., pp. 34–37.

23. Khalid B. Sayeed, *Politics in Pakistan: The Nature and Direction of Change* (New York: Praeger, 1980), p. 43.

24. Former Minister of Education G. M. Shah interview at his home in Islamabad, January 1990.

25. Z. A. Bhutto, "A Development for Democracy?" December 1956, in *PoP,* 1:66ff. Subsequent quotes are from this article, up to p. 73.

26. Firoz Khan Noon, *From Memory* (Lahore: Ferozsons, 1966), pp. 71–72.

27. Ibid., pp. 293–96.

28. UN Deputy Secretary General Yusuf Buch interview in the delegate's lounge of the UN Secretariat, New York, 17 July 1989.

29. Z. A. Bhutto, "Defining Aggression," New York, 25 October 1957, in *PoP,* 1:76.

30. Ibid., p. 79.

31. Ibid., p. 83.

Chapter 5: Apprenticeship to Power

1. Z. A. Bhutto to Major General Iskander Mirza, April 1958, held in BFLA.

2. Mohammad Ayub Khan, *Friends, Not Masters: A Political Autobiography* (London: Oxford University Press, 1967), pp. 54–57. Hereafter cited as Ayub's *Autobiography.*

3. Z. A. Bhutto, "Territorial Sea Limits," address to the First Committee of the United Nations Conference on the Sea, Geneva, 17 March 1958, in *PoP,* 1:94ff.

4. Ibid., pp. 99–102.

5. Begum Sahiba Nusrat Bhutto's interview. Subsequent quotations from Begum Nusrat from the same interview.

6. Benazir's *Autobiography,* p. 30.

7. Ayub's *Autobiography,* p. 57.

8. Firoz Khan Noon, *From Memory* (Lahore, Ferozsons 1966), p. 297.

9. Ayub's *Autobiography,* pp. 70–75.

10. Ibid.

11. Overheard by Professor Muhammad Reza Kazimi of St. Patrick's College,

Karachi, while Zulfi was speaking with J. A. Rahim and other PPP founders in Dacca's Intercontinental Hotel, and kindly reported to the author in Karachi on 10 January 1990.

12. Humayun Mirza to "My dear Zulfikar," from the World Bank, Washington, D.C., 10 March 1977, held in BFLA.

13. Mody, *Zulfi,* p. 61.

14. General Gul Hasan interview at his club in Rawalpindi, 24 January 1990.

15. Edgar A. Schuler and Kathryn R. Schuler, *Public Opinion and Constitution Making in Pakistan, 1958–1962* (Kalamazoo: Michigan State University Press, 1966), p. 24. The next two quotations are also from this source.

16. Ayub's *Autobiography,* p. 79.

17. Note from Minister Bhutto to President Ayub Khan, in *White Paper: The Performance of the Bhutto Regime,* vol. I, *Mr. Z. A. Bhutto, His Family and Associates* (Islamabad: Government of Pakistan, 1979), pp. 2–3.

18. Ayub's *Autobiography,* p. 207.

19. Bhutto PCH.

20. Ambassador Nasim Ahmad interview in the delegate's lounge of the UN Secretariat, New York, 17 July 1989.

21. Z. A. Bhutto, UN speech, 11 November 1959, in *PoP,* 1:105.

22. Ibid., p. 108.

23. Z. A. Bhutto, "On Tied Aid," speech in Karachi, 30 November 1959, ibid., pp. 112ff.

24. Ibid., p. 115.

25. Z. A. Bhutto to Field Marshal Mohammad Ayub Khan, 11 November 1959, in Zulfikar Ali Bhutto, *New Directions* (London: Namara Publications, 1980), pp. 112–14. Also held in BFLA.

26. Z. A. Bhutto to Foreign Minister Manzur Qadir, 11 November 1959, held in BFLA.

27. Lyndon Baines Johnson, *The Vantage Point: Perspectives of the Presidency, 1963–1969* (New York: Holt, Rinehart & Winston, 1971), p. 339.

28. Bhutto, *New Directions,* p. 120.

29. Z. A. Bhutto, "Pakistan-Soviet Oil Agreement," 23 March 1961, in *PoP,* 1:131.

30. Z. A. Bhutto, "Impressions of the United Nations," address to the Pakistan UN Association, Karachi, 22 May 1961, ibid., p. 139.

31. Ibid., p. 142.

32. Z. A. Bhutto, "Disarmament Problems," convocation address, Sind University, 30 March 1962, ibid., pp. 168–69.

33. Bhutto PCH. All quotes in the next paragraph from same source.

34. Z. A. Bhutto, "Role of Political Parties," speech in the National Assembly, 10 July 1962, in *PoP,* 1:170–80. Subsequent quotes from the same source.

35. Husna Sheikh interview at her flat in London, 19 June 1989. Subsequent Husna quotations are from the same interview.

36. Ibid.

37. Ardeshir and Nancy Cowasjee interview at their home on Mary Road in Karachi, 19 April 1989.

Chapter 6: Foreign Minister to the Field Marshal

1. Bhutto PCH.
2. Z. A. Bhutto, "The Sino-Pakistan Boundary Agreement," 26 March 1963, in *PoP,* 1:184–91.
3. Ibid., p. 188.
4. Z. A. Bhutto, "Reply to Nehru," Lahore, 14 July 1963, in *PoP,* 1:192–94.
5. Taseer, *Bhutto,* p. 50.
6. Foreign Minister Bhutto, speech in the National Assembly, concluding the foreign policy debate, 24 July 1963, in *PoP,* 1:195–205, quotation at p. 198.
7. Ibid., pp. 198–201.
8. Ibid., p. 202.
9. Bhutto PCH.
10. Rita Dar interview in New Delhi, 21 January 1990.
11. Z. A. Bhutto, "The UN and World Peace," speech to the Lions Club, Karachi, 21 November 1963, in *PoP,* 1:207–8.
12. Z. A. Bhutto to President of the Security Council, 16 January 1964; reprinted in *The Kashmir Question,* ed. K. Sarwar Hasan and Z. Hasan (Karachi: Pakistan Institute of International Affairs, 1966), pp. 427–31.
13. Z. A. Bhutto interview, BBC, London, 30 January 1964, in *PoP,* 1:211–12.
14. Sheikh Abdullah's statement of 14 June 1953, quoted in Hasan and Hasan, *The Kashmir Question,* p. 416.
15. Ibid., pp. 432–34. The next quote is from the same source.
16. Z. A. Bhutto, speech to National Press Club, Washington, D.C., 27 April 1964, *PoP,* 1:214.
17. Ayub's *Autobiography,* pp. 233–35.
18. Mumtaz Ali Bhutto interview.
19. Nancy Cowasjee interview at her home on Mary Road in Karachi, 19 April 1989.
20. Foreign Minister Z. A. Bhutto to President's Secretariat, 24 September 1964. Bhutto's minute on letter No. HC/TS/101/64/325, dated 19 September 1964, from High Commissioner for Pakistan, New Delhi.
21. See T. V. Kunhi Krishnan, *Chavan and the Troubled Decade* (Bombay: Somaiya Public Private, 1971), pp. 99–115.
22. Kuldip Nayar, *India after Nehru* (Delhi: Vikas, 1975), pp. 23–24.
23. File No. PI(D), 7/44/64, held in BFLA.
24. Hari Ram Gupta, *The Kutch Affair* (Delhi: U.C. Kapura & Sons, 1969), p. 141.
25. Gul Hasan interview.
26. Kuldip Nayar, *Between the Lines* (Bombay: Allied Publishers, 1969), p. 113.
27. Z. A. Bhutto, memo to President Ayub Khan, 27 May 1965, held in BFLA.
28. *Dawn* (Karachi), 17 June 1965, quoted in Gupta, *The Kutch Affair,* p. 294.
29. *Dawn* (Karachi), 23 June 1965, quoted ibid., p. 297.
30. *Dawn* (Karachi), 1 July 1965, quoted ibid., pp. 306–7.
31. Hasan and Hasan, *The Kashmir Question,* pp. 434–36.
32. Z. A. Bhutto, memo to President Ayub, summer of 1965, held in BFLA.
33. Z. A. Bhutto, "top secret" minute, in "India" volume held in BFLA.

34. Ibid.

35. Field Marshal Ayub Khan, 29 August 1965. Annexure G to GHQ letter No. 4050/5/MO-1, 1 August 73, held in BFLA.

36. Krishnan, *Chavan and the Troubled Decade,* p. 141.

37. Z. A. Bhutto, "Response" to Zia *White Papers,* pp. 88–89, held in BFLA.

38. Z. A. Bhutto, secret memo to President Ayub, held in BFLA.

39. Mumtaz Ali Bhutto interview.

40. Z. A. Bhutto, memo in "India" volume held in BFLA.

41. S. M. Burke and Lawrence Ziring, *Pakistan's Foreign Policy,* 2d ed. (Karachi: Oxford University Press, 1990), p. 340.

42. Z. A. Bhutto, "India's Aggression," speech to UN Security Council, in *PoP,* 1:221ff.

43. Ibid.

44. Z. A. Bhutto, "Self-determination and Kashmir," address to UN General Assembly, 28 September 1965, *PoP,* 1:228ff. Subsequent quotes from the same speech.

45. Foreign Minister to Foreign Secretary, Secret Cypher Message 7033, 12 October 1965, held in BFLA.

46. Secret cable No. 7105, Foreign Minister to Foreign Secretary, New York, 14 October 1965, held in BFLA.

47. Zulfikar Ali Bhutto, *The Myth of Independence* (Lahore and Karachi: Oxford University Press, 1969), p. 71.

48. Ibid., p. 76.

49. Z. A. Bhutto, "Cease-Fire Violations by India," address to UN Security Council, 25 October 1965, in *PoP,* 1:286.

50. Mumtaz Ali Bhutto interview.

Chapter 7: Winters of His Discontent

1. Kuldip Nayar, *India: The Critical Years* (Delhi: Vikas, 1971), p. 199.

2. Ibid., p. 207.

3. Taseer, *Bhutto,* p. 71.

4. Bhutto PCH.

5. Ibid.

6. Ibid.

7. Ibid.

8. President Ayub Khan to Prime Minister Indira Gandhi, from Camp Larkana, February 1966, held in BFLA.

9. Z. A. Bhutto, "On Indo-Pakistan War, 1965," address to the National Assembly, 16 March 1966, in *PoP,* 1:287–311. Subsequent quotations on this subject are from that speech.

10. Z. A. Bhutto to Ayub Khan, 11 April 1966, in Bhutto, *New Directions,* pp. 124–26.

11. Z. A. Bhutto to Ayub Khan, 11 May 1966, ibid., pp. 122–24.

12. Ibid., p. 124.

13. Roedad Khan interview at his home in Islamabad, 10 April 1989.
14. Bhutto PCH.
15. Taseer, *Bhutto,* pp. 76–77.
16. Z. A. Bhutto to Kazi Fazlullah, Beirut, July 15 1966, held in BFLA.
17. Bhutto PCH.
18. Pakistan People's Party, *Election Manifesto* (Lahore: Classic, 1970), pp. 5–6.
19. J. A. Rahim to "My dear Bhutto," Paris, 8 October 1966, held in BFLA.
20. Ibid.
21. Z. A. Bhutto, speech to All-Pakistan Students Federation, London, 13 August 1966, in *Politics of the People,* vol. 2, *Awakening the People,* (hereafter cited as *Awakening*) (Rawalpindi: Pakistan Publications, n.d.), pp. 4–15.
22. Ibid.
23. Ibid.
24. Bhutto PCH.
25. Z. A. Bhutto, "Pakistan and Nuclear Proliferation," Larkana, 29 December 1966, in *Awakening,* pp. 19–21.
26. Rounaq Jahan, *Pakistan: Failure in National Integration* (New York: Columbia University Press, 1972). See also Richard Sisson and Leo E. Rose, *War and Secession: Pakistan, India, and the Creation of Bangladesh* (Berkeley, University of California Press, 1990).
27. Z. A. Bhutto, "My Debut in Journalism," 12 January 1967, in *Awakening,* pp. 22–31.
28. Gul Hasan interview.
29. Bhutto, "My Debut in Journalism."
30. Z. A. Bhutto, "Starting with a Clean Slate," address to the Muzaffargarh Bar Association, 17 January 1968, in *Awakening,* pp. 43–50.
31. Pakistan People's Party, *Election Manifesto.*
32. Bhutto, "Starting with a Clean Slate," p. 45.
33. Ibid., p. 46.
34. G. W. Choudhury, *The Last Days of United Pakistan* (Bloomington: Indiana University Press, 1974), pp. 27ff.
35. Bhutto, "Starting with a Clean Slate," pp. 47–48.
36. Z. A. Bhutto, "We Shall Not Be Cowed," speech at Mirpur Khas, 18 February 1968, in *Awakening,* pp. 54–60.
37. That opinion concluded a heavy file from the Law Ministry on charges of "corruption" against "Mr. Z. A. Bhutto, a former Foreign Minister of Pakistan, [who] owned considerable waste land, . . ." 7 April 1968, held in BFLA.
38. Z. A. Bhutto, address to the Nawabshah Bar Association, 21 February 1968, in *Awakening,* pp. 61–70.
39. Ibid., p. 65.
40. Ibid., p. 66.
41. Z. A. Bhutto, "Incompetence Intensified Crisis," address to Khairpur Bar Association, 8 March 1968, in *Awakening,* pp. 71–76.
42. Ibid.

43. Z. A. Bhutto, "A New Class of Landlords," address to the Larkana Bar Association, 12 March 1968, in *Awakening*, pp. 80–88.

44. Ibid.

45. Ibid., pp. 84–85.

46. Z. A. Bhutto, "On Leaving the Government," address to Sind's first Pakistan People's Party convention, Hyderabad, 21 September 1968, in *Awakening*, pp. 119–33. Subsequent quotes are from the same speech.

47. Z. A. Bhutto, "Dictatorship Is Crumbling," Abbottabad, 29 October 1968, in *Awakening*, pp. 164–68. The next quote is from the same source.

48. Z. A. Bhutto, "The Struggle Continues," Peshawar, 5 November 1968, in *Awakening*, pp. 186–97.

49. Z. A. Bhutto, affidavit in Lahore High Court, 5 February 1969, in *Awakening*, pp. 201–29, quotation at p. 221.

50. Ibid., pp. 221–22.

51. *Pakistan Times*, 14 November 1968, held in BFLA.

52. Bhutto affidavit in Lahore High Court, p. 202.

53. Ibid., p. 203. The next quote is from the same page.

54. Mohammad Asghar Khan, *Generals in Politics: Pakistan, 1958–1982* (London and Canberra: Croom Helm, 1983), p. 13.

55. Taseer, *Bhutto*, p. 99.

56. Ibid., p. 100.

57. Quoted in Richard S. Wheeler, *The Politics of Pakistan: A Constitutional Question* (Ithaca: Cornell University Press, 1970), p. 274.

58. Choudhury, *The Last Days of United Pakistan*, p. 41.

59. Asghar Khan, *Generals in Politics*, pp. 14–15.

60. Choudhury, *The Last Days of United Pakistan*, p. 22.

61. Z. A. Bhutto to Raja Mohammad Sarwar Khan, 17 May 1968, held in BFLA.

62. Z. A. Bhutto to Col. Dost Mohammad, 17 May 1969, held in BFLA.

63. Col. Dost Mohammad to Mr. Z. A. Bhutto, KDML/16/A, 19 May 1969, held in BFLA.

64. Z. A. Bhutto to Col. Dost Mohammad, May 25 1969, held in BFLA.

65. Masud Mufti to Z. A. Bhutto, D.O. No. CB/272, 5 August 1969, held in BFLA.

66. Z. A. Bhutto to Masud Mufti, 12 September 1969, held in BFLA.

67. Ibid.

68. Ibid.

69. Husna Sheikh interview.

Chapter 8: Free Elections and the Birth of Bangladesh

1. Z. A. Bhutto, "Launching the Election Campaign," Karachi, 4 January 1970, in *Politics of the People*, vol. 3, *Marching Towards Democracy* (hereafter cited as *Marching* (Rawalpindi: Pakistan Publications, n. d.), pp. 3–8. Subsequent quotes are from the same source. For an excellent account of Bhutto's campaign and the election, see Anwar H. Syed, *The Discourse and Politics of Zulfikar Al: Bhutto* (New York: St. Martin's Press, 1992), pp. 67–87.

2. Z. A. Bhutto to Mr. Karam Elahi, Superintendent of Police, Larkana, 10 January 1970, held in BFLA.

3. Z. A. Bhutto, "The Change in Foreign Policy," speech in Rawalpindi, 17 January 17, 1970, in *Marching*, pp. 9–16.

4. Z. A. Bhutto, "Politics of the People," Peshawar, 18 January 1970, in *Marching*, pp. 17–21.

5. Z. A. Bhutto, "Basic Issues Are Economic," Mardan, 25 February 1970, in *Marching*, pp. 22–25.

6. Z. A. Bhutto, "Socialism Is Islamic Equality," Gjurat, 1 March 1970, in *Marching*, pp. 26–32.

7. Z. A. Bhutto, "India's Attack on Pakistan," Lahore, 8 March 1970, in *Marching*, pp. 33–43, at p. 42.

8. Z. A. Bhutto, "The Incident at Sanghar," Karachi, 12 April 1970, in *Marching*, pp. 44–49.

9. Z. A. Bhutto, *The Great Tragedy* (Karachi: Pakistan People's Party, September 1971), p. 14.

10. Mody, *Zulfi*, p. 106.

11. Choudhury, *The Last Days of United Pakistan*, p. 98.

12. Z. A. Bhutto, "The Main Issue Is Equality," Campbellpur, 19 April 1970, in *Marching*, pp. 50–57, quotation at p. 54.

13. Ibid.

14. Z. A. Bhutto, "A Long March for People's Rights," Abbottabad, 19 April 1970, in *Marching*, pp. 58–64, quotation at p. 61.

15. Z. A. Bhutto, "Constitution Not Final Goal," Kohat, 25 April 1970, in *Marching*, pp. 74–78, quotation at p. 78.

16. Z. A. Bhutto, "Beware of Vote Beggars," Peshawar, 27 April 1970, in *Marching*, pp. 84–89, quotation at pp. 84–85.

17. Z. A. Bhutto, "Inauguration of PPP Office," Quetta, 12 June 1970, in *Marching*, p. 117.

18. Z. A. Bhutto, speech in Quetta, 14 June 1970, in *Marching*, pp. 120–25.

19. Z. A. Bhutto, "Consulting the People," Malir, 5 August 1970, in *Marching*, pp. 130–34.

20. Z. A. Bhutto to Mr. O. M. Qarni, 16 August 1970, held in BFLA.

21. Z. A. Bhutto, "Prepare for a People's Government," Lahore, 14 October 1970, in *Marching*, pp. 135–42, quotation at p. 139.

22. Sheikh Mujibur Rahman, "Bengalees Shall Not Be Allowed to Turn Slaves," speech at Dumni, 20 October 1970, in *The Bangladesh Papers* (Vanguard Books, 1973), p. 100. Next quote from the same speech, pp. 101–2.

23. Z. A. Bhutto, speech in Lahore, 3 November 1970, in *Marching*, p. 147.

24. Z. A. Bhutto, "Towards a New Pakistan," Lahore radio and television speech, 18 November 1970, in *Marching*, pp. 152–60, quotation at pp. 153–54.

25. Ibid., pp. 155–56.

26. Z. A. Bhutto, "Thanking the Voters," Lahore, 12 December 1970, in *Marching*, pp. 161–65, quotation at p. 163.

27. Bhutto, *The Great Tragedy*, p. 20–21.

28. Ibid., p. 22.

29. He later insisted that all he had metaphorically meant, and actually said was that if they went to Dacca to attend a "National Assembly" meeting called by Mujib, they would "have no [political] legs to stand on."

30. Z. A. Bhutto, "Deadlock on the Constitution," Punjab University, Lahore, 22 February 1971, in *Marching*, pp. 166–70.

31. Ibid., pp. 169–70.

32. Ibid.

33. Reported in *The Pakistan Times*, 25 February 1971, Quoted in *The Bangladesh Papers*, pp. 170–77.

34. Z. A. Bhutto, "The Crisis: Two Alternatives," Lahore, 28 February 1971, in *Marching*, pp. 171–76.

35. Ibid.

36. Reported in *Morning News*, 2 March 1971, quoted in *The Bangladesh Papers*, pp. 188–89.

37. Mujibur Rahman, quoted in *The People* (Dacca), 2 March 1971, in *The Bangladesh Papers*, pp. 189–91.

38. Z. A. Bhutto, "Ready for Negotiations," Karachi, 2 March 1971, in *Marching*, pp. 177–79.

39. Mujibur Rahman, 2 March 1971, quoted in *The Bangladesh Papers*, p. 191.

40. Text of Yahya Khan's broadcast, ibid., pp. 214–16.

41. Ibid., pp. 217–18.

42. Z. A. Bhutto, "A Compromise Is Possible," Karachi, 14 March 1971, in *Marching*, pp. 181ff. The next quotes are from Ibid., pp. 184–186.

43. Ibid., p. 189.

44. Ibid., p. 191.

45. Ibid., p. 193.

46. Nawab Akbar Khan Bugti, reported in *Dawn* (Karachi), quoted in *The Bangladesh Papers*, pp. 237–38.

47. Z. A. Bhutto, press conference, Karachi, 15 March 1971, in *Marching*, pp. 194–95.

48. Mian Mumtaz Daultana, quoted in *Dawn* (Karachi), 17 March 1971, quoted in *The Bangladesh Papers*, pp. 249–50.

49. Mujibur Rahman, *Dawn* (Karachi), 16 March 1971, ibid., p. 248.

50. Bhutto, *The Great Tragedy*, p. 38–39.

51. Iqbal Ismay interview at the home of a mutual friend in Karachi, 25 January 1990.

52. Bhutto, *The Great Tragedy*, pp. 40–41.

53. Ibid., p. 41.

54. Ibid., p. 42.

55. Ibid., pp. 42–43.

56. Ibid., p. 43.

57. Ibid., p. 44.

58. Ibid., p. 45.

59. Ibid., p. 46.

60. Mujibur Rahman, quoted in *Dawn* (Karachi), 23 March 1971, *The Bangladesh Papers*, pp. 257–58.

61. Z. A. Bhutto, quoted ibid. (Karachi), 23 March 1971, ibid., p. 260.
62. Yahya Khan, quoted in *The Pakistan Times* (Lahore), 23 March 1971, ibid., p. 261.
63. Mujibur Rahman, quoted in *Morning News,* 23 March 1971, ibid., p. 262.
64. Tajuddin Ahmed, quoted in *Dawn* (Karachi), 25 March 1971, ibid., p. 266.
65. Mujibur Rahman, quoted ibid., pp. 266–67.
66. Bhutto, *The Great Tragedy,* p. 48.
67. Ibid., p. 50.
68. Asghar Khan, *Generals in Politics,* p. 34.
69. Bhutto, *The Great Tragedy,* pp. 50–51.
70. Stanley Wolpert, *Roots of Confrontation in South Asia* (New York: Oxford University Press, 1982), p. 151.
71. Yahya Khan, *Dawn* (Karachi), 27 March 1971, in *The Bangladesh Papers,* p. 277.
72. Transcript of the taped meeting between Z. A. Bhutto and Mujibur Rahman, 27 December 1971, held in BFLA.
73. Bhutto, *The Great Tragedy,* pp. 52–53.
74. Husna Sheikh interview.
75. Kaiser Rasheed to "Dear Mr. Bhutto," 14 April 1971, held in BFLA.
76. Choudhury, *The Last Days of United Pakistan,* pp. 190–93.
77. Bhutto PCH.
78. Minutes of the meeting held on 29 July 1971, in Karachi, between President Yahya Khan and his aides, and Mr. Z. A. Bhutto and his advisers, held in BFLA.
79. Ibid.
80. Z. A. Bhutto to Lt. Gen. Peerzada, 2 September 1971, held in BFLA.
81. Z. A. Bhutto to Gen. Agha Mohammad Yahya Khan, President of Pakistan, Rawalpindi, 8 September 1971, held in BFLA.
82. Z. A. Bhutto, "Power Must Be Transferred," 8 September 1971, in *Marching,* pp. 202–3.
83. Z. A. Bhutto, "Let Democracy Return," Karachi, 11 September 1971, in *Marching,* pp. 204–8.
84. Z. A. Bhutto, press statement, Karachi, 29 September 1971, in *Marching,* pp. 209–18, quotation at p. 216.
85. Z. A. Bhutto, "Designs Against the People," Multan, 8 October 1971, in *Marching,* pp. 225–30, quotation at p. 228.
86. Ibid., p. 229.
87. Quoted in Stanley Wolpert, *A New History of India* (New York: Oxford University Press, 1977), pp. 402–3.
88. "N.S. to Dear Sir," 25 November 1971, typed, held in BFLA.
89. Z. A. Bhutto to General Yahya Khan, Camp, Rawalpindi, *Confidential,* 28 November 1971, held in BFLA.
90. Ibid., p. 2.
91. Z. A. Bhutto to Lt.-Gen. Peerzada, Camp, Peshawar, 2 December 1971, held in BFLA.
92. Z. A. Bhutto to General Ayub Khan, Camp, Rawalpindi, 3 December 1971, held in BFLA.

Chapter 9: President Bhutto "Picks Up the Pieces"

1. Henry Kissinger, *White House Years* (Boston: Little, Brown, 1979), p. 907.
2. Ibid., pp. 907–8.
3. Z. A. Bhutto, speech to UN Security Council, 12 December 1971, in *Marching,* pp. 231–60, quotation at p. 232.
4. Ibid., p. 234.
5. General Niazi, quoted in Robert Payne, *Massacre* (New York: Macmillan, 1973), pp. 121–22.
6. Bhutto, speech to UN Security Council, 12 December 1971, p. 246.
7. Ibid., pp. 251–53.
8. Ibid., p. 256.
9. Ibid., p. 260.
10. Z. A. Bhutto, "My Country Beckons Me," address to UN Security Council, 15 December 1971, in *Marching,* pp. 268–76.
11. Benazir's *Autobiography,* p. 52.
12. Gavin Young, "How Dacca Fell: The Inside Story," *Observer* (London), 19 December 1971, typed out, held in BFLA.
13. Taseer, *Bhutto,* p. 130.
14. Roedad Khan interview.
15. Hafeez Pirzada interview at his home in Karachi, 7 April 1989.
16. Gul Hasan interview.
17. Ibid.
18. Z. A. Bhutto, "Address to the Nation," 20 December 1971, in Z. A. Bhutto, *Speeches and Statements,* vol. 1, *December 20, 1971–March 31, 1972* (Karachi: Government of Pakistan, 1972), pp. 1–16, quotation at p. 1.
19. Ibid., pp. 3–4.
20. Ibid., p. 5.
21. Ibid., p. 7.
22. Ibid., pp. 15–16.
23. Bhutto PCH.
24. Transcribed tape of Z. A. Bhutto's private conversation with Sheikh Mujibur Rahman, 27 December 1971, held at BFLA.
25. Transcript of the tape of their second private meeting in the same house of "detention," held in BFLA.
26. Taseer, *Bhutto,* p. 133.
27. Z. A. Bhutto, "Address to the Nation," 2 January 1972, in Bhutto, *Speeches and Statements,* 1:33–34.
28. Z. A. Bhutto, "Speech delivered at Public Meeting," Karachi, 3 January 1972 ibid., pp. 35–55, quotation at pp. 37–38.
29. Ibid., p. 51.
30. Ibid., p. 45–46.
31. Ibid., pp. 46–47.
32. Z. A. Bhutto, letter to ministers and governors, 6 January 1972, ibid., p. 58.
33. Z. A. Bhutto, "Statement on Convening Provincial Assemblies," 22 January 1972, ibid., p. 59.

34. Z. A. Bhutto, "Address to Businessmen and Industrialists," Karachi, 24 January 1972, ibid., pp. 66–67.

35. Z. A. Bhutto, speech in People's Great Hall, Peking, 1 February 1972, ibid., p. 72.

36. Z. A. Bhutto, "Address to the Nation," 10 February 1972, ibid., pp. 75–78.

37. Z. A. Bhutto to "My dear Brother" Talpur, *Confidential,* 28 February 1972, held in BFLA.

38. Z. A. Bhutto, "Address to the Nation," 1 March 1972, in Bhutto, *Speeches and Statements,* 1:98–106, quotation at p. 98.

39. Ibid., pp. 100–101.

40. Yahya Bakhtiar interview in Karachi, April 1989.

41. Z. A. Bhutto, "Address to the Nation," 1 March 1972, in Bhutto, *Speeches and Statements,* 1:106.

42. Z. A. Bhutto, "Address to the Nation," 3 March 1972, ibid., pp. 108–9.

43. Ibid.

44. General Tikka Khan interview at the Governor's House in Lahore, 17 April 1989.

45. Ibid.

46. Shahid Javed Burki, *Pakistan under Bhutto, 1971–1977,* 2d ed. (London: Macmillan 1988), p. 70.

47. Tikka Khan interview.

48. Z. A. Bhutto, "Address to the Nation," 3 March 1972, in Bhutto, *Speeches and Statements,* 1:110–11.

49. *A Defence Plan for Pakistan,* held in BFLA.

50. Z. A. Bhutto, "Address to the Nation," 3 March 1972, in Bhutto *Speeches and Statements,* 1:112.

51. Ibid., p. 113.

52. Ibid., p. 115.

53. President Z. A. Bhutto to the Chief of Staff, Army, 23 June 1972, held in BFLA.

54. Ibid.

55. Z. A. Bhutto, speech in Lahore, 19 March 1973, in Bhutto, *Speeches and Statements,* 1:137–40.

56. Ibid., pp. 140–141.

57. Z. A. Bhutto, address to the National Assembly, 14 April 1972, in Bhutto, *Speeches and Statements,* vol. 2, *April 1, 1972–June 30, 1972* (Karachi: Government of Pakistan, 1972), pp. 18–48, quotation at p. 47.

58. Z. A. Bhutto, speech at oath-taking ceremony, Rawalpindi, 21 April 1972, ibid., pp. 49–53.

59. President Z. A. Bhutto to "Dear Mir Sahib" (R. B. K. Talpur), 28 April 1972, held in BFLA.

60. President Z. A. Bhutto to "My dear Mumtaz" (M. A. Bhutto), 1 May 1972, No. PSP-4(7)72, held in BFLA.

61. Oriana Fallaci, *Interview with History,* trans. John Shepley (Boston: Houghton Mifflin, 1976), p. 163; following quotation, pp. 188–89.

62. Ibid., pp. 190–91. The following quote is from p. 200.

63. Z. A. Bhutto, address to the nation, 27 June 1972, *Pakistan Horizon* 25 (Third Qtr. 1972):111.

64. Benazir's *Autobiography,* p. 53.

65. Mody, *Zulfi,* p. 146.

66. Ibid., p. 141.

67. Ibid., p. 148.

68. Benazir's *Autobiography,* p. 57.

69. Z. A. Bhutto, speech at Lahore airport, 3 July 1972, *Pakistan Horizon* 25 (Third Qtr. 1972):118–19.

70. Ibid., pp. 119–20.

71. Z. A. Bhutto, address at Rawalpindi airport, 4 July 1972, in Z. A. Bhutto, *Speeches and Statements,* vol. 3, *July 1, 1972–September 30, 1972* (Karachi: Government of Pakistan, 1972), pp. 10–15.

72. Z. A. Bhutto, speech in National Assembly, 14 July 1972, ibid., pp. 20–68, quotation at p. 22. Following quotation from pp. 24–25.

73. Ibid., p. 34.

74. Ibid., p. 35.

75. Husna Sheikh interview.

76. Z. A. Bhutto, address to the National Assembly, 14 July 1972, *Speeches and Statements,* 3:45–46.

77. Z. A. Bhutto, "The Simla Accord," address to Pakistan Institute of International Studies, Karachi, 21 July 1972, *Pakistan Horizon* 25 (Third Qtr. 1972):3–16.

78. President Z. A. Bhutto to Secretary General Defence; Chief of Staff, Army; Chief of Staff, Air; Chief of Staff, Navy, "Top Secret," August 1972, held in BFLA.

79. President Z. A. Bhutto, address to National Assembly, 14 July 1972, *Speeches and Statements,* 3:56–57.

80. Ibid., p. 58.

Chapter 10: Provincial Problems Proliferate

1. Z. A. Bhutto, speech to National Assembly, 14 July 1972, in Bhutto, *Speeches and Statements,* 3:60.

2. Ibid., p. 65.

3. Ibid., p. 66.

4. Ibid., p. 67.

5. Told to me by Justice A. R. Cornelius during interview in Faletti's Hotel, Lahore, 1989.

6. Z. A. Bhutto, address to the nation, 15 July 1972 in Bhutto, *Speeches and Statements,* 3:69–71, quotation at p. 70.

7. Z. A. Bhutto, address to the nation, 16 July 1972, ibid., p. 74.

8. Ibid.

9. Husna Sheikh interview.

10. Z. A. Bhutto, "Address to PPP Workers," Karachi, 20 July 1972, Bhutto, *Speeches and Statements,* 3:83.

11. Z. A. Bhutto, "Address to Senior Officials of Sind Province," Karachi, 22 July 1972, ibid., p. 87.

12. Ibid., p. 89.

13. President Bhutto to Chief Minister Bhutto of Sind, 18 August 1972, held in BFLA.

14. President Bhutto to Chief Minister Bhutto of Sind, 21 August 1972, held in BFLA.

15. President Bhutto to Chief Minister Bhutto of Sind, 2 September 1972 from "CAMP KARACHI," held in BFLA.

16. Z. A. Bhutto, "Message on Quaid-i-Azam's death Anniversary," 11 September 1972, in Bhutto, *Speeches and Statements,* 3:207–8.

17. Home Minister Abdul Qaiyum Khan to President Z. A. Bhutto, 3 May 1972, held in BFLA.

18. Reported by Asrar Ahmad, "Cracks in the Ruling Party," *Outlook,* 23 September 1972, p. 4.

19. Z. A. Bhutto interview with Editor Van Rosmalen of *Elsevier's Magazine* (Amsterdam), in Rawalpindi, 1 October 1972, in Bhutto, *Speeches and Statements,* vol. 4, *October 1, 1972–December 31, 1972* (Karachi: Government of Pakistan, 1973), p. 2.

20. Saied Ahmad Khan to President Z. A. Bhutto, 7 September 1972, held in BFLA.

21. President Bhutto to Mian Anwar Ali, 12 September 1972, held in BFLA.

22. Z. A. Bhutto, address at inauguration of Karachi's nuclear power plant, 28 November 1972, in Bhutto, *Speeches and Statements,* 4:162.

23. Z. A. Bhutto television interview with H. K. Burki and Safdar Qureshi, 20 December 1972, ibid., p. 241.

24. Ibid., p. 243.

25. Z. A. Bhutto, speech at Karachi public meeting, 3 January 1973, in Bhutto, *Speeches and Statements,* vol. 5, *January 1, 1973–March 31, 1973* (Karachi: Government of Pakistan, 1973), p. 7.

26. Ibid., p. 10.

27. Bhutto PCH. The following quotation is from the same source.

28. Z. A. Bhutto, "Proclamation by the President on Baluchistan," 15 February 1973, in Bhutto, *Speeches and Statements,* 5:81.

29. Z. A. Bhutto, address to the National Assembly, 23 February 1973, ibid., p. 9.

30. Z. A. Bhutto, "Pakistan Day Message," 23 March 1973, ibid., p. 144.

31. Z. A. Bhutto, aide memoire, 4 April 1973, in Bhutto, *Speeches and Statements,* vol. 6, *April 1, 1973–August 13, 1973* (Karachi: Government of Pakistan, 1973), p. 15.

32. Ibid., pp. 26–27.

Chapter 11: Foreign Triumphs, Domestic Tragedies

1. Z. A. Bhutto, message on Iqbal Day, 21 April 1973, in Bhutto, *Speeches and Statements,* 6:47.

2. Z. A. Bhutto, address at Pakistan Military Academy, 21 April 1973, ibid., pp. 49–50.

3. Z. A. Bhutto, address to the National Assembly, 9 July 1973, in Bhutto, *Speeches and Statements,* 6:129.

4. Ibid., p. 132.

5. Ibid., pp. 134–36.

6. Ibid., p. 137.

7. Ibid.

8. President Bhutto's "Secret" memo to Governor Khar of Punjab, G. M. Jatoi, and Rafi Raza, 6 July 1973, held in BFLA.

9. Z. A. Bhutto to Nawab Mohammad Akbar Khan Bugti, Governor of Baluchistan, "Confidential: To Be Opened by Addressee Only," 10 May 1973, held in BFLA.

10. President Z. A. Bhutto to Governor Akbar Bugti, 9 June 1973, held in BFLA.

11. "Anonymous" Report on Baluchistan, enclosed by Governor M. A. K. Bugti in his letter of 3 August 1973 to President Z. A. Bhutto, held in BFLA.

12. Z. A. Bhutto, after-dinner speech at the Savoy, London, 24 July 1973, in Bhutto, *Speeches and Statements,* 6:172–73.

13. Z. A. Bhutto, address to the nation, 14 August 1973, in Bhutto, *Speeches and Statements,* vol. 7, *August 14, 1973–December 31, 1973,* (Karachi: Government of Pakistan, 1974), p. 1.

14. Ibid., p. 5.

15. Z. A. Bhutto, address to the nation, 13 September 1973, ibid., pp. 51–54.

16. Z. A. Bhutto, speech in White House, 18 September 1973, ibid., p. 63.

17. Joint Pakistan-U.S. statement, 20 September 1973, ibid., pp. 79–80.

18. Z. A. Bhutto, address to UN General Assembly, 20 September 1973, ibid., pp. 81–91, quotation at p. 84.

19. Ibid., p. 85.

20. Ibid., p. 87.

21. Ibid., pp. 88–89.

22. Z. A. Bhutto, on "Meet the Press," NBC, 23 September 1973, ibid., pp. 102–10, quotation at p. 109.

23. Z. A. Bhutto, at Islamabad Airport, 27 September 1973, ibid., p. 117.

24. Z. A. Bhutto, to POWs on their return to Pakistan, 28 September 1973, ibid., p. 121.

25. Masood Mahmood interview.

26. Z. A. Bhutto, address to service chiefs, 11 October 1973, in Bhutto, *Speeches and Statements,* 7:128–29.

27. Prime Minister Benazir Bhutto interview at the Prime Minister's House in Rawalpindi, 9 April 1989.

28. Bhutto PCH.

29. Z. A. Bhutto to King Faisal, 16 October 1973, in Bhutto, *Speeches and Statements,* 7:152.

30. Z. A. Bhutto, press conference, Karachi, 20 October 1973, ibid., pp. 154–55. The following quote is from Ibid., p. 156.

31. Mir Ali Ahmad Talpur interview, by M. B. Naqvi, in *Outlook,* 27 October 1973, pp. 5–6.

32. "Bugti" to "Dear Mr. Prime Minister," Karachi, 31 October 1973 holograph letter, held in BFLA.

33. Z. A. Bhutto, speech after dinner honoring Dr. Henry Kissinger, at Prime Minister's House, Rawalpindi, 9 November 1973, in Bhutto, *Speeches and Statements,* 7:175–76.

34. Z. A. Bhutto, speech in Larkana, 30 November 1973, ibid., p. 192.

35. Z. A. Bhutto interview with Dr. Satish Kumar of *Hindustan Times,* 28 December 1973, ibid., p. 288.

36. Zeba Bakhtiar interview in Karachi, April 1989.

37. Chief Minister Mumtaz Ali Bhutto of Sind to Prime Minister Z. A. Bhutto, "Top Secret" letter of 16 October 1973, held in BFLA.

Chapter 12: Prime Minister Bhutto at the Peak of His Power

1. Vina Mody interview in Los Angeles, 1989.

2. Z. A. Bhutto to jurists' conference, quoted in *Outlook,* 23 February 1974, p. 7.

3. A. K. Brohi, quoted ibid., p. 9.

4. Bhutto, *New Directions,* pp. 75–92, quotation at pp. 79–80.

5. Ibid., pp. 81–83.

6. Ibid., p. 85.

7. Ibid., pp. 91–92.

8. Pakistan-UAE joint communiqué, 13 March 1974, in *Pakistan Horizon* 27 (Second Qtr. 1974):137.

9. Tehmina Durrani, *My Feudal Lord.* (New Delhi: Sterling Publishers, 1991).

10. Hanif Ramay interview at his home in Lahore, April 1989.

11. Abdul Ghaffar Khan interview with Yusuf Lodi, quoted in *Outlook,* 13 April 1974, p. 4.

12. Vice-Premier Teng, speech at Beijing banquet to honor Prime Minister Bhutto, in *Pakistan Horizon* 27 (Second Qtr. 1974):149.

13. Z. A. Bhutto, speech at Vice-Premier Teng's banquet, 12 May 1974, ibid., p. 150.

14. Z. A. Bhutto, statement on India's nuclear explosion, 19 May 1974, ibid., pp. 131–34.

15. Mujibur Rahman's speech quoted in *Dawn* (Karachi), 28 June 1974, reprinted in *Pakistan Horizon* 27 (Third Qtr. 1974):190–91.

16. Z. A. Bhutto, speech at Dacca banquet, 27 June 1974, ibid., pp. 191–92.

17. Husna Sheikh interview.

18. J. A. Rahim's own account of his assault on 3 July 1974 in Islamabad. The typescript, hand-corrected by Rahim, was given to me by Ardeshir Cowasjee from his personal archives on Mary Road, Karachi, 1989. This account of Rahim's brutal beating was corroborated by his son, Sikander Rahim, who works for the World Bank, and called me from Washington, D.C., in April 1992, confirming from his personal knowledge everything to which his father had earlier attested.

19. Bhutto PCH.
20. Taseer, *Bhutto,* p. 177.
21. Air Marshal (ret.) Zulfikar Ali Khan interview in his home in Islamabad, 1989.
22. Hanif Ramay interview.
23. Prime Minister Bhutto's speech to the National Assembly, 26 August 1974, in *Pakistan Horizon* 27 (Third Qtr. 1974):166–83, quotation at p. 177.
24. Z. A. Bhutto interview with Jack Reynolds of NBC, Pindi, 23 September 1974, in *Pakistan Horizon* 27 (Fourth Qtr. 1974):159.
25. Ibid., pp. 163–64.
26. Z. A. Bhutto, on Baluchistan, 15 October 1974, ibid., p. 168.
27. Z. A. Bhutto, speech in Moscow, 26 October 1974, ibid., p. 180.
28. Z. A. Bhutto, speech at Kissinger banquet, 31 October 1974, ibid., pp. 186–87.
29. Ahmad Raza Kasuri IB report, p. 00336, held in BFLA.
30. Taseer, *Bhutto,* p. 177.
31. Masood Mahmood interview.
32. Zulfikar Ali Bhutto, *My Execution.* (London: Musawaat Weekly International, 1980), pp. 79–80.
33. Taseer, *Bhutto,* p. 178. Reconfirmed to me by Salmaan Taseer at his home in Lahore, January 1990.
34. Z. A. Bhutto, press conference, Foreign Press Association of Pakistan, 20 December 1974, in *Pakistan Horizon* 28 (First Qtr. 1975):149–65, quotation at pp. 157–60.

Chapter 13: From "Leader of Pakistan's People" to "Leader of the Third World"?

1. Typed report of the tribunal's investigation into the Kasuri murder, held in BFLA.
2. Z. A. Bhutto to G. M. Khar, first typed in early January 1975, and "revised" on 30 January 1975, held in BFLA.
3. Abdul Huq to "My dear Prime Minister," Dacca, 16 December 1974, sent through Mahmud Ali, No. PS/MS/8-75, Islamabad, 6 January 1975, held in BFLA.
4. Abdul Malek to "My dear Mr. Bhutto," from Saudi Arabia, 22 January 75, held in BFLA.
5. Z. A. Bhutto, reply to David Rockefeller letter of 24 February 1975, held in BFLA.
6. G. M. Khar to author at Lahore airport VIP waiting room, 1989.
7. Jam Sadiq Ali to PM Z. A. Bhutto, 31 December 74, held in BFLA.
8. Ibid., p. 3.
9. Mumtaz Ali Bhutto to Z. A. Bhutto, Prime Minister, 26 January 1975, held in BFLA.
10. Mumtaz Ali Bhutto to PM Z. A. Bhutto, "TOP SECRET," 20 January 1975, held in BFLA. Following quotation from the same source.
11. Mumtaz Ali Bhutto to PM Z. A. Bhutto, 28 February 75, held in BFLA.

12. Mumtaz Ali Bhutto to PM Z. A. Bhutto, 5 October 75, held in BFLA.

13. A. W. Katpur to PM A. Z. Bhutto ("Jiay Bhutto" PPP, SIND, Karachi), 4 April 1975, held in BFLA.

14. Z. A. Bhutto, statement in *Dawn* (Karachi), 26 February 1975, reproduced in *Pakistan Horizon* 28 (First Qtr. 1975):171.

15. Z. A. Bhutto, press conference, 10 March 1975, in *Pakistan Horizon* 28 (Second Qtr. 1975): 119–20.

16. Ibid., p. 123.

17. Z. A. Bhutto, speech to the National Assembly, 27 March 1975, ibid., p. 144.

18. Z. A. Bhutto, speech at banquet in Karachi, 13 April 1975, ibid., p. 151.

19. Vice-Premier Li Hsien-nien, speech, Lahore, 23 April 1975, in *Pakistan Horizon* 28 (Third Qtr. 1975):129.

20. Z. A. Bhutto interview with David Frost, Rawalpindi, 15 May 1975, ibid., p. 145.

21. Ibid., p. 146.

22. Morarji Desai interview with Oriana Fallaci, 26 June 1975, New Delhi, published in *New Republic,* reprinted ibid., p. 160.

23. Z. A. Bhutto, press conference, 15 June 1975, ibid., p. 155.

24. Attorney General Yahya Bakhtiar, opening address in the Supreme Court of Pakistan, in the Reference by The Islamic Republic of Pakistan on Dissolution of NAP, Rawalpindi, 19, 20, and 23 June 1975.

25. Yahya Bakhtiar's Rejoinder in the Supreme Court of Pakistan to written statement of Mr. Abdul Wali Khan, in Reference by The Islamic Republic of Pakistan on the Dissolution of NAP, Rawalpindi, 27 August 1975, pp. 4–5.

26. Anthony Mascarenhas, *Bangladesh: A Legacy of Blood.* (London: Hodder & Stoughton, 1986), pp. 73–74.

27. Z. A. Bhutto, statement on 15 August 1975, in *Pakistan Horizon* 28 (Third Qtr. 1975):150.

28. Begum Suhrawardy to "My dear Prime Minister Saheb," Karachi, 21 September 1975, held in BFLA.

29. Z. A. Bhutto, interview in Bucharest, Rumania, 24 October 1975, ibid., pp. 225–26.

30. Z. A. Bhutto, national broadcast, 10 November 1975, *Pakistan Horizon* 28 (Fourth Qtr. 1975):227–28.

31. Ibid., p. 230.

32. Z. A. Bhutto, broadcast to the nation, 21 December 1975, in *Pakistan Horizon* 29 (First Qtr. 1976):125–26.

33. Bhutto PCH.

Chapter 14: Prelude to New National Elections

1. Z. A. Bhutto, interview in *Egyptian Gazette and Mail,* 29 January 1976, reprinted in *Pakistan Horizon* 29 (First Qtr. 1976):150–52.

2. Bhutto PCH.

3. Bhutto's interview, 29 January 1976, reprinted in *Pakistan Horizon* 29 (First Qtr. 1976):157.

4. Ambassador Henry Byroade's interview at his home in Maryland, 20 July 1989.

5. Saied Ahmad Khan to Prime Minister Z. A. Bhutto, held in BFLA.

6. General Faiz Ali Chishti, *Betrayals of Another Kind: Islam, Democracy and the Army in Pakistan* (London: Asia Publishing House, 1989), p. 27.

7. Wajid S. Hasan interview in Karachi, 15 April 1989.

8. Gul Hasan interview.

9. Benazir's *Autobiography,* p. 68.

10. Benazir Bhutto interview.

11. Gul Hasan interview.

12. Bhutto's PCH.

13. Z. A. Bhutto, inaugural speech to the opening session of the International Seerat Congress, Rawalpindi, 3 March 1976, in *Pakistan Horizon* 29 (Second Qtr. 1976):201.

14. Z. A. Bhutto, banquet speech for the shah of Iran, 9 March 1976, ibid., p. 175.

15. Bhutto PCH.

16. Z. A. Bhutto, "RCD: Challenge and Response," in *Pakistan Horizon* 29 (Second Qtr. 1976):3–7.

17. Ibid., pp. 11–12.

18. "TOP SECRET" summary for Z. A. Bhutto of his official visit to China, 26–30 May 1976, "Summit-level Talks Between PM Z. A. Bhutto and Premier Hua Kuafeng of PRC," held in BFLA.

19. Ibid.

20. Z. A. Bhutto, speech in Kabul, 10 June 1976, in *Pakistan Horizon* 29 (Third Qtr. 1976):151.

21. Z. A. Bhutto, "Rejoinder" (to Zia's *White Papers*), held in BFLA.

22. Z. A. Bhutto, address to Minorities Conference, 13 June 1976, ibid., pp. 138–39.

23. Ibid., pp. 140–42.

24. Mrs. Dina Wadia to "Dear Mr. Bhutto," New York, 2 August 1976, held in BFLA.

25. Z. A. Bhutto to "Dear Chief Minister," 9 September 1976, held in BFLA.

26. *Bhutto Says,* comp. and ed. Bashir Ahmed (Karachi: Motherland Press, [reprinted] September 1976), p. iii.

27. Z. A. Bhutto, broadcast to the nation, 17 July 1976, in *Pakistan Horizon* 29 (Third Qtr. 1976):144–45.

28. Ibid., p. 146.

29. Ibid., p. 148.

30. Ibid., pp. 148–49.

31. Hanif Ramay interview.

32. Z. A. Bhutto to General M. Zia ul-Haq, 1976, held in BFLA.

33. Z. A. Bhutto to "Dear Dr. Kissinger," 21 November 1976, held in BFLA.

34. "TOP SECRET" draft, minutes of a meeting between Z. A. Bhutto and Mr. MacLellan Canadian ambassador to Pakistan, 28 June 77, held in BFLA.

35. Z. A. Bhutto, message to Prime Minister S. Bandaranaike, 15 August 1976, in *Pakistan Horizon* 29 (Third Qtr. 1976):167.

36. Z. A. Bhutto to Prime Minister S. Bandaranaike, 7 September 1976, held in BFLA.

37. Minister Aziz Ahmad, speech in Mexico City, 16 September 1976, in *Pakistan Horizon,* 29 (Fourth Qtr. '76,) pp. 269–73.

38. Z. A. Bhutto, statement on Mao's death, 9 September 1976, ibid., p. 262.

39. Z. A. Bhutto, message to the nation, 25 December 1976, in *Pakistan Horizon* 30 (First Qtr. 1977):170–71.

40. Z. A. Bhutto, address to the nation, 5 January 1977, ibid., pp. 183–84.

41. Ibid., pp. 187–88.

Chapter 15: New Elections and Their Tragic Aftermath

1. Asghar Khan, *Generals in Politics,* p. 103.

2. Z. A. Bhutto to Chief Minister Sadiq Hussain Qureshi, 2 February 1977, held in BFLA.

3. Henry Byroade interview.

4. DIB Sheik's "SECRET/IMMEDIATE" cover letter to Bhutto, 19 February 1977, held in BFLA. Underlined in original.

5. Rao A. R. Khan to Bhutto, *SECRET,* 23 February 1977, held in BFLA.

6. Ibid.

7. DG ISI Lt.-Gen. Ghulam Jilani Khan to Chief of Army Staff General M. Zia-ul-Haq, Rawalpindi, 3 December 1976, copy to Mr. Afzal Said Khan, PM's Sectt (Pub), PM's House, Rawalpindi, held in BFLA.

8. Z. A. Bhutto, marginalia holograph, ibid.

9. "Zia" to "My dear Prime Minister," GHQ, Rawalpindi, 24 November 1976 (17/3/COAS (P) *SECRET*), held in BFLA.

10. General M. Zia-ul-Haq, 5 February 1977, No. 4852/320/PS-1 (b), held in BFLA.

11. Z. A. Bhutto, inaugural speech to Seerat Conference, Rawalpindi, 19 February 1977, in *Pakistan Horizon* 30 (First Qtr. 1977):190–91.

12. Z. A. Bhutto, statement, 9 March 1977, *Pakistan Horizon* 30 (Second Qtr. 1977):164.

13. Z. A. Bhutto to "My dear Mufti Sahib," 13 March 1977, held in BFLA.

14. Mufti Mahmood, president PNA, to "My dear Bhutto Sahib," Peshawar, 14 March 1977, held in BFLA.

15. Shahinshah Pahlevi to "Dear Mr. Prime Minister," 22 March 1977, held in BFLA.

16. Mufti Mahmud to "My dear Bhutto Sahib," 17 March 1977, in *Pakistan Horizon* 30 (Second Qtr. 1977):165–66.

17. PM Z. A. Bhutto to "My Dear Mufti Sahib," 19 March 1977, ibid., pp. 166–67.

18. Ibid., p. 168.

19. PM Z. A. Bhutto to Justice S. A. Jan, 21 March 1977, ibid., p. 172.

20. Draft minutes of "*TOP SECRET*" meeting between Prime Minister Z. A. Bhutto and Mr. Halil Basol, Turkish commerce minister, 28 March 1977, held in BFLA.

21. PM Morarji Desai to "My Dear Mr. Prime Minister" (Z. A. Bhutto), New Delhi, 18 April 1977, held in BFLA.

22. Former Air Marshal Rahim Khan's wire to Z. A. Bhutto, Madrid, 13 April 1977. I thank Begum Rahim Khan for this wire, a copy of which was given to me during her interview at her home in Maryland, 3 March 1992.

23. Z. A. Bhutto, statement, 20 April 1977, *Pakistan Horizon* 30 (Second Qtr. 1977):92.

24. Draft minutes of PM Z. A. Bhutto's "TOP SECRET" meeting with Sheikh Riyadh Al-Khatib, 20 April 1977, held in BFLA.

25. 21 April 1977 interview granted to Ambassador Al-Khatib, ibid.

26. "SECRET/IMMEDIATE" report of 28 May 1977 public meeting in Birmingham, from Ch. Noor Hussain to "My Dear Prime Minister" (Z. A. Bhutto), 31 May 1977, held in BFLA.

27. Interview granted by PM Bhutto to Ambassador Al-Khatib, 21 April 1977, ibid.

28. "SECRET/MOST IMMEDIATE," Rao A. Rashid Khan to PM Z. A. Bhutto, 21 April 1977, held in BFLA. The following quotation is from the same source.

29. Asghar Khan, *Generals in Politics,* p. 112.

30. Draft minutes of "TOP SECRET" meeting between PM Bhutto and Ambassador Al-Khatib, 30 April 1977, held in BFLA.

31. The original of this letter was shown to me by Begum Rahim Khan and contains the word "violated," which I have underlined, rather than "vitiated," which was used when Asghar Khan's *Generals in Politics* was published five years later (p. 117, last paragraph).

32. Rao A. Rashid Khan to PM Bhutto, 17 May 1977, PM's Secretariat (Public) corresp. 1361, held in BFLA.

33. DIB Rao A. Rashid Khan, "SECRET/IMMEDIATE," 3 June 1977, No. IH-6308-V. (13), held in BFLA.

34. Maulana I. H. Thanvi to PM Z. A. Bhutto, 26 May 1977, ibid.

35. Mian Abdul Hayee to "My dear Zulfi," 11 April 1977, ibid.

36. Mr. Abayakoon, to PM Z. A. Bhutto, 15 June 1977, ibid.

37. S. A. Saeed to "My dear Mr. Bhutto," Lahore, 14 May 1977, ibid.

38. Draft minutes of "*TOP SECRET*" meeting between PM Z. A. Bhutto and Ambassador S. A. Azimov of the Soviet Union, 9 May 1977, ibid.

39. Chishti, *Betrayals of Another Kind,* p. 51.

40. Mufti Mahmood to Mr. Z. A. Bhutto, Sihala Camp Prison, 12 May 1977, held in BFLA.

41. Draft minutes of "*TOP SECRET*" meeting between PM Bhutto and Ambassador Al-Khatib, 26 May 1977, held in BFLA.

42. Draft minutes of "*TOP SECRET*" "interview" to Ambassador Al-Khatib granted by PM Z. A. Bhutto, 27 May 1977, ibid.

43. "*EYES ONLY*" DIB Rao A. Rashid Khan to the Prime Minister, "*TOP SECRET/IMMEDIATE,*" 10 June 1977, ibid.

44. Draft minutes of "*TOP SECRET*" meeting between PM Bhutto and Mr. Umit Bayulken, secretary general of CENTO, 6 April 1977, ibid.

45. Draft minutes of "*TOP SECRET*" meeting between PM Bhutto and Ambassador S. A. Azimov, 16 June 1977, 11:00 A.M., ibid.

46. Ibid.

47. Draft minutes of "*TOP SECRET*" meeting between PM Bhutto and King Khalid in Riyadh, 18 June 1977, ibid.

48. Draft minutes of "*TOP SECRET*" audience granted by PM Bhutto to Mr. Nihat Dinc, Turkish ambassador, 27 June 1977, ibid.

49. Bhutto's "Rejoinder" to Zia's *White Papers,* p. 288, held in BFLA.

50. Benazir's *Autobiography,* p. 80. The following quotes in this paragraph are from Ibid., pp. 81–82.

51. Chishti, *Betrayals of Another Kind,* p. 64.

52. Ibid., p. 74.

Chapter 16: Zulfi's Fall—From Martial Coup to Martyrdom

1. General Zia ul-Haq, address to the nation, 5 July 1977, in *Pakistan Horizon,* 30 (Third and Fourth Qtrs. 1977):210.

2. Ibid., pp. 211–12.

3. Asghar Khan, *Generals in Politics,* pp. 127, 142.

4. Zia interview in *Newsweek,* reprinted ibid., pp. 228–29.

5. Ibid., p. 229.

6. Chishti, *Betrayals of Another Kind,* p. 18.

7. Ibid., p. 74.

8. Bhutto's PCH.

9. Tehmina Durrani, *My Feudal Lord* (New Delhi: Sterling, 1991), p. 238.

10. Zia press conference, 14 July 1977, *Pakistan Horizon* 30 (Third and Fourth Qtrs.):214–15.

11. Ibid., pp. 218–19.

12. Ibid., p. 219.

13. Ibid., pp. 219–20.

14. Ibid., p. 221.

15. Z. A. Bhutto to "My dear Ambassador" (Azimov), 14 July 1977, Murree; in holograph at bottom of the letter, "For security reasons I have asked my man to bring this letter back in original after you have noted its contents." Held in BFLA.

16. Z. A. Bhutto to H. E., ambassador of the People's Republic of China, Murree, 14 July 1977, held in BFLA.

17. Asghar Khan, *Generals in Politics,* p. 129.

18. Masood Mahmood interview.

19. Afzal Saeed Khan's statement, recorded by Deputy Director (FIA) Saad Aharif, 2 September 1977, held in BFLA.

20. Chishti, *Betrayals of Another Kind,* p. 78.

21. Durrani, *My Feudal Lord,* p. 244.

22. Zia, address to the nation, in *Pakistan Horizon* 30 (Third and Fourth Qtrs. 1977):231–32.

23. Ibid., p. 234.

24. Benazir's *Autobiography,* p. 89.

25. Zia, speech of 14 August 1977, in *Pakistan Horizon* 30 (Third and Fourth Qtrs. 1977):237.

26. Ibid., p. 238.

27. The first *White Paper, On The Conduct of the General Elections in March 1977,* was published by the government of Pakistan, in Rawalpindi, July 1978; the second, was *Misuse of Media (20 December 1971–4 July 1977),* published in August 1978; and the third, *The Performance of the Bhutto Regime,* was published in January 1979 and comprised four volumes: vol. 1, *Mr. Z. A. Bhutto, His Family and Associates;* vol. 2, *Treatment of Fundamental Institutions;* vol. 3, *Misuse of the Instruments of State Power;* and vol. 4, *The Economy.*

28. Chishti's *Betrayals of Another Kind,* p. 20.

29. Yahya Bakhtiar interview.

30. French Police conducted a thorough autopsy. Benazir Bhutto, however, argued that "There was a strong possibility, if not probability, that Shah had been killed by agents of the regime." Benazir's *Autobiography,* p. 254.

31. Zia, address to the nation, 6 September 1977, in *Pakistan Horizon* 30 (Third and Fourth Qtrs. 1977):245.

32. Zia interview by Hassan Kaleemi, reprinted from *Dawn* (Karachi), 11 September 1977, ibid., p. 248.

33. Zia, address to political leaders, 13 September 1977, ibid., pp. 249–50.

34. Z. A. Bhutto, chairman of the PPP, to Begum Nusrat Bhutto, acting chairman, PPP, 13 September 1977, held in BFLA.

35. Mumtaz Ali Bhutto interview.

36. Benazir's *Autobiography,* p. 97.

37. Mumtaz Ali Bhutto interview.

38. Zia, press statement, 17 September 1977, in *Pakistan Horizon* 30 (Third and Fourth Qtrs. 1977):251–52.

39. Asghar Khan, *Generals in Politics,* p. 134.

40. Zia, address to the nation, 1 October 1977, in *Pakistan Horizon* 30 (Third and Fourth Qtrs. 1977):253–56.

41. Z. A. Bhutto, Reply—44 (p. 80) in "Rejoinder" filed before the Supreme Court of Pakistan on 10 October 1977 to accusations made by Zia—"Respondent"—and filed on his behalf by A. K. Brohi in written form.

42. "Constitutional Petition" No. I-R of 1977 in the Supreme Court of Pakistan, Original Jurisdiction, Begum Nusrat Bhutto v. The Chief of Army Staff, etc. Rejoinder to the Reply of the Respondent Dated 26 October 1977. Z. A. Bhutto's charges concluded in Kot Lakhpat Jail, Lahore, 31 October 1977, held in BFLA.

43. S. Anwarul Haq, C. J., *Supreme Court Judgement* "on Begum Nusrat Bhutto's Petition, challenging the Detention of Mr. Z. A. Bhutto and others." Lahore, 10 November 1977, pp. 9–11.

44. Ibid., p. 38.

45. Ibid., p. 101.

46. Bhutto, *My Execution,* pp. 31–32.

47. Benazir's *Autobiography,* p. 107.

48. Zia interview by Sandy Gall of ITV, London, 23 March 1978, in *Pakistan Horizon* 31, nos. 2 and 3 (1978):234.

49. Bhutto's PCH.

50. Anwarul Haq interview at his home in Lahore, 1989.

51. A. R. Cornelius interview at his hotel apartment in Lahore, 1989.

52. Bhutto's PCH.

53. Roedad Khan interview.

54. Yahya Bakhtiar's interview.

55. Benazir Bhutto's typed and hand-corrected letter to Prison Authorities, appealing for permission for Dr. Niazi to see her father in his death cell, held in BFLA.

56. Dr. Z. Niazi interview in Karachi, April 1989.

57. Benazir's *Autobiography,* p. 3.

58. Nazar Mohammed, quoted in Benazir's *Autobiography,* p. 12.

59. Ibid., pp. 4–5.

Bibliography

Letters and Documents: Public and Private

The most important primary archive for Z. A. Bhutto's life is preserved in his private library at 70, Clifton, Karachi. The Bhutto Family Library and Archive contains many steel trunks full of copies of public and private documents and letters retained by Prime Minister Bhutto in his Karachi home, as well as the originals of letters sent to him both from abroad and from Pakistan. Most of the letters and documents are in files, some labeled, others unlabeled, though many remain loose in envelopes or manila folders. There is no catalogue or index of Bhutto Family Library and Archive holdings, but the papers and letters alone are extensive, well over 50,000 pages, possibly more than 100,000.

From the time that he entered government service, Z. A. Bhutto kept copies of official papers he drafted and received, a practice he continued throughout his more than half a decade as president and prime minister. After his last incarceration, Prime Minister Bhutto wrote extensively in his prison cell, and the original holograph (Bhutto's Prison Cell Holograph, PCH) is preserved in the Bhutto Family Library and Archive. The library itself contains more than 10,000 volumes, including most of the books Z. A. Bhutto wrote, and his very fine collection of Napoleana.

Khalid Shamsul Hasan, the founding president of the Shamsul Hasan Foundation for Historical Studies and Research in Karachi, has carefully preserved many original Quaid-i-Azam Jinnah letters at his foundation's home, including one of Zulfi Bhutto's first letters, written to Mr. Jinnah in 1945. Khalid Hasan has also collected many rare books on the cultural history of Sind, to which I was granted full access.

Ardeshir Cowasjee's fine library and private archive of Pakistani history at his home on Mary Road in Karachi was also generously made available to me for research whenever I visited Karachi, and there too I was given access to one of Zulfi Bhutto's first postcards.

Thanks to Professor Sharif al Mujahid, then director of the Quaid-i-Azam Academy in Karachi, I had full access to a complete set of I. H. Burney's weekly *Outlook,* as well as to other newspapers and journals published during Bhutto's era in high office. The *Outlook* was begun in April of 1972 and unfortunately forced to stop publication in mid-1974, because of its outspoken critique of government leaders.

Education Minister S. G. M. Shah kindly made his private library with its collection of the *Sind Quarterly,* which he founded in 1973, available to me during my visit to Islamabad. Other important primary source series that I have used are the *Pakistan Horizon,* a quarterly journal published by the Pakistan Institute of International Affairs in Karachi, which reproduces many documents in each issue. I consulted every issue from volume 20 through volume 30. I also used *India News,* published weekly by the Information Service of the Embassy of India, Washington, D.C., volumes 10–14, and the *Indian and Foreign Review,* a weekly published in New Delhi, volumes 7–14.

A "Freedom of Information" letter of inquiry to the U.S. Department of State yielded several hundred recently declassified "Confidential" telegrams and other documents concerning Z. A. Bhutto, from 1968 to 1979. The Central Intelligence Agency refused, however, "either to confirm or deny" the "existence" of any information about him in any of their archives or files, in response to another "Freedom of Information" letter from me.

Zulfikar Ali Bhutto's Books, Articles, Speeches, and Other Published Statements

Peace-Keeping by the United Nations. Karachi: Pakistan Publishing House, 1967.

Political Situation in Pakistan. New Delhi: Veshasher Prakashan, 1968.

The Myth of Independence. Lahore and Karachi: Oxford University Press, 1969.

The Great Tragedy. Karachi: Pakistan People's Party, 1971.

Politics of the People (speeches, statements, and articles). Edited by Hamid Jalal and Khalid Hasan. Vol. 1, *Reshaping Foreign Policy, 1948–1966.* Vol. 2, *Awakening the People, August 1966–December 1969.* Vol. 3, *Marching towards Democracy, January 1970–December 1971.* Rawalpindi: Pakistan Publications, n.d.

Speeches and Statements. Vol. 1, December 20, 1971–March 31, 1972; vol. 2, April 1, 1972–June 30, 1972; vol. 3, July 1, 1972–September 30, 1972; vol. 4, October 1, 1972–December 31, 1972; vol. 5, January 1, 1973–March 31, 1973; vol. 6, April 1, 1973–August 13, 1973; vol. 7, August 14, 1973–December 31, 1973; vol. 8, January 1, 1974–December 31, 1974. Karachi: Government of Pakistan, 1972–75.

Bilateralism: New Directions. Islamabad: Government of Pakistan, 1976.

The Third World: New Directions. London: Quartet Books, 1977.

My Pakistan. New Delhi: Biswin Sadi Publications, 1979.

If I Am Assassinated. . . . New Delhi: Vikas, 1979.

My Execution. London: Musawaat Weekly International, 1980.

New Directions. London: Namara Publishers, 1980.

Published Documents

Election Manifesto of the Pakistan People's Party, 1970. Lahore: Classic, 1971.
The Bangladesh Papers. Vanguard Books, 1973.
The Constitution of the Islamic Republic of Pakistan. Passed by the National Assembly of Pakistan on the 10th April 1973 and Authenticated by the President of the National Assembly on the 12th April 1973.
The State vs Zulfikar Ali Bhutto: Lahore High Court Judgement and Two Supreme Court Judgements. Karachi: Maz, n.d.
Constitutional Petition No.I-R of 1977 in the Supreme Court of Pakistan, Original Jurisdiction, *Begum Nusrat Bhutto v. The Chief of Army Staff,* et al., *Rejoinder to the Reply of the Respondent,* 26 October 1977.

Interviews

Pakistan

Prime Minister Benazir Bhutto
Begum Sahiba Nusrat Bhutto
Governor and Chief Minister (Sind) Mumtaz Ali Bhutto
Begum Manna (née Bhutto) Islam
Foreign Minister Sahabzada Yaqub Khan
Minister Abdul Hafeez Pirzada
Attorney General Yahya Bakhtiar
Chief Minister (Sind) G. M. Jatoi
Chief Minister (Baluchistan) Nawab Mohammad Akbar Bugti
Chief Minister (Baluchistan) Ataullah Mengal
Governor and Chief Minister (Punjab) G. Mustafa Khar
Governor (Sind) Justice Fakruddin Ibrahim
Chief Minister (Punjab) Hanif Ramay
Chief Minister (Sind) Jam Sadiq Ali
Governor (Punjab) General Tikka Khan
Minister Roedad Khan
Minister Rafi Raza
Minister S. G. M. Shah
Minister Dr. Mubashir Hasan
Speaker Malik Meeraj Khalid
Ambassador Abida Hussain
Ambassador Air Marshal Zulfiqar Ali Khan
Ambassador Ehsan Rashid
Chief Justice A. R. Cornelius
Chief Justice Anwarul Haq
Justice Dorab Patel
Justice Riaz Ahmad Sheikh
Minister Syed Fakhar Imam
Justice Naseem Shah
Ambassador Sirdar Izzat Hyat-Khan
Ambassador Mian Mumtaz Daultana
The Hon. Sikandar Hayat Khan
Mayor Hakim M. Ahsan
General Gul Hasan
Admiral S. M. Ahsan
Lieutenant General S. G. M. Peerzada
Lieutenant General Fazle Haq
Brigadier Noor Hussain
Begum Hashmut Hussain
Begum Shaista Suhrawardy Ikramullah
Begum Saida Razi Isa
Khalid Shamsul Hasan
Wajid Shamsul Hasan
Dr. Sajjed Shamsul Hasan
Dr. (Miss) K. F. Yusuf
Khurshid Hasan Meer, Esq.
Salmaan Taseer
Dr. Jami Mehta
Omar Qureshi
Ardeshir Cowasjee
Nancy Cowasjee

Begum Hamida Khuhro
Dr. Zafar Niazi
Begum Z. Niazi
Khalid Butt
Begum Rehana Islam
Gulgee
Yusuf Lodi (Vai ell)
Dr. A. Haye Saeed
Begum Salima H. Agha
Kamal Azfar, Esq.
Begum K. Azfar
Suhail Lari
Yasmeen Lari
Tariq Islam
Rafi Munir
Dr. Riaz Ahmad
Dr. Sikandar Hayat
Omar Kalim
Prof. M. R. Kazimi
Dr. Rahimutullah
Muzaffar Husain
Dr. Badr Siddiqi
Rizvan Kehar
Begum Rizvan (née Bhutto) Kehar
Solicitor Dingomal, Esq.
Professor Afak Haydar
Dr. M. Yusuf Abbasi
Khurshid Hadi
Professor M. Naeem Qureshi
Zeba Bakhtiar
Ambassador Robert Oakley
Consul General Joe Melrose
Consul Helena Finn
Consul Kent Obee

India

The Hon. Piloo Mody
Mrs. Vina Mody
Foreign Minister Inder K. Gujral
Maharaja Krishna Rasgotra
Foreign Minister Raja Dinesh Singh
Khushwant Singh
Patwant Singh
Maharaja Madhukar Shah
Begum Mehru Rahim Khan
Dr. Pran Chopra
Dr. Prabha Chopra
Madame Vijaya Lakshmi Pandit
Mrs. Rita Dar
Chief Minister S. Nijalingappa
Chief Minister Farooq Abdullah
General Jagjit Singh Aurora
Dr. Rajendra Sareen
Minister K. C. Pant
Justice R. S. Narula
Ambassador Shankar Bajpai
Ambassador S. K. Lambah
Ambassador Dr. Karan Singh
Ambassador Kewal Singh
Foreign Minister Swaran Singh
Justice Dr. Nagendra Singh
Ambassador William Clark

London

Husna Sheikh
Sir Cyril Henry Philips
Dr. Z. H. Zaidi

United States of America

Ambassador Jamsheed K. A. Marker
Mrs. Arnaz Marker
Ambassador Nasim Ahmad
Deputy Secretary General Yusuf Buch
Consul General S. Hadi Raza Ali
Ambassador Henry Byroade
Ambassador J. Kenneth Galbraith
Assistant Secretary of State Teresita Schaffer
Professor Howard Schaffer

Deputy Assistant Secretary of State Nancy Ely-Raphel
Professor Ralph Braibanti
Professor Hafeez Malik
Professor Holden Furber
Professor Anwar H. Syed
Dr. Shahid Javed Burki
Dr. Leo Rose
Sikander Rahim
Elizabeth Horowitz, Esq.
Dr. Leili Bakhtiyar
Begum Mehru Rahim Khan
Vina Mody
Joyce Hundel
Masood Mahmood
Professor Shafik Hashmi
Jehangir Jal Mugaseth
Al Cechvala
Mary Ellen Denton
Joe Phillips
Inder Chabra
Dr. Craig Baxter
William A. Codus

Books and Articles

Abdullah, Farooq, and Sati Sahni. *My Dismissal.* Delhi: Vikas, n.d.

Afzal, M. Raffque. *Political Parties in Pakistan, 1947–1958.* Islamabad: National Commission on Historical and Cultural Research, 1976.

Ahmad, Aijaz. "Democracy and Dictatorship in Pakistan." *Journal of Contemporary Asia* 8, no. 4 (1978):477–88.

Ahmed, Akbar S. *Discovering Islam.* London: Routledge, 1989.

Ahmed, Manzooruddin, ed. *Contemporary Pakistan.* Karachi: Royal Book, 1980.

Akbar, Ghulam. *He Was Not Hanged.* Lahore: Midas, 1989.

Akbar, M. J. *India: The Siege Within.* Harmondsworth: Penguin, 1985.

Akhtar, Jamnadas. *Political Conspiracies in Pakistan.* Delhi:Punjab Pushtak Bhandar, 1969.

Ali, Chaudhri Muhammad. *The Emergence of Pakistan.* New York, Columbia University Press, 1967.

Ali, Tariq. *Pakistan: Military Rule or People's Power?* Harmondsworth: Penguin, 1983.

Aman, Akhtar. *Pakistan and the Challenge of History.* Lahore: Universal, n.d.

Azad, Maulana Abul Kalam. *India Wins Freedom: The Complete Version.* New Delhi: Orient, Longman, 1989.

Azfar, Kamal. *Pakistan: Political and Constitutional Dilemmas.* Karachi: Pakistan Law House, 1987.

Batra, J. C. *The Trial and Execution of Bhutto.* Delhi: Kunj, 1979.

Bazaz, Nagin. *Ahead of His Times: Prem Nath Bazaz.* New Delhi: Sterling, n.d.

Bazaz, Prem Nath. *The History of the Struggle for Freedom in Kashmir.* Islamabad: Government of Pakistan, 1954.

Beg, Aziz. *Battle of Ballot or War of Attrition: Perils of Polls in Pakistan.* Islamabad: Babur and Amer, 1978.

Bhatia, H. S. *Portrait of a Political Murder.* New Delhi: Deep & Deep, 1979.

Bhutto, Benazir. *Daughter of the East: An Autobiography.* London: Hamish Hamilton, 1988.

————. *The Way Out.* Karachi: Mahmood, 1988.

Braibanti, Ralph. *Research on the Bureaucracy of Pakistan.* Durham: Duke University Press, 1969.
Brodie, Fawn. *The Devil Drives: A Life of Sir Richard Burton.* New York: Norton, 1967.
Broomfield, John H. *Elite Conflict in a Plural Society.* Berkeley and Los Angeles: University of California Press, 1968.
Burke, S. M., and Lawrence Ziring. *Pakistan's Foreign Policy: An Historical Analysis.* 2d ed. Karachi: Oxford University Press, 1990.
Burki, Shahid Javed. *Pakistan under Bhutto, 1971–1977.* 2d ed. London: Macmillan, 1988.
Burki, Shahid Javed, and Craig Baxter, with contributions by Robert LaPorte, Jr., and Kamal Azfar. *Pakistan under the Military: Eleven Years of Zia ul-Haq.* Boulder: Westview Press, 1989.
Butani, D. H. *The Future of Pakistan.* New Delhi: Promilla, 1984.
Channo, Sahib Khan. *The Movement for Separation of Sind From the Bombay Presidency, 1847–1937.* Ph.D. Thesis, University of Sind, 1983.
Chishti, Lt. Gen. Faiz Ali. *Betrayals of Another Kind.* London: Asia, 1989.
Chopra, Pran. *India's Second Liberation.* Delhi: Vikas, 1973.
Chopra, P. N., chief ed. *Towards Freedom, 1937–47,* vol. 1. New Delhi: Indian Council of Historical Research, 1985.
————, ch. ed., assisted by Prabha Chopra. *The Collected Works of Sardar Vallabhbhai Patel,* vols. 1–3. Delhi: Konark, 1990–93.
Chaudhari, Mohammed Ahsen. *Pakistan and the Great Powers.* Karachi: Council for Pakistan Studies, 1965.
Choudhury, G. W. *Pakistan's Relations with India, 1947–1966.* London: Pall Mall Press, 1968.
————. *The Last Days of United Pakistan.* Bloomington: Indiana University Press, 1974.
Clark, Ramsay. "Trial of Ali Bhutto and the Future of Pakistan." *Nation,* 19 August 1978, pp. 136–40.
Duncan, Emma. *Breaking the Curfew: A Political Journey through Pakistan.* London: Arrow, 1990.
Durrani, Tehmina. *My Feudal Lord.* New Delhi: Sterling, 1991.
Election Manifesto of the Pakistan People's Party, 1970. Lahore: Classic, 1971.
Fallaci, Oriana. *Interview with History.* Translated by John Shepley. Boston: Houghton Mifflin, 1976.
Feldman, Herbert. *Revolution in Pakistan.* London: Oxford University Press, 1967.
————. *From Crisis to Crisis, 1962–1969.* London: Oxford University Press, 1972.
————. *The End of the Beginning: Pakistan, 1969–1971.* London: Oxford University Press, 1975.
Goodnow, Henry Frank. *The Civil Service of Pakistan.* New Haven: Yale University Press, 1964.
Gopinath, Meenakshi. *Pakistan in Transition.* New Delhi: Manohar, 1975.
Gulraj, J. P. *Sindh and Its Sufis.* London, 1979.
Gupta, Hari Ram. *The Kutch Affair.* Delhi: U. C. Kapura & Sons, 1969.

Gupta, Sisir. *Kashmir: A Study in India-Pakistan Relations*. Bombay: Asia, 1966.
Hasan, Khalid. *The Umpire Strikes Back*. Lahore: Vanguard Books, n.d.
Hasan, Khalid Shamsul. *Quaid-i-Azam's Unrealised Dream*. Karachi: Shamsul Hasan Foundation for Historical Studies and Research, 1991.
Hasan, K. Sarwar and Z. Hasan, eds. *The Kashmir Question*. Karachi: Pakistan Institute of International Relations, 1966.
Haq, S. Anwarul, C. J. *Supreme Court Judgement on 'Begum Nusrat Bhutto's Petition, Challenging the Detention of Mr. Z. A. Bhutto and others.'* Lahore, 10 November 1977.
Jahan, Rounaq. *Pakistan: Failure in National Integration*. New York: Columbia University Press, 1972.
Jalal, Ayesha. *The State of Martial Rule*. Cambridge: Cambridge University Press, 1990.
Jinnah, M. A. *Speeches of Quaid-i-Azam Mohammad Ali Jinnah as Governor General of Pakistan*. Karachi: Sind Observer Press, 1948.
Johnson, Lyndon Baines. *The Vantage Point: Perspectives of the Presidency, 1963–1969*. New York: Holt, Rinehart & Winston, 1971.
Kak, B. L. *Z. A. Bhutto's Notes from the Death Cell*. New Delhi: Rada Krishna Prakashan, 1979.
Kamal, K. L. *Pakistan: The Garrison State*. New Delhi: Intellectual, 1982.
Kaushik, S. N. *Pakistan under Bhutto's Leadership*. New Delhi: Uppal, 1985.
Kennedy, Charles H. *Bureaucracy in Pakistan*, Karachi: Oxford University Press, 1987.
Kissinger, Henry. *White House Years*. Boston: Little, Brown, 1979.
Khan, Mohammad Asghar. *Generals in Politics: Pakistan, 1958–1982*. London and Canberra: Groom Helm, 1983.
Khan, Mohammad Ayub. *Friends, Not Masters: A Political Autobiography*. London: Oxford University Press, 1967.
Khan, Rahmatullah. *Kashmir and the United Nations*. New Delhi: Vikas, n.d.
Khan, Air Marshal Zulfiqar Ali. *Pakistan's Security: The Challenge and the Response*. Lahore: Progressive, 1988.
Krishnan, T. V. Kunhi. *Chavan and the Troubled Decade*. Bombay: Somaiya, 1971.
Kumar, Satish. *The New Pakistan*. New Delhi: Vikas, 1978.
LaPorte, Jr., Robert. "Pakistan in 1972: Picking up the Pieces." *Asian Survey* 13 (January–June 1973):187–98.
Lari, Yasmeen. *Traditional Architecture of Thatta*. Karachi: Heritage Foundation, 1989.
Malik, Hafeez, ed. *Iqbal: Poet-Philosopher of Pakistan*. Washington, D.C.: Public Affairs Press, 1963.
———. *Muslim Nationalism in India and Pakistan*. Washington, D.C.: Public Affairs Press, 1963.
Mankekar, D. R. *Homi Mody: A Many Splendoured Life*. Bombay: Popular Prakashan, 1968.
Mansingh, Surjit. *India's Search for Power*. New Delhi: Sage, 1984.
Mascarenhas, Anthony. *Bangladesh: A Legacy of Blood*. London: Hodder & Stoughton, 1986.

Mody, Piloo. *Zulfi, My Friend*. Delhi: Thomson Press, 1973.
Moinuddin. *Sind: Land of Legends*. Karachi: National Book Fund, 1975.
Mujahid Sharif, al. *Ideological Orientation of Pakistan*. Islamabad: National Book Fund, 1976.
————. *Quaid-i-Azam Jinnah: Studies in Interpretation*. Karachi: Quaid-i-Azam Academy, 1978.
Mukerjee, Dilip. *Zulfiqar Ali Bhutto: Quest for Power*. New Delhi: Vikas, 1972.
Mullik, B. N. *My Years with Nehru, 1948–1964*. Bombay: Allied, 1972.
Munir, Muhammad, C. J. (ret.). *From Jinnah to Zia*. Lahore: Vanguard Books, 1980.
Nayar, Kuldip. *Between the Lines*. Bombay: Allies, 1969.
————. *The Critical Years*. Delhi: Vikas, 1971.
————. *India after Nehru*. Delhi: Vikas, 1975.
Noon, Firoz Khan. *From Memory*. Lahore: Ferozsons, 1966.
Palmer, Norman D. "The Two Elections: A Comparative Analysis." *Asian Survey* 17 (July 1977):648–66.
Payne, Robert. *Massacre*. New York: Macmillan, 1973.
Rahman, Sheikh Mujibur. *Bangladesh, My Bangladesh*. New Delhi: Orient, Longman, 1972.
Rahman, General M. Attiqur. *Back to the Pavilion*. Karachi: Ardeshir Cowasjee, 1990.
Rasa, Mian Sayed Rasul. *The Architect of New Pakistan*. Rawalpindi: Sarhad, 1977.
Sahni, Naresh Chander. *Political Struggle in Pakistan*. Jullunder: New Academic, 1969.
Sareen, Rajendra. *Pakistan—The India Factor*. New Delhi: Allied, 1984.
Sayeed, Khalid B. *Pakistan: The Formative Phase*. Oxford: Oxford University Press, 1968.
————. *Politics in Pakistan: The Nature and Direction of Change*. New York: Praeger, 1980.
Schuler, Edgar A., and Kathryn R. Schuler. *Public Opinion and Constitution Making in Pakistan, 1958–1962*. Kalamazoo: Michigan State University Press, 1966.
Shah, Sayid Ghulam Mustafa. *The British in the Sub-Continent*. Karachi: Sindhi Kitab Ghar, 1989.
Sham, Mahmud. *Larkana to Peking*. Karachi: National Book Foundation, 1976.
Sharma, B. L. *The Kashmir Story*. Bombay: Asia, 1967.
Sengupta, Jyoti. *Bangladesh in Blood and Tears*. Calcutta: Naya Prokash, 1981.
Schofield, Victoria. *Bhutto: Trial and Execution*. London: Cassell, 1979.
Singh, Karan. *Sadar-i-Riyasat: An Autobiography*. Vol. 2, *1953–1967*. Delhi: Oxford University Press, 1985.
Singh, Khushwant. *Ranjit Singh: Lion of Punjab*. London, 1962.
Singhal, Damodar P. *Pakistan*. Englewood Cliffs, N.J.: Prentice-Hall,
Sisson, Richard, and Leo E. Rose. *War and Secession: Pakistan, India, and the Creation of Bangladesh*. Berkeley & Los Angeles: University of California Press, 1990.
Syed, Anwar H. *Pakistan: Islam, Politics, and National Solidarity*. New York: Praeger, 1982.

———. *The Discourse and Politics of Zulfikar Ali Bhutto*. New York: St. Martin's Press, 1992.

———. "Pakistan in 1976: Business as Usual." *Asian Survey* 17 (February 1977):181–90.

———. "Pakistan in 1977: The 'Prince' Is under the Law." *Asian Survey* 18 (February 1978):117–125.

———. "Z. A. Bhutto's Self-Characterizations and Pakistani Political Culture." *Asian Survey* 18 (November 1978):125–66.

Taseer, Salmaan. *Bhutto: A Political Biography*. London: Ithaca Press, 1979.

Weinbaum, M. G. "The March 1977 Elections in Pakistan. Where Everyone Lost." *Asian Survey* 17 (July 1977):599–618.

Wheeler, Richard S. *The Politics of Pakistan*. Ithaca: Cornell University Press, 1970.

White Paper: The Conduct of the General Elections in March 1977. Rawalpindi: Government of Pakistan, July 1978.

White Paper: Misuse of Media (20 December 1971–4 July 1977). Rawalpindi: Government of Pakistan, August 1978.

White Paper: The Performance of the Bhutto Regime: vol. 1, *Mr. Z. A. Bhutto, His Family and Associates;* vol. 2, *Treatment of Fundamental Institutions;* vol. 3, *Misuse of the Instruments of State Power;* and vol. 4, *The Economy*. Islamabad: Government of Pakistan, January 1979.

Wilcox, Wayne A. *Pakistan: The Consolidation of a Nation*. New York: Columbia University Press, 1963.

Wolpert, Stanley. *Jinnah of Pakistan*. New York: Oxford University Press, 1984.

———. *A New History of India*. 4th ed. New York: Oxford University Press, 1993.

———. *Morley and India, 1906–1910*. Berkeley and Los Angeles: University of California Press, 1967.

———. *Massacre at Jallianwala Bagh*. New Delhi: Penguin, 1988.

———. *Roots of Confrontation in South Asia: Afghanistan, Pakistan, India and the Superpowers*. New York: Oxford University Press, 1982.

Wriggins, Howard, ed. *Pakistan in Transition*. Islamabad: Quaid-i-Azam University Press, 1975.

Yusuf, Hamid. *Pakistan in Search of Democracy, 1947–77*. Lahore: Afrasia, n.d.

Zaman, Fakhar, and Akhtar Aman. *Z. A. Bhutto: The Political Thinker*. Lahore: People's, 1973.

Ziaullah, Syed, and Samuel Baid. *Pakistan: An End without a Beginning*. New Delhi: Lancer International, 1985.

Ziring, Lawrence. *The Ayub Era: Politics in Pakistan, 1958–1969*. Syracuse: Syracuse University Press, 1971.

———. "Pakistan: The Campaign before the Storm." *Asian Survey* 17 (July 1977):581–98.

———. "Pakistan and India: Politics, Personalities and Foreign Policy." *Asian Survey* 18 (July 1978):709–27.

Ziring, Lawrence, Ralph Braibanti, and Howard Wriggins, eds. *Pakistan: The Long View*. Durham: Center for Commonwealth and Comparative Studies, 1977.

Books in Urdu

Ahmad, Ghafoor. *Phir Martial Law Agiya* (Then Came Martial Law). Lahore: Jung, 1988.

Munir, Ahmad. *Siyasi Uttar Charhao* (Political Ups and Downs). Lahore: Jung, N.d.

Niazi, Maulana Kausar. *Aur Line Kat Ga'i* (And the line was cut). Lahore: Jung, 1987.

Rashid, Rao Abdul. *Jo Main Ney Daikha* (What I Have Seen). Lahore: Atishfashan, 1985.

Index